Harpers Ferry Armory and the New Technology

THE CHALLENGE OF CHANGE

Harpers Ferry Armory and the New Technology

THE CHALLENGE OF CHANGE

MERRITT ROE SMITH

Cornell University Press   ITHACA AND LONDON

Cornell University Press gratefully acknowledges a grant from the Andrew W. Mellon Foundation that aided in bringing this book to publication.

First published 1977 by Cornell University Press.
Published in the United Kingdom by Cornell University Press Ltd.
2–4 Brook Street, London W1Y 1AA.

First Printing, Cornell Paperbacks, 1980

International Standard Book Number (cloth) 0-8014-0984-5
International Standard Book Number (paper) 0-8014-9181-9
Library of Congress Catalog Number 76-28022
Printed in the United States of America
Librarians: Library of Congress cataloging information appears on the last page of the book.

For Wilson N. Smith (1906–1962),
father and friend

Acknowledgments

L ike most authors, I owe many debts. I am deeply grateful to my
family for longstanding encouragement and support; to Philip
S. Klein for first kindling my interest in the history of technology;
to Albert Blair and Sara D. Jackson for leading me to many un-
tapped resources at the National Archives; to Agnes B. Cadwell,
Martin R. Conway, B. Franklin Cooling, Benjamin Davis, Rodney
A. Pyles, Stephen T. Riley, Judith A. Schiff, Hilda E. Staubs, and
Juliette Tomlinson for similar favors during the course of research;
to Harold J. Bingham for kindly allowing access to the James T.
Ames Papers at his home; to Edwin A. Battison for patient tutorage
and incisive commentary on many points of technical detail; to
Charles B. Dew, Eugene S. Ferguson, Hugo A. Meier, Glenn
Porter, and Nathan Rosenberg for reading the manuscript at
various stages and offering invaluable recommendations; to Charles
Christenson for help in constructing the tables; to Bronwyn M.
Mellquist for perceptive criticism and countless hours of fine
editorial assistance; to Jane Dieckmann for further aid in the
preparation of the manuscript; to Les Benedict, John C. Burnham,
June Z. Fullmer, Myron Hedlin, W. David Lewis, Allan R. Mil-
lett, and Warren Van Tine for friendly counsel and innumerable in-
sights into the historian's craft. All these people and more have
helped immeasurably to shape my thoughts and fashion this
presentation.

This book contains material, considerably revised and expanded,
from my earlier essays in the *Virginia Magazine of History and
Biography* 81 (1973), *Technology and Culture* 14 (1973), and *Techno-*

logical Innovation and the Decorative Arts, Winterthur Conference Report 1973 (Charlottesville: University Press of Virginia, 1974). I am indebted to the editors of these publications for permission to reprint certain passages in the present text. Lastly, I wish to acknowledge the generous support of the Smithsonian Institution, which made eighteen months of uninterrupted research possible.

MERRITT ROE SMITH

Columbus, Ohio

Contents

Illustrations

Abbreviations

AAR	Records of the Allegheny Arsenal (Record Group 156, National Archives)
AGO	Records of the Adjutant General's Office (Record Group 94, National Archives)
GAO	Records of the United States General Accounting Office (Record Group 217, National Archives)
HFNP	Harpers Ferry National Historical Park
HR	Records of the United States House of Representatives (Record Group 233, National Archives)
JAG	Records of the Office of the Judge Advocate General (Record Group 153, National Archives)
NPS	Records of the National Park Service (Record Group 79, National Archives)
OCO	Records of the Office of the Chief of Ordnance (Record Group 156, National Archives)
OIG	Records of the Office of the Inspector General (Record Group 159, National Archives)
OQG	Records of the Office of the Quartermaster General (Record Group 92, National Archives)
OSW	Records of the Office of the Secretary of War (Record Group 107, National Archives)
PGW	Papers of George Washington (Library of Congress)
RPO	Records of the Patent Office (Record Group 241, National Archives)
SAR	Records of the Springfield Armory (Record Group 156, National Archives)
USS	Records of the United States Senate (Record Group 46, National Archives)

Harpers Ferry Armory and the New Technology

THE CHALLENGE OF CHANGE

Introduction

The labouring classes [of America] are comparatively few in number, but this is counter-balanced by, and indeed may be regarded as one of the chief causes of, the eagerness with which they call in the aid of machinery in almost every department of industry. Wherever it can be introduced as a substitute for manual labour, it is universally and willingly resorted to. . . . It is this condition of the labour market, and this eager resort to machinery wherever it can be applied, to which, under the guidance of superior education and intelligence, the remarkable prosperity of the United States is mainly due.

These comments by the noted British engineer and machine builder Joseph Whitworth typified European opinion of American industrial development in the mid-nineteenth century. By 1854, the date of Whitworth's special report to Parliament, many observers had detected similar trends in the American economy. What fascinated them was not so much that the people of the United States had adopted the new machine technology, but rather that they had done it with such little difficulty. Whitworth, whose memory of labor riots and machine breaking in Regency England remained vivid, expressed amazement at the propensity of Americans to favor the adoption of machinery. "The workmen hail with satisfaction all mechanical improvements," he wrote, "the importance of which . . . they are enabled by education to understand and appreciate." In Whitworth's judgment a liberal system of public education coupled with a scarcity of labor in the United States fostered unusually harmonious relations between factory masters and mechanics which benefited all members of society.[1]

Other visitors witnessed the same phenomenon, but advanced different explanations. To the Frenchman Alexis de Tocqueville, by no means an uncritical observer, America's industrial potential rested with her democratic institutions and inbred utilitarian attitudes. Charles Dickens detected similar traits, while C. L. Fleischmann, a German bureaucrat who spent several years in the United States as an employee of the Patent Office, and his compatriot Francis Grund ascribed industrial progress to the country's wealth of natural resources and "enterprising spirit" of its citizenry. Others found the seeds of industry deeply implanted in the lure of the wilderness, the majesty of personal freedom, native genius, disdain for liberal science, economic necessity, a compulsive search for national identity, vanity, love of change, and an unabated eagerness to "get on." Yet, whatever the reason, everyone agreed that mechanization had found a most favorable environment in North America.[2]

By the 1850s the United States no longer remained the relatively simple, homogeneous society once exalted by Thomas Jefferson and other disciples of agrarian democracy. Having achieved a high degree of mechanical proficiency and having entered a period of sustained economic growth, the young republic had relinquished its abject dependency on European technology and no longer stood in awe of the Old World's industrial prowess. So thorough was this transition that European enterprisers, impressed by American man-

1. Whitworth's report, originally published in the British *Parliamentary Papers* (1854), has been reprinted in *The American System of Manufactures* edited with a lengthy introductory essay by Nathan Rosenberg (Edinburgh: Edinburgh University Press, 1969), pp. 387–388. Labor scarcity, the most frequently-cited reason for American advances in the mechanical arts during the early nineteenth century, has recently been the subject of an intense but unresolved debate among practitioners of the "new economic history"; see, for example, Edward Ames and Nathan Rosenberg, "The Enfield Armory in Theory and History," *Economic Journal* 78 (1968):827–842; and Paul J. Uselding, "Technical Progress at the Springfield Armory, 1820–1850," *Explorations in Economic History* 9 (1972):291–316.

2. In addition to the original works of de Tocqueville, Dickens, Fleischmann, and Grund, see Henry T. Tuckerman, *America and Her Commentators* (New York: Scribner's, 1864); John G. Brooks, *As Others See Us* (New York: Macmillan, 1908); and Jane L. Mesick, *The English Traveller in America, 1785–1835* (New York: Columbia University Press, 1922).

ufacturing expertise, began to import Yankee machinery. One of the first industries to receive such recognition was the railroad. As early as 1837, American engine builders were exporting steam loco- motives to European customers, and in 1842 Russia's czarist govern- ment commissioned a Philadelphia firm, Winans, Eastwick & Harri- son, to supply the rolling stock for a 460-mile railroad connecting Moscow and St. Petersburg.[3]

Even more indicative of the changes taking place was the impres- sive debut of American metalworking technology at the 1851 Crys- tal Palace Exhibition in London. Here for the first time little- known manufacturers impressed visitors with the precision design and functional simplicity of their machine-made products. Exciting considerable comment and admiration were the unpickable pad- locks of Alfred C. Hobbs of New York, Samuel Colt's revolving pistols, and six completely interchangeable rifles made by Robbins & Lawrence, an obscure firearms and machine company in Wind- sor, Vermont. All three firms won prizes at the exposition, but more significantly the excellence of their wares prompted the Brit- ish government to send Joseph Whitworth and five other commis- sioners to New York's Crystal Palace Exhibition and, from there, on a fact-finding tour of the northeastern section of the country during the summer of 1853. The following spring Her Majesty's Ordnance Board, eager to take advantage of the so-called "Amer- ican System" of interchangeable production, dispatched another three-man "Committee on Machinery" to the United States for the express purpose of introducing similar improvements at the Enfield armory near London. By the time the investigators returned home in August 1854, they had placed orders for over $105,000 worth of machinery with seven different firms, five of which were located in New England. Within four years several other foreign governments sent similar investigatory teams to the United States. Although little noticed at the time, these events signaled America's coming of age as an industrial power.[4]

3. Victor S. Clark, *History of Manufactures in the United States*, 3 vols. (Wash- ington: Carnegie Institution, 1929), 1:362–363, 508.
4. Rosenberg, *American System*, pp. 180–192; U.S., Congress, Senate, *Military Commission to Europe in 1855 and 1856: Report of Major Alfred Mordecai*, 36th Cong., 1st sess., 16 June 1860, Senate Ex. Doc. No. 60, p. 159; Colonel Henry K. Craig to Henry W. Clowe, March 6, 1856, Letters Sent to Ordnance Officers, OCO.

A fact that escaped Whitworth and his contemporaries was that mechanization, in spite of its many practical benefits, had exacted a price. The very process of factory innovation had been a trying experience in which the wage of progress for countless individuals had been mental anguish, physical debilitation, and oftentimes death. Being transient and little acquainted with the country's bucolic past, most foreign observers failed to appreciate fully the impelling changes that attended the growth and assimilation of the new technology in America. Industrialization had wrenched the country out of a rural age and catapulted it into a different epoch. Dominated by what Daniel Webster called "this mighty agent, Steam," the coming of industry unleashed new and unfamiliar forces that made life more complex and uncertain. Hardly anything—politics, family life, values, religion, the economy—remained untouched by the momentum of technological change. "The old universe was thrown into the ash heap," historian Henry Adams ruefully commented on his childhood years during the 1840s. "He [Adams] and his eighteenth-century troglodytic Boston were suddenly cut apart—separated forever."[5]

Adams, the son of a diplomat and the descendant of two presidents, longed to return to the idyllic days of his forefathers, and other members of society echoed his expressions of loss and frustration. No one felt the impact of industrialization more directly than the common man, who, in an ironic way, became a symbol of the age. The factory, the railroad, the steamboat—indeed all the elements of the new civilization—required greater discipline, increased regimentation of daily routines, and, above all, a heightened consciousness of time. In most communities the factory and station clock, the most precise of all machines, came to control time.

While this seemingly elementary need of business management enforced a greater standardization of daily life on nearly all Americans, it operated with particular efficacy on the working population. Craft-trained artisans, the earliest recruits of industrialism, increasingly found themselves isolated from employers, irritated by the growing intensity of production, and threatened by the ma-

5. *The Education of Henry Adams* (Boston: Houghton Mifflin, 1918), p. 5.

chine. The division of labor, an adjunct of mechanization and a critical component of the factory system, not only bastardized their crafts but also weakened their bargaining power. Most ominous of all, conformity supplanted individuality in the productive process, an experience that sapped morale, widened class distinctions, and, in many cases, hampered creativity.

While craftsmen felt abused by the factory system, the greatest difficulties confronted those without skills. Since their tasks could easily be performed by others, few opportunities existed for wage increases and other job-related emoluments. Deprived of promotions and other avenues of mobility, unskilled workers often became sullen, careless, and troublesome victims of impersonal hierarchical organizations. All these factors introduced elements of tension and insecurity in American life that had never been there before.[6]

The subject of this book, the national armory at Harpers Ferry, Jefferson County, Virginia, well illustrates the anxieties and misgivings that attended the rise of large-scale manufacturing in the early nineteenth century. The story of Harpers Ferry, most notably the efforts of its inhabitants to preserve accustomed life styles and practices in the wake of accelerating technology, presents a microcosmic view of the industrial revolution which is perhaps more suggestive of America's bittersweet relationship with the machine than many historians have heretofore recognized. In addition to specifying the organizational and technological changes that occurred at Harpers Ferry between 1798 and 1861, this study seeks to identify those who initiated new ideas, evaluate the significance of their work in a national context, and determine how the community at large responded to the emergence of novelty in thought and action. The scenario is both subtle and complicated. It pictures an isolated rural society floundering between the two worlds of agrar-

6. Only a handful of scholars, mostly cultural historians, have addressed themselves to this question. Particularly noteworthy are Leo Marx, *The Machine in the Garden* (New York: Oxford University Press, 1964); Douglas T. Miller, *The Birth of Modern America, 1820–1850* (New York: Pegasus, 1970); Charles L. Sanford, "The Intellectual Origins and New-Worldliness of American Industry," *Journal of Economic History* 18 (1958):1–16; and Sanford, ed., *Quest for America* (New York: New York University Press, 1964).

ian pastoralism and industrial progress. Mechanization is held suspect and the imperatives that accompany technological advance are tolerated only as long as workers are allowed to retain certain rights and privileges associated with pre-industrial traditions. In such a milieu elements of scorn, ridicule, and apprehension are present. Yet, interestingly enough, there also exists a certain degree of wonderment, even tantalizing fascination with the changes taking place. To ignore these contradictory impulses is to slight the tremendous complexity of the community's industrial experience.

The Harpers Ferry story diverges sharply from oft-repeated generalizations that "most Americans accepted and welcomed technological change with uncritical enthusiasm."[7] Such stereotypes are based primarily on studies treating the economic development of New England, which in many respects was atypical. The realities of the situation at Harpers Ferry also contrast with the published statements of the British Committee on Machinery, which, upon touring the country in 1854, extolled "the extreme desire manifested by masters and workmen to adopt all labour-saving appliances," while ranking the Virginia armory among the five most progressive manufacturing establishments of its type in the United States.[8] Like Whitworth, the members of the committee did not perceive the many difficulties that attended building the factory and adopting new techniques. Nor did they appreciate the endless troubles management encountered in getting workers to follow an industrial regimen. Given an ignorance of the country's cultural heritage as well as the natural desire of factory masters to put their best foot forward in the presence of foreign dignitaries, this is understandable. Nonetheless, enough instances of labor discord and discontent exist during the antebellum period to impugn the credibility of well-meaning but impressionable foreign observers who reported on the American's abiding love affair with the machine.[9]

7. Edwin T. Layton, Jr., ed., *Technology and Social Change in America* (New York: Harper & Row, 1973), p. 1.

8. Rosenberg, *American System*, p. 128.

9. Herbert G. Gutman, "Work, Culture and Society in Industrializing America, 1815–1919,"*American Historical Review* 78 (1973):531–587; David Montgomery, "The Working Class of the Preindustrial American City, 1780–1830," *Labor*

This book suggests that Harpers Ferry's response to industrialization was hesitant and equivocal. If further studies corroborate my finding, long-standing beliefs about the eagerness with which Americans have embraced new technologies will stand in serious need of revision. Equally important, popular litanies lauding work, progress, and industrialism as basic tenets of the national creed will also require searching revaluation. Could it be that only a small segment of the population held these views and tried to inculcate them through various agencies of economic and social control? Thus far scholarly inquiry has only scratched the surface of this intriguing question. Before any definite pronouncements can be made about the impact of the machine in the United States, a number of "grass roots" investigations must be undertaken at local and regional levels. Until this is done, our view of the historical consequences of industrialism in American civilization is likely to remain as ambivalent and uncertain as were the feelings of people who first encountered the phenomenon during the early nineteenth century.

History 9 (1968):1–22; Montgomery, "The Shuttle and the Cross: Weavers and Artisans in the Kensington Riots of 1844," *Journal of Social History* 5 (1972):411–446; Stephen Thernstrom, *Poverty and Progress: Social Mobility in a Nineteenth Century City* (Cambridge, Mass.: Harvard University Press, 1964).

CHAPTER I

Regional Interests and Military Needs: Founding the "Mother Arsenal," 1794–1801

T he myth of the garden, so ably delineated by Leo Marx, Henry Nash Smith, and other twentieth-century scholars, occupies a dominant place in American thought and feeling. Central to this myth is the idea of a regenerate nation, an agrarian republic situated in an undefiled "middle landscape" where nature and civilization exist in harmonious balance. Since the earliest settlements the image of such a pastoral society, at once free and equal, virtuous and comfortable, and at peace with itself, has "defined the promise of American life."[1]

Of all those who have espoused the pastoral ideal, no one has had a greater impact in fixing the concept in American consciousness than Thomas Jefferson. In his *Notes on the State of Virginia*, first published in 1785, Jefferson isolated the very essence of the American dream, giving it a timeless and enduring quality that no writer before or since has captured with such gusto and sense of immediacy. The secret of Jefferson's success, as Marx points out, lay in his ability to snatch the pastoral ideal from the realm of literary imagination and clothe the concept with economic and political reality. In developing and elaborating the theme of a new Eden, the sage of Monticello drew upon what he knew best, the cultural and physical character of his native state, and filled his notebooks with incisive

1. Henry Nash Smith, *Virgin Land: The American West as Symbol and Myth* (Cambridge, Mass.: Harvard University Press, 1970), p. 123; Marx, *Machine in the Garden*, pp. 5–26.

commentaries on the civil and natural history of the new world environment.[2]

The rustic beauty of Jefferson's Virginia contrasted sharply with the depravity of Old Europe. In the American countryside happy yeomen tended their crops, meadows rustled in the wind, and cattle grazed under the sunlit sky of an infinite universe. Everything existed in proper relationship to nature. In Jefferson's world there was little room for artificiality. Conspicuously absent were cities and the dehumanizing "rules" of industrial civilization. "Let our workshops remain in Europe," he admonished. "While we have land to labour then, let us never wish to see our citizens occupied at a work-bench, or twirling a distaff."[3]

Throughout the *Notes* Jefferson skillfully mixed personal feeling with objective experience. Few things escaped his discerning eye. He described Virginia's Natural Bridge in Rockbridge County, for example, as though he were evaluating a work of art. "It is impossible for the emotions arising from the sublime, to be felt beyond what they are here," he observed. "So beautiful an arch, so elevated, so light, and springing as it were up to heaven, the rapture of the spectator is really indescribable!" By tempering romantic impulse with rational sensibility, Jefferson ingeniously ascribed engineering beauty to natural form, thereby achieving a delicate balance in his pastoral scheme. Equally eloquent was his commentary on the picturesque gorge formed by the confluence of the Potomac and Shenandoah rivers at Harpers Ferry: "The passage of the Patowmac through the Blue ridge is perhaps one of the most stupendous scenes in nature. You stand on a very high point of land. On your right comes up the Shenandoah, having ranged along the foot of the mountain an hundred miles to seek a vent. On your left approaches the Patowmac, in quest of a passage also. In the moment of their junction they rush together against the mountain, rend it asunder, and pass off to the sea."

Harpers Ferry typified Jefferson's "middle landscape." At once the scenery was as "placid and delightful" as it was "wild and

2. Marx, pp. 74, 116–144.
3. Jefferson, *Notes on the State of Virginia* (Philadelphia: R. T. Rawle, 1801), p. 325.

tremendous." The settlement lay at the tip of a triangular tract situated between the two rivers. Behind this point and roughly parallel to the rivers, a narrow belt of land wound around the base of a steep hillside (see Map). On this slope Jefferson recorded his impressions of the vicinity. To the east the waters of the Potomac and Shenandoah, now joined, flowed through a glassy vale of jagged rocks and sharply defined plateaus toward the Chesapeake. Across the Shenandoah and to the right was Loudoun Heights (Figure 20), an imposing mountain that seemed to hover over the village and to cast its shadow on the rapids and Virginius Island below. To the left on the far side of the Potomac in a more distant northeasterly direction stood the rocky precipice of Maryland Heights, rising majestically hundreds of feet above the river to provide a panoramic view of Harpers Ferry and the rich grain lands lying westward in the Valley of Virginia. To behold these sights, Jefferson wrote, "is worth a voyage across the Atlantic."[4]

Although Jefferson never completely abandoned his pastoral vision, economic and political realities during the early national period forced him to compromise the theory. Many of his fellow countrymen were strongly committed to making the new republic economically self-sufficient. This meant, in effect, accommodating commerce and manufactures, the two things he wished most to quarantine in Europe. Apprehensive that too large a dosage of change and modernity might upset the delicate balance of the "middle landscape," Jefferson wanted to proceed with caution. Whenever he acquiesced to forces molding the new America, he did so with "painful anxiety." Jefferson, to be sure, did not oppose progress. As Marx indicates, a more devoted student of science and technology could not be found in America. He simply did not want to jump headlong into a frenzied program of national development at the expense of what mattered most—the preservation of values associated with a rural society.[5]

Jefferson's uneasy reflections about the future of the agrarian republic differed considerably from the thought of George Washington. Like Jefferson, Washington envisaged "the Garden of Amer-

4. Jefferson, pp. 34–36, 45.
5. Marx, pp. 135, 140.

ica," but his conception of the pastoral life was more idyllic than real and certainly was never intended as a serious guide to social policy. First and foremost, Washington was a businessman, a firm believer in the ethos of capitalism and an unabashed prophet of economic progress. Since the French and Indian War he had been an avid land speculator, acquiring property whenever he thought it might turn a profitable dollar. By the time the *Notes on the State of Virginia* was published, he had become deeply involved in the affairs of the Potowmack Company, a joint stock venture aimed at improving navigation along the Potomac River and attracting trade to the ports of Alexandria and Georgetown. His collected papers are filled with references to Mount Vernon, western lands, commercial development, and, after 1790, the building of the nation's capital in the District of Columbia. Along with his official duties as president of the United States, these matters engrossed Washington's personal attention during the 1790s. Each of them indicates the scope and regional orientation of his private affairs. Combined, they related to a larger scheme of things often described as "the favorite object of his heart"—the economic development of the Potomac Valley with the new "Federal City" as its focal point.[6]

Washington's certainty that commerce and industry would be a stabilizing influence on the new nation strengthened his determination to make the Federal City "the great emporium of the United States."[7] That the growth of trade and manufacturing might destroy the bucolic simplicity of life in the region never seemed to enter his mind. In this conviction he was joined by numerous

6. Washington to Sir John Sinclair, December 11, 1796, *The Writings of George Washington*, ed. John C. Fitzpatrick, 39 vols. (Washington: GPO, 1931–1944), 35:324–331; Tristram Dalton to John Adams, March 26, 1797, Letters Received (quotations from the Adams Papers are from the microfilm edition, by permission of the Massachusetts Historical Society). For Washington's private business affairs see Raymond G. Peterson, Jr., "George Washington, Capitalistic Farmer" (Ph.D. diss., The Ohio State University, 1970), and Marcus Cunliffe, *George Washington: Man and Monument* (New York: New American Library, Mentor Books, 1960), p. 59.

7. Washington to Sinclair, December 11, 1796, *Writings*, 35:329. The material in this paragraph and the following section is more fully documented in Merritt Roe Smith, "George Washington and the Establishment of the Harpers Ferry Armory," *Virginia Magazine of History and Biography* 81 (1973):415–436.

persons on both sides of the Potomac as far west as Cumberland, Maryland. From a former governor of Maryland to a store owner in Shepherdstown, they anxiously awaited the day when the Federal City would become the economic as well as the political hub of America. As landowners and businessmen they all stood to gain handsomely by the growth of population and trade along the Potomac River. Such enthusiastic boosterism set the stage for the establishment of the Harpers Ferry armory.

Prior to 1794 all arms furnished to United States troops had been purchased either from importers or private manufacturers who contracted with the government for their delivery at various depots located in Massachusetts, New York, Pennsylvania, and Virginia. All sorts of deceptions and forfeitures had occurred under the contract system and, as president, Washington was determined to remedy the situation. He therefore sponsored and Congress approved a bill in April 1794 "for the erecting and repairing of Arsenals and Magazines." The legislation provided $81,865 for the establishment of as many as four national armories and gave the president wide discretionary powers in executing the order. He held the options of not only deciding the number of arsenals and armories to be built but also choosing their locations and using the appropriations in almost any lawful manner he saw fit. Furthermore, he was given the authority of appointing (or dismissing) superintendents and master armorers at will, an important source of patronage which in later years became a subject of bitter contention in local politics.[8]

The selection of Springfield, Massachusetts, as the site of the first national armory came as no surprise. In a letter to the president dated December 14, 1793, Secretary of War Henry Knox had strongly recommended this small river town in the western section of his native state as the only spot already owned by the United States that possessed adequate buildings, water power, and transportation facilities. Apparently Washington had no objection to Knox's proposal and, upon passage of the armory bill, immediately approved the measure.[9]

8. U.S. *Statutes at Large*, 1:352.
9. *American State Papers: Military Affairs*, 1:44.

The decision on the location of arsenals in the middle and southern states involved more sensitive issues. Considering the meager amount of money appropriated for the purpose, Washington had two alternatives. One was to use the funds to rebuild pre-existing Revolutionary War magazines already owned by the government and located at Philadelphia and Carlisle, Pennsylvania, West Point, New York, and New London, Virginia. If three of these were rebuilt, as many as four arsenals were feasible. The other option was to purchase land at a new location and build an arsenal with the hope of developing it into a full-fledged factory at some later date. Only one such arsenal could be constructed because of the extra expenditures involved. Although Henry Knox and his successor at the War Department opposed the idea, Washington decided to build one large arsenal rather than rebuild three smaller ones. Furthermore, he resolutely determined to build the new installation at Harpers Ferry.

Washington's decision received warm endorsement from a group of down-river merchants headed by Tobias Lear of Georgetown and George Gilpin of Alexandria. Closely knit by common objectives, this small but influential clique zealously supported any project that tended to channel hinterland trade into the Potomac basin. All of them held stock in the Potowmack Canal Company and, at one time or another, had lobbied in the Maryland and Virginia legislatures in order to persuade those states to continue sponsorship and financial support of the venture. As directors of the Potowmack Company, Lear and Gilpin had visited Harpers Ferry on numerous occasions and had circulated glowing reports of its potential as a millsite. As close friends of the president, both men would play an important role in arranging the government's land purchases at the settlement.[10]

Lear especially perceived the economic benefits of locating a large government factory at Harpers Ferry. The continuous consumption of various necessities (iron, coal, grindstones, files, tools)

10. Corra Bacon-Foster, "Early Chapters in the Development of the Potomac Route to the West," *Records of the Columbia Historical Society* 15 (1912):167, 177, 179, 184; Lear to Washington, February 13, 1795, Washington to Lear, November 2, 1795, PGW (microfilm).

would provide lucrative contracts for mercantile houses in the region; incoming shipments of raw materials and outgoing shipments of finished firearms would increase the alarmingly low toll revenues of the Potowmack Company; employment of a large labor force would bring in more people to populate the Valley and thereby increase the demand for agricultural and store-bought goods; finally, the existence of an arms factory would enhance the value of property in the immediate vicinity. Lear viewed the arsenal's function in regional terms as a stimulus to business enterprise along the Potomac. It was just the tonic the Valley needed, and for these reasons he became totally committed to seeing the plant constructed at Harpers Ferry.[11]

Between 1794 and 1797, Lear and Gilpin, acting as special agents of the president, worked assiduously to complete the land transactions at Harpers Ferry. Much to Washington's irritation, however, many delays occurred, not the least of which was caused by the hesitancy of the John Wager family, descendants of the earliest settler of the area, to sell their centrally located holdings along the Potomac and Shenandoah rivers and by the existence of an unsettled lawsuit over the ownership of a 230-acre "sawmill tract" contiguous with the Wager property.[12] Equally disquieting, the president faced opposition within his own cabinet.

Knox had never been favorably disposed toward building an arsenal at Harpers Ferry. In his mind, other more advantageously located sites existed in Pennsylvania, Maryland, and Virginia. His successor at the War Department, Timothy Pickering, felt even

11. Lear to Washington, October 19, 1795, PGW.

12. Three parties owned land at Harpers Ferry. The Wager family owned the indispensable 125-acre "Ferry Tract" at the junction of the Potomac and Shenandoah rivers; Burges Ball, a close friend and relative of Washington, held a seventy-year lease on 600 acres of woodland owned by Henry Lee; and Thomas Rutherford, Jr., the son-in-law of General William Darke, owned a 230-acre "sawmill" plot which ran from river to river and lay between the Wager and Ball properties. At this time, the tract was in suit between Rutherford, the Wagers, and a man named Bready who operated a sawmill on the Shenandoah side of Harpers Ferry. See Washington to Secretary of War, September 16, 29, 1795, and October 16, 1797, *Writings*, 34:307–308, 318; 36:54; Lear to Washington, October 19, 1795, May 30, 1796, and January 17, 1797, Washington to Lear, January 13, 1797, PGW.

stronger about the issue, arguing that the appropriations provided by Congress "would be inadequate for a single new establishment." Instead of an arsenal, he hoped Washington would be content to build a depot at Harpers Ferry "where, afterwards, the works necessary in the formation of all implements of war might be erected as the requisite funds could be provided." In the meantime, Pickering recommended expanding operations at Springfield, where sufficient land, labor, and buildings already existed. At this arsenal, he argued, invaluable experience could be accumulated and guidelines formed for the erection of others.[13]

In an effort to deflate the exaggerated claims advanced by Lear and Gilpin and at the same time persuade the president to reconsider his decision, Pickering commissioned Colonel Stephen Rochfontaine, a French-born military engineer, to re-examine different sites along the Potomac "fit for the establishment of an arsenal." After conducting the survey in the spring of 1795, Rochfontaine submitted a report that did not even mention Harpers Ferry as a viable site. When called upon for an explanation, the Frenchman replied that there had been no oversight or neglect on his part. Rather, in his opinion, Harpers Ferry did not warrant serious consideration since "there was no ground on which convenient buildings could be placed" at reasonable expense and, more important, "no water work would be safe there" because of the settlement's susceptibility to floods. Instead of convincing the president to modify his plans, Rochfontaine's report upset him. Washington remained more adamant than ever, insisting that Harpers Ferry was the best possible site, even demanding that Rochfontaine return to the settlement—this time in the company of Lear and Gilpin —and revise the report so that it coincided with his expectations.[14]

In the end it took nearly three years to overcome disinterest and opposition at the War Department and consummate the land purchases at Harpers Ferry. By the time the final deeds reached Phila-

13. Octavius Pickering and Charles W. Upham, *The Life of Timothy Pickering*, 3 vols. (Boston: Little, Brown, 1873), 3:140–141.

14. U.S., *Annals of Congress*, 6:2569–2570; Washington to Burges Ball, March 2, 1795, to Lear, November 30, 1795, *Writings*, 34:129, 381; Lear to Washington, October 19, 1795, Washington to Lear, November 2, 1795, PGW.

delphia in the spring of 1797, Washington had relinquished the presidency to John Adams and had retired to Mount Vernon. Even then, owing primarily to lack of funds and the noncommittal attitude of Adams' secretary of war, James McHenry, the proposed arsenal remained in abeyance, nothing more than a paper project. Not until an undeclared war with France in 1798 and Washington's return to public life as commander-in-chief of the specially created "provisional army" were serious efforts made to activate the factory.[15]

Of all the persons connected with its establishment, the Harpers Ferry arsenal owed its existence most to George Washington. With single-mindedness he chose the site and ordered its purchase against the advice of a military engineer and two secretaries of war; after two years of neglect he redirected the Adams administration's attention to the project in 1798; and, though he did not live to see the arsenal's completion, he heightened the strategic value of the site by ordering three regiments under General Charles Cotesworth Pinckney—nearly one fourth of the provisional army—to Harpers Ferry in 1799.[16] Throughout these proceedings, Washington remained at one mind with Lear, Gilpin, and other members of the Potowmack Company. As businessmen whose fortunes depended primarily on the continued development of up-river trade, they eagerly sought to cultivate new "sure-pay" customers, particularly those providing large orders for supplies on a long-term basis. The Harpers Ferry project appeared as a juicy plum nearly ripe for the picking, and they mustered all their influence to ensure that the tasty morsel would not be snatched out of their hands by equally hungry competitors.

While championing Harpers Ferry as an arsenal site, Washington repeatedly stated that the place not only possessed an "inexhaustible supply of water" and "centrality among Furnaces and

15. Washington to Lear, January 13, 1797, Lear to Washington, January 17 and February 10, 1797, PGW; Washington to Secretary of War, May 6 and July 5, 1798, *Writings*, 36:253–254, 318.
16. Washington to Colonel Thomas Parker, September 28, 1799, to Alexander Hamilton, September 29, 1799, PGW.

Forges" but also was removed far enough inland "to be perfectly secure" against foreign invaders. Strategic considerations, he maintained, required an arsenal near the new Federal City.[17] The locality's defensive position was, in fact, questionable, but even apart from this, he failed to perceive that the military advantages of the area augured ill for industrial development. Even under the best of circumstances the building of a factory, particularly one designed to manufacture products as complex as firearms, was a difficult and trying experience. A successful effort required a trained labor force, special tools, tested managerial skill, easy access to raw materials and sources of technical information, and, most important of all, a social milieu adaptable to change and regimentation. In all these categories Harpers Ferry was sadly deficient. Although Knox, Pickering, and Rochfontaine had recognized these shortcomings and tried to dissuade the president from implementing his plan, their warnings about inadequate facilities, the danger of freshets, and unnecessary expenditures fell on deaf ears. Washington refused to be swayed from his purpose, insisting that less fault-finding and more diligence on the part of government officials would achieve the desired results.

What had inspired Jefferson in 1785 became a severe handicap thirteen years later. Harpers Ferry's greatest liability was its isolated, even frontier-like position. In 1798 the village was little more than a trading outpost occupied by a handful of residents. Although scores of settlers had crossed the Potomac at the Ferry, almost all of them were transients who continued westward either to homesteads in the Shenandoah Valley or to more prosperous villages at Charlestown, Winchester, and Strasburg. The closest towns of any size were Frederick, Maryland, some twenty miles to the east, and Hagerstown, Maryland, twenty-five miles north. Since neither possessed large mercantile houses, armorers tools and other essential manufacturing supplies had to be procured in cities along the eastern seaboard and shipped to Harpers Ferry at considerable expense. Most often these provisions were purchased in Philadelphia and Baltimore and then hauled overland in Conestoga

17. Washington to Secretary of War, May 6, 1798, *Writings*, 36:253; *American State Papers: Military Affairs*, 1:44.

wagons pulled by five-horse teams. Depending on the point of origin, the weight of the consignment, and the condition of the roads, such trips required anywhere from three to thirteen days. Goods coming upriver from Alexandria and Georgetown were equally dependent upon weather conditions, the density of traffic, and the temperament of crews. Whatever the means of conveyance, existing avenues of transportation were both slow and uncertain, making the problem of planning and coordinating factory production infinitely more difficult. Even bar and pig iron, two essential commodities that Washington considered easily procured at Harpers Ferry, had to be teamed nearly a hundred miles from furnaces and forges located in central Pennsylvania. Viewed in this light, transportation factors alone placed Harpers Ferry at a great disadvantage in comparison with Springfield. Standing on the banks of the Connecticut River, secure from floods, and easily accessible to the ports of Hartford, Boston, and New York, the Massachusetts armory seemed to possess all the advantages that the Potomac site lacked.[18]

From a social and economic perspective Harpers Ferry was extremely provincial in 1798. News traveled slowly—sometimes days, at other times weeks behind current affairs of the outside world, depending on who happened to pass through the neighborhood. Except for a small sawmill and a country store which catered to the needs of local farmers on a partly cash, partly barter basis, the settlement contained no business establishments. The only other structures consisted of a cluster of houses, several stables, and some sheds used for storing goods awaiting shipment downriver to Alexandria and Georgetown. Since no churches, schools, or other communal institutions existed at the Ferry, everyday life assumed a decidedly rural character in accordance with the agrarian ways of the surrounding countryside. At the center of this society were a number of gentry families who, through landed wealth, inheri-

18. Jefferson, *Notes on Virginia*, pp. 34–36; Thomas, Lord Fairfax, *Journey from Virginia to Salem, Massachusetts, 1799* (London: Privately printed, 1936), pp. 28–29; Payrolls and accounts, Harpers Ferry armory, 1816–1818, Second Auditor's Accounts, GAO; Joseph Barry, *The Strange Story of Harper's Ferry*, 3d ed. (Shepherdstown, W.Va.: The Shepherdstown Register, 1959), pp. 5–14.

tance, and local prominence, dominated the politics and economy of the region. Of lesser rank but larger in numbers were small free-holders who owned from forty to one hundred acres of land, some livestock, and perhaps a slave or two. Whether one worked a farm or a plantation, the tempo of life was set by the seasons. In such an environment measured time, a precious commodity in the business world, had little meaning. Like country dwellers in old England, these Virginians tended to arrange their daily tasks in accordance with the cycles of nature, frequently punctuating intense bouts of labor with hunting expeditions, barbecues, visits to neighbors, and attendance at court days, militia musters, and other festive occasions. Such diversions served to alleviate the drudgery and isolation of rural existence while reinforcing a romanticized myth of Southern leisure. Although travelers received hospitable treatment, outsiders who settled in the area were not accepted until they had proven their worth as good citizens and neighbors. Those who transgressed established norms and traditions quickly found themselves ostracized from the mainstream of social life and respectability.[19]

As part of a larger rural culture, the early residents of Harpers Ferry assigned more importance to their own neighborhood, its agrarian institutions, traditions, and interests, than to those of the nation at large. Given these provincial attachments, they viewed the proposed arsenal as a regional project sponsored in name only by the federal government. This perspective did not appreciably change once manufacturing operations began in 1802. Indeed, as the years passed, a majority of the local populace tended to construe the arsenal's function in even narrower terms. In their eyes, the factory existed not as an efficient producer of military ordnance but as a convenient pork barrel of jobs, contracts, and political patronage for those who inhabited the inner reaches of the Potomac Valley. As late as 1841 the civilian appointees who managed the works clearly mirrored these sentiments. At the same time they deeply resented any attempts by federal officials in Washington to

19. Cf. E. P. Thompson, *The Making of the English Working Class* (New York: Vintage Books, 1963), pp. 15–447. These points will be fully documented in Chapters 2, 6, and 9.

challenge their authority by altering administrative procedures, enforcing common regulations, or changing personnel at the arsenal. These attitudes greatly complicated the process of building a factory, developing a technological capacity, and instilling work discipline at Harpers Ferry during the early nineteenth century.[20]

The decision to begin construction at Harpers Ferry and to expand the proposed works from a small arsenal to a full-sized manufacturing armory occurred during the fall and early winter of 1798. Secretary of War McHenry made the announcement after numerous consultations with fellow cabinet members as well as with the commander-in-chief of the provisional army, George Washington. While he was apprehensive about the cost of the undertaking, President Adams seemed willing to abide by whatever McHenry thought best. Pickering and Secretary of Treasury Oliver Wolcott, however, opposed the measure, the former because he continued to believe that the production of private contractors augmented by that of the Springfield armory could meet all necessary demands for small arms, the latter because he considered it an unnecessary expenditure of government revenues. On the other hand, Attorney General Charles Lee and Secretary of the Navy Benjamin Stoddert favored the idea. Both urged that the works at Harpers Ferry "should be pushed with all possible vigor" and completed before the War Department embarked on any new ventures.[21]

Of all the members of Adams' cabinet, Stoddert appeared most enthusiastic about Harpers Ferry's prospects. Washington had previously written to him suggesting that "the establishment of a public foundery and Armory at the junction of the Potomac and Shenandoah, will afford no small advantage in arming the ships" of the newly created navy.[22] As a former Georgetown merchant and stockholder in the Potowmack Company, Stoddert needed very little

20. See, for example, Michael C. Sprigg to Colonel George Bomford, June 29, 1829, Major John Symington to Captain William Maynadier, July 12, 1849, Letters Received, OCO.

21. Bernard C. Steiner, *The Life and Correspondence of James McHenry* (Cleveland: The Burrows Brothers Co., 1907), pp. 394–395, 400–402; Washington to Secretary of War, October 1 and December 13, 1798, *Writings*, 36:477–479; 37:44–45.

22. Washington to Secretary of Navy, September 26, 1798, *Writings*, 36:465.

convincing. His response to one of McHenry's questionnaires was clear and to the point:

I think your submission contained a query, whether the Work begun at Harpers Ferry should be finished—or whether addl. works should be made there. . . . It is most clearly my opinion, that the Arsenal at Harpers Ferry, should there be a difference in point of magnitude in the three [proposed arsenals], should be the more important—the Mother Arsenal. It is without comparison the most convenient of the three to the Western Country. It is more convenient than either of the other places [Springfield and Rocky Mount, South Carolina] to all parts of the States. It is nearly in the centre.[23]

Despite the assurances of Stoddert and Lee, McHenry continued to doubt the wisdom of their advice throughout 1798 and 1799. He repeatedly stated his misgivings and questioned associates about the necessity of building a large arms factory on the Potomac. Nevertheless, he made preparations in August 1798 by appointing John Mackey as paymaster and storekeeper at Harpers Ferry and Joseph Perkin as the arsenal's first superintendent.[24]

Other than his foreign birth, outspoken devotion to Hamiltonian Federalism, and friendship with McHenry and other party dignitaries, little is known about Mackey's personal life prior to his appointment as paymaster. He probably had served in some subordinate capacity to Samuel Hodgdon, the Intendant General of Military Stores in Philadelphia before moving to Harpers Ferry in mid-September 1798. Although he was well educated and had aspirations of becoming a well-to-do gentleman, chronic financial difficulties prevented him from assuming a higher station in society. The thought of being confined to the status of a secondary civil servant plagued Mackey and heightened his sense of frustration and insecurity. This, in turn, generated a mercurial temperament as complex as it was implacable.[25]

23. Stoddert to McHenry, August 1, 1799, quoted in Steiner, pp. 401–402.

24. Perkin received his appointment on August 6, 1798; Mackey on August 30. See William Simmons to Mackey, December 20, 1799, Letter Book F, First Auditor's Accounts, GAO.

25. For information on Mackey, see Hodgdon-Pickering Papers, 1795–1801, AGO.

To superiors Mackey was hardworking and intensely loyal, though at times a bit too ingratiating. Among friends he was gracious, urbane, and witty. To subordinates he was arrogant, ill tempered, and contemptuous of the rights of others. Colleagues considered him vain, overly ambitious, and unwilling to share authority. To Jeffersonians he was an unscrupulous and vindictive political foe. Though denied the fruits of wealth and social position, he nonetheless retained a strong attachment to elitist principles while professing an equally strong aversion toward the unwashed masses.

Compared with Mackey, Perkin came from an unpretentious, even austere, background. An Englishman by birth and a Moravian by conviction, he had served an apprenticeship in the Birmingham gun trade before emigrating to America around 1774. With the onset of the Revolutionary War he worked for John Strode, a well-known businessman and manager of the Rappahannock Forge near Falmouth, Virginia. Recognizing Perkin's ability as a "usefull ingenious man," Strode "kept him chiefly at Gunlocks," a task assigned only to the most skilled artisans.[26] With the termination of hostilities in 1781, Perkin left Falmouth and moved to Philadelphia, where he set up a small gun shop in partnership with Samuel Coutty. There he remained, making arms for private as well as public patrons and serving as a part-time inspector for the Intendent of Military Stores, until called upon in 1792 to assume supervision of the New London arsenal in Campbell County, Virginia. When the government abandoned the New London installation six years later, Secretary of War McHenry transferred Perkin to Harpers Ferry.[27]

In a letter of instruction dated August 6, 1798, McHenry ordered Perkin to the Potomac to conduct a preliminary survey of the "Waters and grounds for erecting the necessary buildings and works for the establishment of an Armoury." At the same time the

26. Strode to Henry Dearborn, February 12, 1807, Letters Received, OSW.
27. Perkin to Hodgdon, January 5, 1793, Seth B. Wigginton to Hodgdon, August 10, 1797, Hodgdon-Pickering Papers, AGO; Accountant's Office, War Department, Letter Book G, First Auditor's Accounts, GAO; Henry J. Kauffman, *The Pennsylvania-Kentucky Rifle* (Harrisburg, Pa.: Stackpole Books, 1960), p. 319.

secretary acquired the services of another Englishman, James Brindley, to accompany and assist the new superintendent. Allegedly a nephew of the celebrated builder of the Bridgewater and Grand Trunk canals and a noted engineer in his own right, Brindley had come to America in 1786 at the request of Robert Morris to oversee the construction of an inland waterway connecting the Schuylkill and Susquehanna rivers in Pennsylvania. For more than a decade he had worked for several other river development companies in a similar capacity. He was already familiar with Harpers Ferry, having visited the village in 1786 as a consultant to the Potowmack Company. At that time Washington had opined that "no person in this country has more practical knowledge" of canal navigation than Brindley.[28]

Perkin and Brindley arrived at Harpers Ferry in mid-August 1798. After spending several weeks going over the land at the junction of the two rivers, they concluded that the most promising spot for waterworks lay on the Potomac side of the settlement. Completing the survey, they returned to Philadelphia to draw up a report and discuss their findings with the secretary of war.

Although McHenry received their estimates and proposals in September, he did not issue immediate orders to proceed. Apparently Brindley had recommended that an experienced engineer be commissioned to build the armory canal and milldam, but Paymaster Mackey effectively discouraged the idea at the War Department. Convinced that the armory required "nothing more than a common Water course of about 15 feet wide and 3 feet deep," he felt certain that he could complete the project without the added expense of hiring a professional engineer.[29] However, he did not object to engaging an engineer to survey the canal along the Potomac and build the framework of a dam above it at Harpers Ferry. This was just as well since McHenry had already assigned the task to Brindley. By the time these arrangements were completed the winter season was approaching. Because Mackey had yet to as-

---

28. Accountant's Office, War Department, Letter Book G, GAO; Bacon-Foster, "Potomac Route West," pp. 167–169; Daniel H. Calhoun, *The American Civil Engineer* (Cambridge, Mass.: The M.I.T. Press, 1960), pp. 11–12.

29. Mackey to Hodgdon, March 12, 1799, Document File, HFNP.

semble a labor force, set up housing, or procure necessary tools and supplies, the project was postponed until the following spring. Meanwhile, Perkin—freed of the responsibility of excavating the canal—remained in Philadelphia making final arrangements for the move to Harpers Ferry.

When Perkin returned to Harpers Ferry in the fall of 1798, he had assembled a small contingent of armorers who had formerly plied their trade at New London, Virginia. Since no armory buildings had yet been erected, he set them to work converting an old frame warehouse into a temporary workshop for reconditioning damaged arms received from depots in Carlisle, Philadelphia, and New London. Other than assisting an artillery officer named Major Louis Tousard conduct experiments with local iron ores for casting cannon, Perkin spent most of his time supervising the repair of arms, interviewing prospective suppliers of raw materials, and devising plans for the construction of permanent armory buildings. Even so, very little had been accomplished by February 1799, because McHenry had not issued orders determining the number of buildings or the size of the permanent work force at Harpers Ferry. The disposition of these questions loomed important since little could be done on the physical plant until adequate housing was provided for the armorers and their families.[30]

Another serious obstacle to progress was the lack of a suitable person to design the new workshops. While Mackey had charge of their construction, Perkin retained the prerogative of designating the types of structures best suited for manufacturing purposes. At this juncture the first of several confrontations occurred between the paymaster and superintendent in which political partisanship as well as personal animosity played a part. The two men disagreed not so much over the style of architecture as over the choice of the architect.

Long before McHenry gave final sanction to an armory at Harpers Ferry, Benjamin Henry Latrobe had expressed interest in designing the factory in a letter to Jefferson dated March 28, 1798.

30. Mackey to Hodgdon, September 17, 25, 1798, January 28, 1799, HFNP; Mackey to Secretary of War, February 10, 1799, Hodgdon-Pickering Papers, AGO; McHenry to Samuel Dexter, May 24, 1800, Misc. Letters Sent, OSW.

Apparently Latrobe and Perkin discussed the matter in some detail after Perkin had returned to Philadelphia from his first visit to Harpers Ferry in August 1798. He seemed eager to enlist Latrobe's services, but the secretary of war adamantly resisted the idea because Latrobe was "known to be in the closest sympathy with all the Virginia democrats." This turn of events pleased Mackey for two reasons. First, he felt perfectly capable of laying out the armory buildings without the help of an architect, least of all a Jeffersonian. Second, it served as a rebuff to Perkin, whom he had disliked intensely ever since the superintendent's arrival at Harpers Ferry. With Latrobe disposed of and Perkin put in his place, Mackey had a relatively free hand not only in designing the workshops but also in engaging whomever he pleased to do the building and supply the raw materials.[31]

Much to Mackey's chagrin, actual construction of the factory and public waterway at Harpers Ferry did not begin until the second week in May 1799. Two months earlier, on March 12, he had expressed eagerness to start the project: "All the materials for the buildings have been secured by contracts. Carpenters, Brickburners, Bricklayers are engaged—in a word every thing preparative to our operations is done, but a beginning cannot be made until the course of the Canal be accurately marked, for by this the site of the buildings is to be determined." Mackey repeatedly "urged the necessity of sending Mr. Brindley forward" in several letters to the War Department. But prior commitments in the east prevented Brindley's arrival at Harpers Ferry until April 28. Once there, the Englishman fulfilled his obligations in a relatively short period of time. He finished staking out the limits of the armory canal within ten days and left again for Philadelphia. Returning to the site on August 9, he began laying the foundation for the Potomac dam. With the help of a small work crew he completed the partial masonry structure in sixty-four days. On October 22 he booked final

31. Talbot Hamlin, *Benjamin Henry Latrobe* (New York: Oxford University Press, 1955), pp. 25, 255–256; Dearborn to Perkin, September 29, 1802, Misc. Letters Sent, OSW; Mackey to Hodgdon, September 17, December 31, 1798, February 21, March 1, 12, 1799, William Small to Hodgdon, March 1, 1799, HFNP.

passage to Philadelphia, thus ending a short but successful association with the War Department.[32]

Construction of the factory buildings presumably did not begin until after May 6, when Brindley completed his survey. Once started, the builders made rapid progress. In less than two months Mackey notified his superiors that "the [smith's] shop is built, the Mill house and factory half built, and the materials for the Arsenal nearly in readiness. Everything goes on well." Six months later, on December 26, 1799, he triumphantly reported the completion—exclusive of machinery—of the main armory building. His description is one of the earliest made of the government works at Harpers Ferry:

The Smith shop and Factory are finished. The Arsenal is built but not entirely finished. Two upper floors are yet to be laid. All, even my numerous Democratic Enemies, agree that the Buildings are elegant.

The Smith Shop is 80 by 26 ft. clear of the walls; The Factory 120 by 26 feet, and the Arsenal 120 by 27 feet within. The [smith] shop contains 10 forges, one of which is designed for a Tilt-hammer; The ground floor of the factory is designed for the boring, grinding & polishing machinery; the filers and stockers will occupy the upper floor. The Garret will receive Gun Stocks lumber etc. The Mill is almost built, but will not be put together until the Water is in the Canal. The Arsenal has three floors. This building stands within the confluence of [the Potomac and Shenandoah] Rivers.[33]

Unfortunately work on the armory canal and dam did not fare nearly so well. From the outset Mackey had insisted that digging the canal required "no ingenuity" and could easily "be perfected by Men of Industry" residing along the Northern Neck. By doing so, he argued, the government would avoid the added cost of hiring "professional Men, who commonly make their employers pay for the name." This strategy evidently appealed to the secretary of war because the measure was approved early in 1799.[34]

32. Mackey to Hodgdon, February 21, March 12, 1799, HFNP; Mackey to Hodgdon, March 6, April 3, 19, May 5, 1799, Hodgdon-Pickering Papers, AGO; Simmons to Perkin, June 1, 1799, Letter Book F, First Auditor's Accounts, GAO; Accountant's Office, War Department, Letter Book G, GAO.

33. Mackey to Hodgdon, July 10, December 26, 1799, HFNP.

34. Mackey to Hodgdon, February 21, March 12, 1799, HFNP.

Having assumed personal responsibility for building the canal, Mackey promptly engaged three men from nearby Shepherdstown to supervise the project. As superintendent of construction he chose Abraham Shepherd, a distinguished Revolutionary War veteran, well-to-do businessman, and community leader. Other than his staunch Federalism, local influence, and friendship with Mackey, Shepherd's sole qualifications were that he owned shares in the Potowmack Company and had built two flat boats for the river trade in 1785. Because of his age and frail health, Shepherd made only occasional visits to the site. Mackey compensated for this by employing Robert Whittel, a local dry goods merchant, as assistant superintendent. Whittel more than likely took the job to assure his getting contracts to supply tools and provisions for the workers. Shepherd and Whittel received some instructions from Brindley, but neither possessed the knowledge required to build a canal. Yet, ironically, both men received wages commensurate to those paid the Englishman. Whittel was assisted by John Tully as overseer of laborers and on-the-job foreman. Tully may have had some prior experience working on the Potowmack Canal, but he was little more than a common laborer and certainly not well versed in the technics of heavy construction. From the beginning the undertaking appeared top-heavy with managers, none of whom were qualified to engineer the project.[35]

Finding reliable workers to excavate the canal posed another serious problem. While Mackey acquired the services of carpenters, masons, and other skilled builders with little difficulty, the supply of unskilled labor proved highly unpredictable at Harpers Ferry. Despite the promise of good wages, room, board, and a daily ration of whiskey, only the most hard-pressed individuals would consent to long hours and back-breaking labor along a hot, muddy river bank. Since conditions were poor and disease prevalent, keeping a full crew at work during the summer and fall of 1799 became extremely difficult. At one point Mackey complained, "Such is the habitual laziness of the poor of this county that nothing but absolute want can drive them from home." Local day laborers, chafing

35. Mackey to Hodgdon, May 2, 1799, HFNP; Accountant's Office, War Department, Letter Book G, GAO.

under the discipline and regularity of the work, abandoned their jobs for more appealing and lucrative employment on nearby farms. Itinerants who stopped to work on the canal often went on drunken sprees or gathered their belongings and moved on to other settlements once they had accumulated a few dollars. With such a high turnover rate, it is doubtful whether Tully ever had more than forty hands working simultaneously at the canal site during the summer of 1799. Each episode accentuated the problem of adapting a rural population to new notions of discipline, punctuality, and uninterrupted toil.[36]

As much as Mackey needed unskilled laborers to excavate the canal, he rejected the idea of using slaves. Several considerations made the strategy unfeasible. For one thing, the Potowmack Company's canal-building experience only a few years earlier had suggested many difficulties in recruiting and retaining a sufficient number of bondsmen. Related to this was the reluctance of many masters to place their slaves under strange overseers and unaccustomed working conditions. Here the owners' concern about accidents, health impairment, harsh treatment, and runaways far outweighed monetary considerations, especially when more agreeable situations could be found for their slaves throughout the countryside. Given the limited supply of slave labor, Mackey's only alternative would have been to mix black and white workers on the canal, thus not only striking at basic social conventions within the community but also sowing volatile seeds for racial discord. Even after manufacturing operations began, white workers never faced the prospect of slaves' filling skilled positions at the armory. Throughout the antebellum period an unwritten code existed at Harpers Ferry that carefully segregated white from black labor. No one, not even the most objective managers, ever seriously considered altering the practice. Although a handful of slaves were later employed at the works as carpenters, water carriers, cartmen, and the like, they were usually the wards of armory officers or their friends, working

36. Mackey to Hodgdon, July 17, 1799, Hodgdon-Pickering Papers, AGO; Mackey to Hodgdon, July 10, 1799, HFNP; Samuel Annin to Simmons, April 15, 1807, Letters Received by the Accountant for the War Department, First Auditor's Accounts, GAO.

either in separate gangs or at special tasks set by white supervisors. Slaves, in short, played a very marginal role in building and maintaining the armory; they played virtually no role at all in its manufacturing operations.[37]

Periodically Mackey made progress reports on the canal, but their contradictory nature tended to confuse rather than edify the secretary of war. On July 10, 1799, for instance, he notified the department, "We have at this time finished fully 1/2 of the Canal." A week later, without any word of explanation, he reported "about 1/3 of the Canal is dug, and should we be able to collect & retain 100 [men] for three months, there is little reason to fear for the completion of this business this year." Had Mackey made good his prediction perhaps McHenry would have ignored the apparent inconsistencies in his reports. But the result of six months' labor proved extremely disappointing. When the project closed for the winter of 1800, the paymaster wrote apprehensively to his friend and benefactor, Hodgdon: "The Bank and Walls of the Canal on *the land* 1/4 mile in length are nearly done, but 3/4 of the whole is yet to be puddled. A third of the dam is done. 100 men in 3 months will complete the *whole*. . . . If I should continue here (which is very doubtful) I shall renew our operations in the spring, so that we may have the business done about the middle of July."[38]

Intent upon expanding his administrative authority, Mackey had sought and accepted full responsibility for the construction of the physical facilities at Harpers Ferry. He had also assured superiors in Philadelphia that the entire plant would be completed and ready for business by the end of 1799. Since the armory depended upon water conveyed by the canal to power its machinery, no comprehensive manufacturing operations could be conducted until the canal was finished. When this did not happen on schedule the entire installation came to a standstill.

Lack of progress on the canal added to a growing dissatisfaction

37. Proceedings of the Board of the President of the Potowmack Company, NPS; Mackey correspondence, 1799–1801, First Auditor's Accounts, GAO; Payrolls and accounts, Harpers Ferry armory, 1816–1852, Second Auditor's Accounts, GAO.

38. Mackey to Hodgdon, July 10, December 26, 1799, HFNP; Mackey to Hodgdon, July 17, 1799, AGO.

with Mackey's performance as paymaster. By December 1799 it had become apparent to residents of Harpers Ferry that his conduct left much to be desired. While harassing those who would not do his bidding, Mackey exhibited favoritism by awarding friends lucrative supply contracts. His haughty treatment of both skilled and unskilled workmen had caused severe labor problems, including a strike at the armory in February 1799.[39] He had not only clashed with Perkin and his armorers over the payment of wages and the distribution of rations, but had become involved in a mud-slinging political feud with local supporters of Jefferson. The latter confrontation occurred after Mackey had fired a worker named William Small and confiscated some government property at Shepherdstown. Well aware of his unpopularity, the neurotic paymaster attempted to defend his actions by arguing that a massive conspiracy was afoot. "Never was a Man more universally hated than I am at

---

39. At issue were complaints that Mackey was making a handsome profit at the armorers' expense by supplying them with rotten rations and pocketing the additional money that good food would have cost. Antagonism grew to such an extent that on February 22, 1799, the workers walked off their jobs in protest. Although they had Perkin's support, Mackey adamantly refused to submit to their demands for fresh food. Referring to the armorers as "wretches who are destitute of every thing but vicious habits," he vividly described the incident in a letter to Hodgdon on March 1, 1799:

> In my last letter to you, I have taken notice of some improper conduct of people at this place: the seeds of sedition which were then sown, have since sprouted into open rebellion. The Mechanics have on friday 22 deserted the workshop on a signal being given for that intent, and retired to the Tavern, where after having farcically imitated a Committee, they drew up a declaration addressed to me, that they must have Cash in lieu of rations. As this was done at a time when 10,000 lb of Beef and pork were on hand; and further as it was promoted by the contrivance, and for the convenience, of about 6 men, but especially as it was done in violation of the spirit and letter of their agreement, were I a party in that contract, their ringleaders, Ja. Nicholson the principal Engine of Sedition, Charles Williams and Henry Martin alias Bewzer should have been instantly secured and lodged in the County prison. But because I had no power of that kind, necessity directed me to the use of means less efficacious for the coertion of them. They have been literally starved into a promise of drawing rations till the first of June 1799; and have on the Tuesday following returned to their occupations.

See Mackey to Hodgdon, February 21, March 1, 12, 1799, Small to Hodgdon, March 1, 1799, HFNP; Testimonial of J. Mackey, March 4, 1799, Mackey to Hodgdon, May 5, 1799, Hodgdon-Pickering Papers, AGO.

this moment throughout the Country," he wrote the Intendent General of Military Stores in July 1799:

Being an European, having taken the military Stores at Shepherdstown from the disaffected, and having dismissed Small are the great sources of malice, defamation and injustice with which I have been persecuted. The levity of Small made him a suitable Engine in the hands of Democrats for embarrassing me. He first had me bound by a Bench of those Gentlemen in a penalty of 1000 Dollars to keep the peace tho' I had never violated it. He then circulated reports that I had beaten a pregnant woman to abortion, shot a negro Man, attempted to ravish his own wife Mrs. Small, and had filched a poor Man somewhere in the Western Counties of all his money.[40]

Though Mackey vehemently denied these charges, irreparable damage had been done to his character. In less than a year he had gained the reputation among armorers and townspeople as a reckless, self-aggrandizing administrator who placed petty interests and party loyalties above all else at Harpers Ferry. Even Perkin, an undemonstrative person who usually refrained from open confrontations, joined the opposition. As popular disenchantment increased, he pressed the War Department for an official investigation of the paymaster's conduct. "He can bar no check on his ambition," the superintendent cautioned, "and I am the ondley one that stands in his way."[41]

Criticism of Mackey also came from members of his own political party, including the commanding general of the provisional army, Washington. In October 1799 the ailing leader complained about the paymaster's failure to make adequate preparations for billeting three regiments at Harpers Ferry. The sternest censure of Mackey, however, came from William Simmons, the ill-humored accountant for the War Department. Throughout the summer and fall of 1799 Simmons scolded Mackey for neglecting to discharge the duties of his office with efficiency and punctuality. He repeatedly

40. Mackey to Hodgdon, July 10, 1799, HFNP. See also Mackey to Hodgdon, August 29, 1799, HFNP; Hodgdon to Mackey, September 2, 1799, Letters Sent, OQG.
41. Perkin to Hodgdon, November 26, 1799, Hodgdon-Pickering Papers, AGO.

urged the paymaster to keep more systematic records and make monthly reports of his activities to the War Department. Mackey simply ignored these requests and continued to submit poorly prepared and incomplete accounts. Lacking certified vouchers, receipts, and inventories, his abstracts of expenditures became almost impossible to audit. After months of futile endeavor to effect reform, Simmons, like Perkin, called for an investigation of the paymaster's affairs. Whether Mackey knew of the impending inquiry or whether it influenced his decision to retire is not known. Whatever the case, he submitted his resignation around January 23, 1800, pleading ill health. Two months later he relinquished his post at Harpers Ferry and left for Philadelphia to close his accounts. With his departure the first tumultuous period in the armory's history came to an end. Although the next few years proved relatively tranquil, personal controversy and political confrontation continued to be an integral part of what may be termed the Harpers Ferry syndrome.[42]

On April 22, 1800, McHenry named Samuel Annin, a former comrade in arms from New Jersey, as the new paymaster. Like his predecessor, Annin had charge "of the Stores, together with the Superintendence of the erection of all the buildings, Dams & other business" at the armory. During his tenure in office "other business" covered a multitude of activities ordinarily assigned to the superintendent. They ranged from the purchase of supplies to, at times, the hiring and firing of workers. Before 1815 the War Department made no attempt to formulate a clear definition distinguishing the duties of the paymaster and the superintendent. In both official and unofficial correspondence the paymaster was frequently addressed as "the superintendent of the public works" at Harpers Ferry. Thus, with respect to the governance of the armory, the jurisdiction of Annin and Perkin was nebulous and overlapping at

42. Washington to Parker, October 27, 1799, PGW; Simmons to Mackey, July 16, October 17, December 20, 1799 (Letter Book F), March 20, April 1, 1800 (Letter Book G), First Auditor's Accounts, GAO; Hodgdon to Mackey, December 21, 1799, February 1, 1800, to Perkins, July 30, 1800, to Annin, September 24, 1800, Letters Sent, OQG.

many points. Fortunately, however, relations between the two men remained cordial and were not marred by the discord that befell the works during Mackey's administration.[43]

Upon arriving at Harpers Ferry on May 1, Annin immediately turned his attention to the uncompleted armory canal. In a letter of instructions the secretary of war had left it up to him to determine whether Shepherd should be retained as superintendent of the project. After making some inquiries, Annin decided to continue Shepherd but at a substantial reduction in salary. This decision upset Shepherd, who refused to "take a cent less per day than he had rec'd from Mr. John Mackey." Immediately after this episode Annin reviewed Mackey's accounts and discovered that Shepherd had defrauded the government of $185 by falsifying pay vouchers for thirty-seven days during the summer of 1799. When this disclosure reached the War Department, Shepherd's fate was sealed. Simmons vigorously denounced "the impositions practised against the United States" and demanded that Shepherd make restitution of every penny he had illegally drawn from the government coffers. Annin, unable "to find a suitable person to take the charge," had no choice but to assume personal supervision of the canal.[44]

Once he had assessed the situation, Annin sought to ensure a steady supply of laborers on the canal by proposing what Washington had previously suggested, that is, the employment of the federal troops stationed at Harpers Ferry. But Washington had since died and General Pinckney grumbled that "soldiers should not dig a canal but study tactics." McHenry personally liked the idea "from motives of economy" but thought it best first to obtain direct presidential authorization for the measure rather than to alienate Pinckney. After some delay John Adams issued an order on May 22, 1800, directing "that as many soldiers shall be employed on the Canal . . . as the good of the service will permit." Each man

43. Simmons to Annin, May 14, 1800, Letter Book G, GAO; Annin to William Eustis, June 28, 1810, Letters Received, OSW; James Stubblefield to Bomford, October 5, 1822, Letters Received, OCO.

44. Annin to Eustis, June 28, 1810, OSW; Dearborn to Annin, May 11, 1801, Misc. Letters Sent, OSW; Simmons to Annin, July 24, 1800, June 1, 1801, Letter Book G, GAO.

who volunteered for the work detail received sixteen cents a day and double rations. Initially fifty soldiers and several sergeants reported for duty at the excavations. By June 1 the work force had grown to one hundred men and was making rapid headway puddling and banking the canal. More than likely, the waterworks would have been finished during the summer of 1800 if this number had been maintained. But, with the disbanding of the provisional army on June 14, Annin lost his guaranteed labor supply and again had to resort to private workers. As a result, progress slackened and the canal was not completed until 1801.[45]

By the time the sluice gates were opened on the canal, Perkin and his men had occupied all but one of the armory buildings over a year. Completing the complex was a mill for heavy forging, erected late in 1800. Like the others, it was constructed of brick masonry and powered by a breast-type water wheel with a fall of twelve feet. In all, the factory had five water wheels averaging 4.2 feet in breadth. The largest was an eight-foot wheel at the forging mill which supplied power to a tilt hammer for working iron and preparing barrel skelps for hand welders. Yet, even though the shops were finished on schedule, the quality of their construction appears to have been shoddy. After less than two years of use, Perkin reported that they were badly in need of repair.[46]

Since these structures represented only the outer shells of the armory, the interiors had to be fitted with millwork and machinery before continuous operations could begin. Making these fittings, in turn, depended largely upon the successful operation of the armory canal, for water wheels were required to power forge hammers, grinding wheels, and other cumbersome equipment. Even after the waterworks and machinery were completed, however, the canal leaked so badly that very little heavy work could be performed on gun barrels and other iron components. Until this situation was

45. Steiner, p. 424; McHenry to Dexter, May 24, 1800, Dearborn to Annin, May 8, 1801, Misc. Letters Sent, OSW; Orderly Book Division Orders, May 22, 29, 1800, JAG.

46. N. King and L. Harbaugh, "Plan of the proposed junction of the Canal . . . at Harpers Ferry," February 1803, Map File, NPS; "Mr. Brindley's Sketch of the Wheels . . . Necessary for the Potomac Works," August 4, 1801, HFNP; Dearborn to Perkin, October 26, November 10, 1801, Misc. Letters Sent, OSW.

remedied, Perkin and his armorers necessarily spent more time making preparations and solving technical problems than actually practicing their trade.[47]

47. J. Newman to Annin, August 4, 1801, Dearborn to James Green (Greer?), March 21, 1803, to Perkin, May 25, 1804, Misc. Letters Sent, OSW; Perkin to Dearborn, July 31, 1801, August 8, 1803, Registers of Letters Received, OSW.

CHAPTER 2

The Craft Origins of Production, 1798–1816

It is difficult to determine exactly when the Harpers Ferry armory became sufficiently manned and equipped to launch full-scale production. Some writers infer from "manufacture of arms" accounts that the factory may have been turning out large numbers of weapons as early as 1798. These accounts, however, comprise expenditures not only for arms actually made at Harpers Ferry but also for the *repair* of arms manufactured elsewhere. Also included are outlays for the purchase of raw materials and the payment of labor. As such, they are officially referred to in War Department annals as "expenditures for the manufacture and repair of arms." Just because Treasury records reveal expenditures at Harpers Ferry from these accounts does not mean that the armory was actually producing arms in 1798, 1799, or even 1800. Hundreds of man hours and thousands of dollars first had to be spent on tooling up the works before manufacturing operations could begin.

On May 24, 1800, James McHenry informed Samuel Dexter, his successor at the War Department, that "the commencement of the fabrication of Arms, has been some time made" at Harpers Ferry while noting at the same time, "much however yet remains to be done, to carry it to maturity." Whether this may be interpreted as signaling the beginning of production is again doubtful. The term "fabrication" had several connotations during the early nineteenth century and, in common usage, could mean one of three things: to assemble, to repair, or to make new. An early inventory indicates that, except for some woodworking tools, very few armorers' supplies had been accumulated at Harpers Ferry by March 1800; nor as late as October had a beginning been made on the construction of machinery. Because power facilities were limited at that time, the "fabrication" must have been restricted to the piecemeal repair and assembly of damaged arms, not the production of new ones. Perkin himself admitted as much when he informed the secretary of war in August 1801, that his men were still engaged "chiefly on old arms."[1]

Although collectors readily concede that smooth-bore muskets bearing Harpers Ferry markings earlier than 1808 are difficult to find, the existence of three specimens dated 1800, 1802, and 1803 has led to the conclusion that the armory "regularly produced" muskets of the French Charleville pattern with minor variations until 1816. To a certain extent this is true. However, smooth-bore muskets did not constitute a "regular" product at Harpers Ferry until 1808. While the armory undoubtedly made several thousand muskets between 1801 and 1808, output was sporadic and certainly not determined to any extent by specific orders from the War Department. Over a four-year period from 1804 through 1807, for instance, deliveries to Samuel Annin at the U.S. arsenal totalled only 342 muskets. During the same period the Springfield armory finished 14,811 muskets. The reason for this meager record is that the armory devoted its resources to the manufacture of a new and more exciting military product—the "short" rifle.[2]

A small quantity of full-stocked rifles may have been made at Harpers Ferry on an experimental basis in 1800 or 1801, but the War Department did not issue an official directive for the preparation of arms of that type until May 25, 1803. On that date Secretary of War Henry Dearborn informed Perkin of his dissatisfaction with the weapons currently issued to regular troops and of his decision

1. *American State Papers: Military Affairs*, 2:481; McHenry to Dexter, May 24, 1800, Misc. Letters Sent, OSW; Perkin to Henry Dearborn, August 4, 1801, Register of Letters Received, OSW; Harpers Ferry armory, March 20, 1800, Misc. War Department Statements, GAO; Samuel Hodgdon to Perkin, August 13, 1800, to Samuel Annin, October 10, 1800, Letters Sent, OQG.
2. Cf. Stuart E. Brown, Jr., *The Guns of Harpers Ferry* (Berryville, Va.: Virginia Book Co., 1968), pp. 13–14, 18; *American State Papers: Military Affairs*, 2:478, 481; 1:679. See Table 1.

"to have a suitable number of judiciously constructed Rifles manufactured at the Armory under Your direction." Dearborn
specifically stipulated that the new model should be shorter in
length, half-stocked, and of a heavier caliber than the ordinary
common rifle, adding that "I have such convincing proof of the
advantage the short rifle has over the long ones (commonly used) in
actual service as to leave no doubt in my mind of preferring the
short rifle, with larger calibers than the long ones usually have."[3]

Upon receipt of these instructions Perkin and several other armorers set to work designing what eventually became known as the
U.S. Model 1803 "short" rifle (Figure 1). Late in November they
submitted several patterns to the War Department for inspection,
one of which was approved with only "very trifling alterations."
Orders were received to make 2,000 rifles; at the same time Dearborn complimented the superintendent for having submitted such
"an excellent pattern." Once production began he seemed so impressed with the performance of the "iron-ribbed" rifles that he
increased the original parcel to 4,000 and in November 1805 asked
Perkin to design and eventually produce a horseman's pistol which
in many details resembled a scaled-down version of the new model
rifle.[4]

Initially Dearborn and Perkin had planned to deliver 2,000 rifles
a year and fill the order by December 1805. Chronic mechanical
difficulties coupled with additional handwork necessary to complete the highly finished weapons, however, tended to slow the rate
of production. In addition, outbreaks of malaria during the summers of 1805 and 1806 caused a labor shortage that seriously taxed
armory operations. Many workmen were stricken with the dreaded
disease or abandoned their jobs to escape to a healthier climate. As
a result, the last parcel of rifles was not completed and turned into
store until February 1807, two months after Perkin himself had
fallen victim to the so-called "bilious fever."[5]

3. Quoted by James E. Hicks, *Notes on United States Ordnance*, 2 vols. (Mt.
Vernon, N.Y.: By the author, 1940), 1:25.

4. Dearborn to Perkin, December 2, 1803, November 1, 1804, November 13,
1805, February 26, 1806, Misc. Letters Sent, OSW; Perkin to Dearborn, June 6,
1803, March 6, 1804, Registers of Letters Received, OSW.

5. *American State Papers: Military Affairs*, 1:199; Annin to Dearborn, February
23, 1807, Letters Received, OSW; Willhelmina Perkin to Secretary of War, December 2, 1806, Registers of Letters Received, OSW.

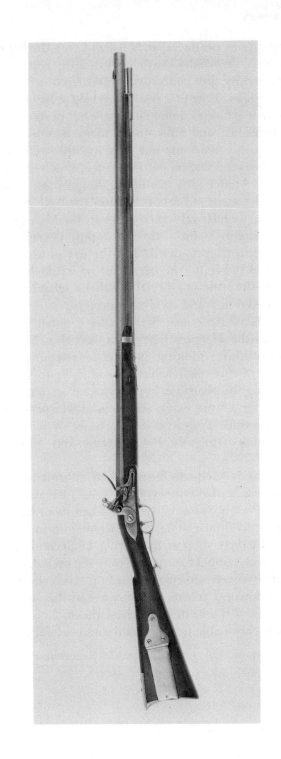

Figure 1. U.S. Model 1803 rifle. Photograph courtesy of the Winchester Gun Museum.

Of all the weapons produced at Harpers Ferry during its sixty-three-year history, the Model 1803 rifle is in many respects unique. It has been prized by gun collectors, popularized in folklore, and studied by historians as the first regulation rifle to be manufactured at the government armories; much literature has been devoted to its ballistics, dimensions, and subsequent usage by military forces. Since Perkin and his assistants not only incorporated Dearborn's ideas but added several features of their own, much of the interest generated by the Model 1803 lies in the design itself. The simple but elegant employment of brass furniture, the well-defined comb of the butt stock, its relatively narrow wrist, the high cheek piece, the large brass patchbox, the wide and slightly curved butt plate, the ornate semigrip trigger guard, and the use of a retaining pin instead of a band to secure the barrel to the stock—these embellishments give the rifle a special aesthetic character directly reflecting the preferences and style of its makers.

Many students of American firearms have noted striking similarities between the Harpers Ferry rifle and the "Pennsylvania-Kentucky" gunsmithing tradition. No one has attempted to explore the implications of this relationship, however, or to suggest what seems to be a highly plausible hypothesis for it, namely, that a significant number of the early armorers at Harpers Ferry had either plied their trade as independent artisans or served long apprenticeships to craftsmen in Pennsylvania and Maryland gun shops.

To test this thesis adequate biographical information about the men who worked at the armory during its formative period is needed. Joseph Perkin was a master craftsman who, by 1792, had an established reputation among Philadelphia gun makers for fine workmanship. In their shop at the corner of Second and Spruce streets, Perkin and Samuel Coutty not only repaired arms but also, according to their advertisements, carried on "Gun and Pistol making in all its branches, where gentlemen may be supplied with Guns and Pistols of the neatest and best quality, on the shortest notice and most reasonable terms."[6] Although several pistols and a

6. *Pennsylvania Gazette*, May 2, 1781, quoted by Kauffman, *Pennsylvania-Kentucky Rifle*, p. 319.

long-barreled fowling piece bear Perkin's signature, no rifle has yet been identified as his work. He probably made rifles for private clients and, perhaps, even for the western trade, however. The general format of the Model 1803 amply shows that he and his assistants were well acquainted with the design and construction of Pennsylvania rifle guns.

From its establishment in 1798, the Harpers Ferry armory profited from the approximately fifteen skilled workmen Perkin brought with him from the federal arsenal in New London, Virginia. With the exception of three or four native Virginians, all had at one time or another resided in Philadelphia and been recruited there. Some, similar to the superintendent himself, had "been breed up among and . . . conducted some of the Principle Gun & Pistol Manufactorys in Great Britain." Others, such as George Metts, Charles Williams, and Perkin's son, Henry, had trained under Philadelphia gun makers. Together they represented Perkin's "principal dependance," a core group upon whom he relied heavily in laying the foundations of the Harpers Ferry armory.[7]

With the start of construction in 1799 and the gradual onset of manufacturing, Perkin sought to expand his labor force. After little success in finding qualified armorers along the Northern Neck, he naturally began to look elsewhere for recruits. As in the past, Philadelphia offered the best prospects. Its reputation as a great craft center and disembarkation point for newly arrived immigrants made it a natural repository of skilled gunsmiths. Moreover, having resided in the city over ten years, Perkin was thoroughly familiar with its commercial environment and had many established contacts there. Relying on this knowledge as well as the assistance of Samuel Hodgdon, the Intendant General of Military Stores, he succeeded in attracting about ten workmen to Harpers Ferry between 1799 and 1801. Among them was Thomas Anneley, a "compleat artificer" in every respect and first master armorer at Harpers Ferry; William Gardner, a highly skilled engraver; Miles Todd, an

7. John Miles to Hodgdon, July 26, 1794, Perkin to Hodgdon, April 27, 1794, Armorers Rough Articles, November 22, 1798, Hodgdon-Pickering Papers, AGO; William Simmons to John Mackey, December 20, 1799, Letter Book F, First Auditor's Accounts, GAO.

experienced barrel forger; Frederick Oswan, an able gunsmith; and James Greer, a highly recommended machinist whom Perkin lured from Robert McCormick's Globe Mills armory in the Northern Liberties section of the city.[8]

While Philadelphians undeniably exerted a strong influence at Harpers Ferry between 1798 and 1816, other Pennsylvania-trained artisans were employed at the government works during this period. Many of the highest paid and most skillful workmen came from towns and hamlets lying beyond the western reaches of the Quaker City. In fact, a collective profile of early armory personnel reveals the existence of a migratory line extending along the "Great Wagon Road" in a southwesterly direction from Philadelphia through Lancaster, York, Hanover, and Frederick to the mouth of the Shenandoah River and the fertile valley beyond. Just as scores of settlers moved westward over this busy trade route throughout the eighteenth century, armorers and their families made their way along this main thoroughfare to Harpers Ferry during the opening decades of the following century.

To narrate the personal history of each worker who followed the Great Wagon Road to Harpers Ferry would entail a tedious biographical roll call and needless repetition. Suffice it to say that in 1816, the first year for which there are adequate records, thirty-three of eighty-four armorers earning monthly wages over forty dollars had either worked or served apprenticeships in Pennsylvania and Maryland gun shops prior to their employment at the national armory. Although this represents 39 percent of the skilled labor force at Harpers Ferry, it is an extremely conservative estimate. First, the figure includes only those armorers who are actually known to have come from the Pennsylvania and Maryland arms industry. Second, it excludes at least twenty-five men who may well have come from similar craft backgrounds but, for lack of positive evidence, cannot be included in the estimate at this time.

8. Simmons to Mackey, December 20, 1799, GAO; Hodgdon to Annin, September 1, October 10, 1800, Letters Sent, OQG; Dearborn to Perkin, May 20, 1801, Misc. Letters Sent, OSW; Greer to Colonel Decius Wadsworth, September 24, 1818, Letters Received, OCO; Kauffman, p. 319; Arcadi Gluckman and Leroy D. Saterlee, *American Gun Makers* (Harrisburg, Pa.: The Stackpole Co., 1953), p. 8; Robert E. Gardner, *Small Arms Makers* (New York: Bonanza Books, 1963), p. 7..

Even taken alone, however, the 39 percent figure is impressive, particularly because it includes four of the five shop foremen at Harpers Ferry in 1816. Though subordinate to the superintendent and master armorer, these men played an increasingly important role in armory affairs after 1808, not only overseeing production but often ruling on the adoption of new techniques.[9]

The roster of early armorers at Harpers Ferry reads like a "Who's Who" of Pennsylvania gun making. Michael Gumpf, for example, was a highly paid stocker who came from a family of eight Lancaster County gunsmiths. George Zorger learned the trade while working under his father, Frederick, one of York's outstanding artisans. Jacob and George Hawken, John Kehler, and George Nunnemacker came from similar family backgrounds. Others such as Martin Breitenbaugh, Philip Hoffman, and Christian Kreps had served apprenticeships under prominent Pennsylvania and Maryland craftsmen to become masters in their own right. In short, the careers of all these men correspond to the extent that they reveal a preponderant Pennsylvania arms-making influence.[10]

Probably the most notable armorer to work at Harpers Ferry before 1816 was Marine T. Wickham. Born and raised in Frederick County, Maryland, he had "served a regular apprenticeship to the business of Gun Making" prior to taking a job at the armory in the summer of 1804. Though Wickham was a product of the so-called

9. Payrolls and accounts, Harpers Ferry armory, 1816, Second Auditor's Accounts, GAO.

10. Dearborn to James Stubblefield, October 5, 1808, Misc. Letters Sent, OSW; Simmons to Annin, October 30, 1810, Letter Book Q, First Auditor's Accounts, GAO; Tobias Lear to Lloyd Beall, October 15, 1815, Letter Book 7, GAO; Stubblefield to Colonel George Bomford, August 6, 1816, July 11, 1829, Philip Hoffman to John C. Calhoun, April 4, 1823, Nahum W. Patch to Bomford, January 25, 1826, William Malleory to Secretary of War, July 13, 1835, Martin Breitenbaugh to Colonel George Talcott, April 15, 1845, Letters Received, OCO; "Proceedings of a Board of Officers: Inspection of Harpers Ferry armory, February 5 to 22, 1842," Reports of Inspections of Arsenals and Depots, OCO; Payrolls and accounts, Harpers Ferry armory, 1816–1817, Second Auditor's Accounts, GAO; Joe Kindig, Jr., *Thoughts on the Kentucky Rifle in Its Golden Age* (Wilmington, Del.: George N. Hyatt, 1960), pp. 51, 285–286, 337–338, 438–439, 443; Henry J. Kauffman, *Early American Gunsmiths, 1650–1850* (Harrisburg, Pa.: The Stackpole Co., 1952), p. 12; Kauffman, *Pennsylvania-Kentucky Rifle*, pp. 246–248, 274, 367; Gluckman and Saterlee, p. 90.

"Emmitsburg School" of gunsmiths, we do not know the name of his early mentor. It may have been Jacob Metzger, a Lancaster-trained rifle maker who moved to Frederick sometime between 1777 and 1788, or perhaps John Armstrong, a highly skilled gun-smith from Emmitsburg, Maryland. Armstrong definitely had con-nections at Harpers Ferry because his eldest son, William, later served as an inspector at the armory during the War of 1812.[11]

Equally adept at forging, filing, stocking, and engraving firearms, Wickham possessed great talent as a gunsmith. More im-portant, he exhibited a rare ability to manage men effectively while at the same time retaining their admiration and respect. These qualities did not go unnoticed, for in 1808 the secretary of war selected Wickham to succeed Perkin's long-time associate, Charles Williams, as master armorer at Harpers Ferry.[12] In this capacity he played a key role in planning a large expansion of the government works between 1808 and 1810. Yet, after less than three years as master armorer, Wickham became disenchanted with his position at the armory and late in 1810 announced his "intention of remov-ing to the Western Country." Before departing, however, Wickham accepted another government position in eastern Pennsylvania, of-fered by the secretary of war.[13]

Residing in Philadelphia, Wickham served as an inspector of contract arms and general troubleshooter at the national armories between 1811 and 1816. During these years he gained a widespread reputation among associates as a "superior artist" and played a prominent part in designing the new Model 1812 regulation mus-

11. Stubblefield to Dearborn, June 20, 1808, Letters Received, OSW.

12. Williams, one of the first workmen at Harpers Ferry, had succeeded An-neley as master armorer on May 20, 1801. After Perkin's death in 1806, he applied for the superintendency at Harpers Ferry. Although the secretary of war ap-pointed James Stubblefield of Stevensburg, Virginia, instead, Williams continued as master armorer until December 22, 1808, when he went to Philadelphia as an inspector of contract arms under Tench Coxe, the purveyor of public supplies. He held this position until April 1814, when he resigned to enter private business. Williams was succeeded by his assistant, Henry J. Perkin, the son of Joseph Perkin.

13. Stubblefield to Eustis, February 25, March 7, 1811, Letters Received, OSW; Eustis to Stubblefield, February 13, March 1, 1811, to Wickham, March 15, 1811, Letters Sent Relating to Military Affairs, OSW.

ket. But, as so often is the case with men of ability in government service, Wickham decided to enter the world of private business after the War of 1812. Upon resigning his inspectorship in 1816, he took over an existing musket contract with the United States and later that year signed his first regular contract for an additional 4,000 arms at $14 a stand. The following year he opened a mercantile house in Philadelphia for the importation of "every European Article and Material used in the public and private manufacturies of Arms." Not surprisingly two of his largest accounts were the Harpers Ferry and Springfield armories. Both ventures prospered and Wickham continued to garner wealth from them until his death in 1834. His rise from an obscure country workshop to a leading arms manufacturer and businessman in Jacksonian America was a truly impressive achievement.[14]

Like Frederick, the village of Hagerstown lay along the Great Wagon Road and became an important recruiting ground of skilled labor for Harpers Ferry. Although many Hagerstown natives worked at the armory during the early nineteenth century, none overshadowed John Resor in terms of lineage and training. As the seventh child born to Peter and Catherine Welshans Resor in 1789, his family roots went to the very heart of Pennsylvania craftsmanship. His paternal grandfather, Matthias Roesser, was one of the earliest gunsmiths in Lancaster County. His maternal grandfather, Joseph Welshans, was a well-to-do mill owner and gunsmith from York. On both sides of the family there were uncles and cousins who made firearms in the finest Pennsylvania tradition.

Like his elder brother, Jacob, John Resor had already served an apprenticeship in his father's shop at Hagerstown before going to Harpers Ferry in 1809. For nearly five years he worked mainly as a gunstocker alongside several other Pennsylvania-trained artisans.

14. Wickham to Roswell Lee, February 10, March 24, 1817, Letters Received, SAR; Wickham to Eustis, April 1, 2, 1811, Letters Received, OSW; Callender Irvine to Bomford, May 13, 25, July 29, 1815, Lieutenant James Baker to Captain John Morton, November 8, 1815, Letters Received, OCO; Contract Books and Statements of Contracts, 1816–1829, OCO; Payrolls and accounts, Harpers Ferry armory, 1818–1820, Second Auditor's Accounts, GAO; "Notes of a Tour of Inspection Commencing 10 December 1822, Armory Harpers Ferry," Inspection Reports, OIG.

As someone "well acquainted with the construction of rifles" he was one of three men chosen in 1814 to introduce similar techniques at the Springfield armory. Upon his return to Virginia in April 1815, the superintendent promoted him to the salaried position of foreman and inspector of finished gunstocks. Several years later he relinquished these duties to assume responsibility for the manufacture and inspection of forged gun barrels, one of the most demanding as well as frustrating jobs at the armory. Resor continued in this capacity until his resignation in December 1831. While his career lacked the spectacular successes of Wickham's, it is of particular interest because Resor's lengthy tenure at the armory indicates how Pennsylvania arms-making traditions became so deeply infused at Harpers Ferry during the opening decades of the nineteenth century.[15]

In addition to those armorers who trained in Pennsylvania and Maryland shops prior to their arrival at Harpers Ferry, other workmen reflected the same traditions though in a more subtle manner. In their case the vehicle of transmission was the armory's apprentice system; through it Pennyslvania practices flowed from one generation to another and were thereby perpetuated at the works. In 1807, James Stubblefield, Perkin's successor as superintendent, provided a capsule description of the system in a letter to the secretary of war: "The condition upon which they have been taken heretofore has been for them to serve from the time they are bound untill of age, and they receive the compensation of $12 per month during their apprenticeship which sum is intended to Board and Cloath them, and as soon as they are able to file a lock, in order to encourage them to be industrious they have a task of 17 Locks per Month, and whatever they complete more than that number they are paid a full price. for.[16] While the relatively novel idea of incen-

15. Kauffman, *Pennsylvania-Kentucky Rifle*, pp. 329–332, 359–360; Kauffman, *Early American Gunsmiths*, pp. xii-xiii; James Monroe to Annin, October 3, 1814, Letters Sent Relating to Military Affairs, OSW; Annin to Monroe, October 8, 1814, Letters Received, OSW; Stubblefield to Bomford, April 7, 1815, Letters Received, OCO; Resor to Major William Wade, May 14, 1825, Correspondence and Reports Relating to Experiments, OCO; Lear to Beall, October 10, 1817, Letter Book 6, First Auditor's Accounts, GAO; Payrolls and accounts, Harpers Ferry armory, 1816–1832, Second Auditor's Accounts, GAO.
16. Stubblefield to Dearborn, May 22, 1807, Letters Received, OSW.

tive pay for extra work must have appealed to the parents of many a young lad, apprentice training at Harpers Ferry remained essentially the same as that practiced throughout the United States during the early nineteenth century. In 1800, for example, a fourteen-year-old boy named Daniel Johnson was bound out by his father as an apprentice to Perkin. The articles of agreement stated that Johnson would serve his new master faithfully and remain at the armory until he reached his twenty-first birthday. Perkin, on his part, agreed to teach Johnson "the Art or Mystery of a Gunsmith," provide for his basic education in the three R's, and present him with a new suit of clothes upon completion of the indenture. Unfortunately Perkin did not live to fulfill the contract, and Johnson later complained to the secretary of war that the government had "entirely neglected" his education and had failed to give him a new suit of clothes at the expiration of his apprenticeship. When Johnson received his freedom in 1807, he had the option of either staying at the armory as a regular workman or going elsewhere as a journeyman gunsmith. Apparently the young man did not feel too badly treated because he not only chose to remain at Harpers Ferry as a lock filer but continued on the payroll until 1828.[17]

Between 1800 and 1813 at least thirty boys served apprenticeships at Harpers Ferry. Depending upon circumstances, the length of an indenture usually was from five to seven years. During that time the youth worked alongside his master and, through practical experience, learned the art of forging, filing, stocking, and finishing a well-crafted arm. George Malleory, for instance, served a six-year indenture under the watchful eyes of his father, William, and Perkin. John Avis, on the other hand, spent only five years as an apprentice to Perkin's successor, Stubblefield. Unlike young Malleory, however, he left the armory. On his twenty-first birthday in 1813, he went to Philadelphia as an inspector of contract arms for the commonwealth of Pennsylvania.[18]

17. Annin to Stubblefield, December 21, 1807, Letters Received, OSW; Wadsworth to Calhoun, March 31, 1820, Letters Sent to the Secretary of War, OCO; Daniel Johnson to Bomford, April 24, 1828, Letters Received, OCO.
18. George Malleory to Henry Clay, January 18, 1841, John Avis to Bomford, June 7, July 18, 1816, Stubblefield to Bomford, July 11, 1816, Letters Received, OCO; Bomford to Lewis Cass, July 2, 1836, Letters Sent to the Secretary of War, OCO.

Around 1809, along with the enlargement of the factory and the introduction of piece-rate accounting procedures, the formal apprentice system gave way to a noncontractual training program at Harpers Ferry. This change denoted a fundamental shift away from traditional craft patterns, indicating that with new methods of organization workmen so well versed in every branch of gun making were no longer required. Under the new system established armorers working as inside contractors assumed responsibility for selecting boys as helpers. Other than the payment of wages, no obligation existed on the armorer's part to feed, clothe, or educate his assistants. Nor, for that matter, was he obliged to teach them the secrets of the trade. For the most part, the degree of skill acquired under the new system depended upon the willingness of the master to impart it. In many cases, therefore, the development of expertise remained minimal, severely restricting a youth's mobility from one job to another.[19]

That more than one interested armorer took time to instruct his helpers on various facets of gun making is illustrated by the careers of Zadoc Butt, John H. King, and Jerome Young. All three began as armorer's assistants in the 1820s, labored diligently, and scaled the ladder of workmanship to become masters. Often, however, a lad's family connections proved more essential to success than his aptitude or ability as an armorer. Boys whose fathers worked at the armory, for instance, stood a far better chance of learning the trade and improving their lot than did those who came from other backgrounds. Indeed to accomplished gunsmiths such as Nunnemacker and Zorger, skill represented a legacy as well as a means of livelihood and in many instances was the sole heritage one could impart to his male heirs. Accordingly, armorers hired their sons as helpers and, at the same time, taught them skills so they could one day secure places in the armory as full-fledged workmen. In this sense the apprentice system never really ended at Harpers Ferry. Through it George Zorger, Jr., John Malleory, Alexander Nun-

19. For descriptions of inside contracting see Felicia J. Deyrup, *Arms Makers of the Connecticut Valley*, Smith College Studies in History, vol. 33 (Northampton, Mass., 1948): 101–102, 161–162; and John Buttrick, "The Inside Contract System," *Journal of Economic History* 12 (1952):205–221.

nemacker, and many other young men got their start as gunsmiths. Moreover it was the means by which craft traditions as well as extended ties of kinship became firmly embedded at the armory during the early decades of the nineteenth century.[20]

Instead of undermining rural customs and values at Harpers Ferry, early labor practices reinforced them. Whether one came from the guilds of Birmingham, the Pennsylvania gun trade, or the armory's own apprentice program, certain habits persisted which mixed factory work with other pursuits in many different ways. Here E. P. Thompson's distinction between "time-oriented" and "task-oriented" labor provides valuable insight. In the former, life and work are clearly divorced from one another whereas in the latter, "Social intercourse and labour are intermingled—the working-day lengthens or contracts according to the task."[21] Much like their English contemporaries, the armorers at Harpers Ferry followed rhythmic rather than regular patterns of toil. They could do so because almost every phase of the productive process was highly individualized and subject to few managerial constraints. Control over the tempo of work, in other words, rested with the armorers themselves. In this scheme, strict adherence to clocked time, unvarying routine, and regimented behavior had little importance. Although the factory opened at sunrise and closed at sunset, artisans were free to come and go as they pleased. So long as each individual adequately performed his task, it mattered little when he worked or what he did with his free time. The speed and dexterity with which an armorer could file a lock or forge a barrel therefore determined the length of his work day. In effect, this meant that younger apprentices and journeymen generally spent a greater amount of time at their benches than older more experienced

20. Payrolls and accounts, Harpers Ferry armory, 1816–1830, GAO; "Proceedings of a Board of Officers," February 1842, OCO. It was not unusual for a father and son to work together at the armory prior to 1840. For example, nineteen families had at least two male members working there in October 1816. The Steadman family led all others with seven, followed by the Malleorys with five, the Stipes with four, and the Bests with four.

21. E. P. Thompson, "Time, Work-Discipline, and Industrial Capitalism," *Past and Present* 38 (1967):60. Similar themes are perceptively developed by Sidney Pollard in *The Genesis of Modern Management* (Cambridge, Mass.: Harvard University Press, 1965), pp. 160–208.

hands. Often skilled artisans would achieve their monthly quota in two or three weeks and could devote the remainder of their time to other pursuits such as farming, operating a small business, or making weapons for private customers in the armory shops. Others preferred to alternate between farming and arms making by allocating five or six hours to each task daily. Harvest time abbreviated the factory schedule even further, although armorers usually compensated for this by great bursts of activity before and after the season. For the most part, hard work prevailed at Harpers Ferry; a systematic work regimen did not. The distinction is important, for the fusion of the two formed the backbone of the new industrial discipline.[22]

Preindustrial culture manifested itself in a variety of ways at Harpers Ferry. Local folklore with its spellbinding tales of "Peacher's Wheel," "Wizard Clip," ghostly apparitions, and other instances of the occult reveals the extent to which primitive morality, superstition, and a sense of mystery pervaded daily life. All sorts of diversions punctuated the working day. Armorers habitually suspended their chores to share a cup of whiskey and engage in conversation. Periodically they would throng to the armory yard to watch dogfights and cockfights, bloody fisticuffs between co-workers, and less belligerent wrestling matches. Invariably onlookers placed wagers on the outcome of the contests. On other occasions armorers would leave their stations to listen to itinerant evangelists and stump orators and examine wares being hawked by peddlers. Interspersed throughout the work year were holidays, barbecues, and celebrations. Election days, the hustings, the fourth of July, and other seasonal observances fostered a carnivallike spirit reminiscent of Bartholomew Fair and "St. Monday" in England. Periodic visits by foreign dignitaries and government officials also prompted much fanfare and excitement. These were festive occasions, a time when the entire community turned out to join in general merriment. Often the drinking, dancing, gaming, and visiting went on until the small hours of the morning.

22. Annin to John Armstrong, March 20, 1815, Letters Received, OSW; Payrolls and accounts, Harpers Ferry armory, 1816–1841, GAO; Charles W. Snell, "A Comprehensive History of the Construction, Maintenance, and Numbers of Armory Dwelling Houses, 1796–1869" (Research report, HFNP, 1958), p. 136.

Rather than opposing or discouraging these celebrations, armory officers were often their most enthusiastic supporters. Such activities provided an emotional release and strengthened morale among the workers. As political appointees and former craftsmen themselves, the superintendent and his minions well understood the importance of maintaining harmonious relations within the community. Holiday festivities afforded an excellent opportunity to curry favor with the armorers by mixing with their families, making flattering remarks, and occasionally treating them to free food and drink. Just as life and work were intermixed at Harpers Ferry, so too were employer-employee relations. A conscious sense of reciprocity governed the superintendent's actions. While he demanded a certain degree of deference and obedience from the workmen, they in turn expected to be coddled and not interfered with. Since they were extremely sensitive about their rights and privileges as skilled artisans, particular care had to be taken not to treat them with condescension. No man worth his salt would stand at command or submit to even the most perfunctory regulations unless he was accorded the dignity and freedom that his skilled status deserved.[23]

Over the years such thoroughly inbred and highly individual work habits served to hinder rather than encourage innovation at Harpers Ferry. Because so many armorers and supervisors had been reared according to the conventions of the craft ethos, they found it extremely difficult to adjust to the increasingly specialized demands of industrial civilization. Except for using commonly known forging, grinding, polishing, boring, and rifling machines, they relied mainly on the dexterous use of hand tools to perform their work. Unlike their more flexible contemporaries in New England, they possessed a high degree of manual skill and saw no need to compensate for what little they lacked by cultivating mechanical know-how. Above all, they considered themselves artisans, not machine tenders, and, as such, believed in the dictum that an armorer's task consisted in making a complete product—lock, stock,

---

23. See, for example, Barry, *Strange Story of Harper's Ferry*, pp. 30, 178–200; Major John Symington to Captain William Maynadier, July 12, 1849, Letters Received, OCO; Colonel James W. Ripley to Colonel Henry K. Craig, April 14, 1859, Reports of Inspections of Arsenals and Depots, OCO.

and barrel. Ideologically speaking, then, they had neither the preparation nor, it seems, the inclination to introduce new techniques at the armory. In a sense they were too skilled, too artistically inclined, too satisfied with the way things were to sanction change or contemplate seriously its possible benefits.[24]

24. The voluminous correspondence of the Ordnance Department and the Springfield armory amply corroborate this point. For an interesting comparison of differences between New England and Pennsylvania arms contractors, see Talcott to Joel R. Poinsett, November 4, 1839, Letters Sent to the Secretary of War, OCO. Summary discussions of New England's penchant for adopting new mechanical techniques are found in Daniel J. Boorstin, *The Americans: The National Experience* (New York: Random House, 1965), pp. 26–34; and Deyrup, pp. 87–99, 117 ff.

CHAPTER 3

Production, Labor, and Management, 1801–1816

Productivity at Harpers Ferry between 1801 and 1806 revealed few signs of growth. After a relatively auspicious start in 1802, output leveled off to about 1,700 arms a year. Lack of reliable water power, poor health conditions, and innumerable problems with the design and manufacture of the Model 1803 rifle considerably retarded armory operations. But perhaps the most telling reason for the lack of productivity gains was the restrictive fiscal program pursued by President Thomas Jefferson during his first administration. Deeply comitted to reducing the public debt, Jefferson's Secretary of Treasury, Albert Gallatin, executed a policy of strict economy in all matters of government spending. Military appropriations particularly became a target of Gallatin's fiscal pruning. With the president's approval and Henry Dearborn's full cooperation, the frugal-minded secretary succeeded in cutting substantially the War Department's budget, thereby reducing the size of the army and limiting the flow of funds to the national armories. Faced with this situation, Joseph Perkin had no choice but to restrict his labor force and to table plans for expanding production.

During Jefferson's second administration, however, renewed turmoil in Europe precipitated a crisis that necessitated a change in government policy. In May 1803, Great Britain had declared war on Napoleonic France and by 1805 war enveloped the European continent. Decrees and counterdecrees issued by the antagonists not only interfered with United States commerce but aroused much popular resentment as well. Most galling to the young republic's pride were stepped-up search, seizure, and impressment practices. At length these repeated violations of American neutrality compelled the Republican administration to consider war with England. Realizing that the country was poorly prepared for any kind of armed conflict, Jefferson decided against a direct military confrontation. Instead, he embarked upon a policy of economic retaliation embodied in the Nonimportation Act of 1806 and the Embargo Acts of 1807 and 1808. At the same time he began making defensive preparations by approving outlays for expanding productive facilities at the national armories and, in April 1808, signed a bill that provided an annual appropriation of $200,000 for arming state militias.[1]

The first indication that the scope of operations at Harpers Ferry would be enlarged came less than a week after the enactment of the first Nonimportation Act. Writing to Perkin on April 23, 1806, Secretary Dearborn revealed the War Department's intention to reintroduce the manufacture of muskets at the factory and asked the superintendent for an estimate of how many armorers would be required to produce 4,000 stands annually. Perkin replied that 120 additional men could do the job "without any considerable additional buildings."[2] Evidently this estimate surprised Dearborn who candidly stated his misgivings in a letter to the superintendent dated May 23, 1806. In particular he thought it odd that the Springfield armory could produce annually 4,000 muskets with sixty-eight men while Harpers Ferry required more than twice that number to achieve the same end. Convinced that something was wrong, Dearborn let Perkin know that his job as well as the armory's reputation were at stake:

Although the Arms made at Springfield are not quite so nicely finished, as those made at Harper's ferry, yet they are generally considered, as very excellent, and as superior to any European Soldier's musket;—unless therefore the number annually manufactured at Harper's ferry, can be

1. See *American State Papers: Military Affairs*, 1:199; William D. Grampp, "A Re-Examination of Jeffersonian Economics," *Southern Economic Journal* 12 (1946):278; Leonard D. White, *The Jeffersonians* (New York: The Free Press, paperback edition, 1965), pp. 2–9, 134–147, 426–432, 532–533.
2. Dearborn to Perkin, April 23, May 3, 1806, Misc. Letters Sent, OSW; Perkin to Dearborn, May 12, 1806, Registers of Letters Received, OSW.

considerably increased, in proportion to the number of hands employed, it will be improper to increase the number of workmen;—and I fear the comparison between the two Armouries; as to the number of Arms annually manufactured, will become a source of uneasiness. You will therefore see the necessity and propriety of making every exertion in your power, to augment the annual Return of Arms manufactured under your direction.

In the same letter the secretary begrudgingly authorized Perkin to hire eight or ten new workmen to begin work on extra tools and equipment needed for making muskets. Yet, other than a token product of 136 stands, the armory throughout 1806 continued to concentrate most of its resources on the manufacture of short rifles and horsemen's pistols.[3]

Although musket work continued to move slowly, the year 1807 marked the beginning of a new era at Harpers Ferry. Following Perkin's death on December 1, 1806, Dearborn initiated a search for a new superintendent. Having previously broached the subject to Eli Whitney, he asked the versatile New Englander on December 10 to take the post. While Whitney had expressed more than passing interest in the position, the request came at an inopportune moment. With lawsuits concerning his cotton gin still pending in the Georgia Circuit Court, he thanked Dearborn for his confidence but declined the offer because, he said, "my present situation is such that I cannot discharge the duties of that important office without a ruinous neglect of my private concerns."[4]

No sooner had Whitney's response reached Washington than Dearborn received a communication from John Strode, a long-time friend and fellow Republican from Culpeper, Virginia, endorsing a request that Perkin's master armorer, Charles Williams, be appointed superintendent. Apparently the secretary had serious doubts about Williams' political sympathies because he asked the Culpeper businessman to "recommend another for the appointment." Strode suggested the name of James Stubblefield, a resident of Stevensburg, Virginia.[5] He did not know Stubblefield be-

3. Dearborn to Perkin, May 23, 1806, Misc. Letters Sent, OSW.
4. Whitney to Dearborn, January 22, 1807, Letters Received, OSW.
5. Dearborn to Strode, February 18, 1807, Strode to Dearborn, February 12, 24, 1807, Letters Received, OSW.

fore making the nomination but, after a personal interview, assured Dearborn:

He appears to be a sensible discreet man about thirty years of Age. . . . In point of ingenuity He does not appear inferior to any I ever convert with; His Moral character is without blemish, and seems to possess strong intellectual faculties, His industry has procured Him a very handsome estate—real & personal, He has & works Two Large productive plantations, and carries on a very extensive trade in the Gunsmith Line, if His knowledge be deficient in any of the branches of business at that place it must be that of making the best sort of Sword blades.[6]

To supplement this report, Strode forwarded a "very private" letter addressed to him by General James Williams of Soldiers Rest near Stevensburg. An old friend of the Stubblefield family, Williams characterized the nominee as a sober and honest man with "as correct habits as any I know." "As to his fitness to superintend the works at Harpers Ferry," Williams concluded, "I can only give my opinion as he has never been ingaged in any works on a large scale he might at first be some what at a loss, tho I have no doubt on my mind that a little time and experience would remove every difficulty, And that he would conduct the business with Honor to himself and advantage to the United States."[7] Impressed by these testimonials, Dearborn invited Stubblefield to Washington for a formal interview. Evidently the meeting went well, for on March 25 he offered Stubblefield the position at an annual salary of $1050, including rations. Three weeks later the War Department officially issued the commission and notified Samuel Annin of the appointment.

Stubblefield had been trained in the craft tradition and his experience as a gun maker did not extend beyond a small country shop. Before assuming the post at Harpers Ferry, he therefore went to the Virginia state armory in Richmond "for the purpose of acquiring some information as to the Mode of Carrying on the business" of a large factory. Dearborn made no attempt to discourage the trip because he felt the experience would be beneficial. Consequently

6. Strode to Dearborn, February 27, 1807, Letters Received, OSW.
7. Williams to Strode, March 1, 1807, Letters Received, OSW.

the new superintendent and his family did not arrive at the Ferry until May 14, 1807.[8]

Once settled at the armory, Stubblefield played a distinctly secondary role to Samuel Annin. For no explicit reason other than length of government service, the War Department continued to recognize the paymaster as the senior administrative officer at Harpers Ferry. Whenever the secretary issued important directives, he addressed them to Annin. In fact, in several letters already alluded to he forthrightly acknowledged Annin as superintendent. The paymaster's duties of conducting all business related to accounts, buildings, and purchases were so thoroughly carried out by Annin that Stubblefield exercised no fiscal power at all at the armory. On at least one occasion the paymaster also took over the superintendent's prerogative of hiring new workmen. There is no indication, however, that Stubblefield ever clashed with Annin over the delegation of authority. For nearly eight years their relationship remained generally friendly and trouble-free.

Only once, in 1810, did Stubblefield attempt to wrest from the paymaster exclusive control over armory purchases. His request was firmly denied. "As the whole of your time will be required in superintending the works and workmen under your charge," the secretary of war responded, "it is considered expedient that Mr. Annin should, as heretofore, continue to make the necessary purchases for the armoury on your requisitions."[9] Clearly the War Department intended Stubblefield to supervise production and little more. Even then he did not act independently of the paymaster, for every important administrative decision—whether it concerned the selection of patterns or the appointment of subordinate officers—had to be approved by Samuel Annin before it could be implemented. All things considered, Stubblefield's charge seemed to be more fitting a master armorer than a duly appointed administrative official.

Even so, the new superintendent brought about some important changes during his first years in office. Under his direction the last

8. Stubblefield to Dearborn, May 5, 15, 1807, Letters Received, OSW.
9. William Eustis to Stubblefield, November 22, 1810, Letters Sent Relating to Military Affairs, OSW.

parcels of rifles and pistols were completed, and the production of Charleville pattern muskets was begun on large scale. At the same time he embarked on an extensive building program which lasted over three years. In order to expand the work area to accommodate 100 men for musket production, Stubblefield asked permission to make "an addition of thirty five feet to the Armoury and Twenty to the Smith's shop; also an additional water wheel, and such other apparatus as may be necessary." Without hesitating, Dearborn approved the measure on November 11, 1807; as usual, Annin took charge of construction.[10]

The paymaster had barely completed the assignment when Dearborn, apparently without solicitation, issued another directive for even greater changes at the armory. In a letter dated June 3, 1808, he notified Stubblefield,

It has been determined to enlarge the Armoury Establishment both at Springfield & Harpers Ferry, & I have given Mr. Annin directions to commence the erection of the necessary buildings, water works, machinery & Apparatus without delay & to have the whole performed on such dimensions & in such manner, as he and you shall agree on generally. For the machinery etc. you will from time to time furnish him with sketches, drawings or directions which he will follow.

It is desirable that the buildings & machinery should be such as may be sufficient for as many workmen, as would be necessary for the manufacture of from 15 to 20,000 muskets annually & a due proportion of Rifles, Pistols & Swords.[11]

In contrast with the War Department's earlier policy of rigidly controlling estimates and expenditures at Harpers Ferry, money seemed to be no obstacle in 1808. Haunted by the specter of war in Europe and the threat of its reaching American shores, even Gallatin recognized that a balanced budget would be practically impossible to maintain, since "all the resources of the country must be called forth to make it efficient" in time of crisis. To be sure, the Republican administration had become more concerned with increasing production at the national armories than with freezing

10. Stubblefield to Dearborn, October 1807, Letters Received, OSW; Dearborn to Annin, November 11, 1807, April 12, 1808, Misc. Letters Sent, OSW.
11. Dearborn to Stubblefield, June 3, 1808, Misc. Letters Sent, OSW.

their appropriations. As a result, total expenditures at Harpers Ferry jumped from $40,631 in 1807 to $104,953 in 1808, $158,835 in 1809, and $145,042 in 1810. During the next decade disbursements for construction and manufacturing would continue to grow, averaging over $176,000 by 1820.[12]

At the time of Stubblefield's appointment in 1807, the Potomac armory comprised 6 buildings staffed by 67 armorers. By the end of his third year as superintendent in May 1810, the facilities had expanded to 12 workshops, 15 dwelling houses, and a work complement of 197 men. Twelve years later, in 1822, the armory boasted 21 shops and annexes, 89 dwellings, and a work force of 234 men. Yet, even though the number of armorers and buildings had more than tripled between 1808 and 1822, expected gains in productivity did not materialize. When Dearborn sanctioned the expansion program of 1808, he had received assurances that the armory would be producing at least 15,000 muskets annually by 1810. The best Stubblefield and his men could do that year, however, was 9,573 musket equivalents. In fact, prior to 1822 the closest Harpers Ferry ever came to achieving the 15,000 quota was in 1818 when it turned into store 9,892 muskets, 2,700 rifles, 17,550 appendages, and 17 pattern pieces (or 13,883 musket equivalents).

Between 1810 and 1821 total outlays for buildings, millwork, dams, canals, and other structures at Harpers Ferry amounted to $127,271; similar improvements at Springfield cost only $37,659. During the same period expenditures for the manufacture and repair of arms exceeded those at the New England installation by $164,728. Although the two armories were nearly of equal size, Springfield continually led in production at costs substantially less per musket.[13]

When asked to account for higher costs and lower output at Harpers Ferry, government authorities seized upon a number of reasons. One explanation advanced time and again was that the

12. White, p. 144; *American State Papers: Military Affairs*, 2:478–481.

13. Annin to Eustis, January 1, 1810, Letters Received, OSW; Benjamin Moor to Colonel George Bomford, October 12, 1835, Letters Received, OCO; *American State Papers: Military Affairs*, 1:679; 2:481–483. For production figures at Harpers Ferry and an explanation of the term "musket equivalent," see Table 1. Production data for Springfield may be obtained from Deyrup, *Arms Makers*, pp. 229–245.

factory produced a greater variety of weapons than Springfield and therefore could not be expected to achieve the same economies of scale realized there. The armory's disappointing performance was also attributed to managerial shortcomings, craft traditions, harsh environmental conditions, bizarre local customs, and the baneful influence of several families who owned and controlled the town of Harpers Ferry. Besides, a serious shortage of skilled labor continually hampered manufacturing activities between 1808 and 1816.

Although Stubblefield had little trouble finding a sufficient number of workers during his first year as superintendent, the labor situation worsened in 1808 and continued to plague him for the next eight years. Indeed, he had trouble not only in recruiting armorers but also in retaining them once they had arrived. The problem became particularly critical during the War of 1812 when the demand for skilled artisans in the firearms industry far outstripped the available supply.

One point of general dissatisfaction was the unhealthiness of the climate at Harpers Ferry. Over the years many workmen left the factory to escape the dreaded "bilious fever" epidemics which occurred almost annually, leaving scores of victims in their wake. Each outbreak of the fever invariably reduced the labor force to a skeleton crew and sometimes even forced closure of the armory for weeks at a time. Not even the promise of wages 12 to 20 percent higher than those paid at Springfield could induce some armorers to return to their jobs after the misery and suffering of a sick season. An officer in 1831 aptly expressed the feelings of many who had worked at the armory a decade or two earlier. Reflecting upon the town's susceptibility to floods in the spring, sweltering heat in the summer, and slack water in the fall, he could not understand what had prompted the government to build at Harpers Ferry in the first place. "About this place," he wrote a friend in Pittsburgh, "there seems to be an atmosphere which may be emphatically termed the Atmosphere of Harpers Ferry."

Every way considered—there are customs and habits so interwoven with the very fibers of things—as in some respects to be almost hopelessly remitless. And in conjunction with this—is associated the ineligibleness of this Station for manufacturing. The prevalence of disease—has undermined the constitution of numbers. The Doctors fees—has operated— with other circumstances to cut out those substances and dry up their unsound. The operations of the Armory at times is much impeded 'by sickness—and renders the place more like a hospital or Quarintine—than advocations—and the rattling of work shops, of the healthful and industrious mechanic.[14]

Lack of adequate housing coupled with the high cost of living constituted another reason why many armorers—especially those with families—refused to accept employment or remain for long at the government works. In a word, living conditions were pitiful. Until 1815 or thereabouts most workers were expected to provide their own lodgings. Because of wide discrepancies in earning power, the quality of residences naturally varied according to the level of one's station. Many armorers, unable to acquire credit except at usurious rates, resorted to putting up one- or two-room shacks in which their wives and children huddled. Damp, poorly lighted, and unventilated, such dwellings became the breeding ground of vermin. Owing to the lack of adequate drainage facilities, the town's drinking water also became contaminated with sewage, a factor that, coupled with the armorers' thirsty disposition, accounted for the unusually high consumption of alcoholic beverages in the community. Conditions such as these not only made for unsanitary living but often resulted in disease and death.[15]

Although they could do little about personal hygiene, Annin and Stubblefield did attempt to clean up the debris-strewn public waterworks and to enhance the armory's appeal by erecting seventy-four dwelling houses on government property between 1810 and 1821.[16] Evidently this strategy, plus a less urgent demand for armorers after the War of 1812, improved the situation. Despite the recurrence of epidemics, there seemed to be relatively little problem in attracting either skilled or unskilled workmen to the armory by

14. Moor to Major Rufus L. Baker, May 5, 1831, Letters Received, AAR.

15. Stubblefield to Dearborn, April 4, 1808, "Friend" to Eustis, July 26, 1809, Colonel John Whiting to Secretary of War, March 1810, Annin to Eustis, December 2, 1812, to John Armstrong, March 17, November 2, 1813, July 26, 1814, Letters Received, OSW.

16. Annin to Eustis, January 1, 1810, OSW; *American State Papers: Military Affairs*, 2:483; Snell, p. 136.

1817. In fact, the labor situation completely changed at Harpers Ferry during the 1820s and 1830s. With high wages and adequate housing virtually assured, competition for positions became so keen that a man's getting a "chance" to work depended as much on whom he knew as on what he knew. While foreigners and outsiders were met with suspicion, local residents received preferential treatment whenever job openings occurred during the antebellum period. Prior to 1817 labor shortages did impede production; after that date, however, even an ample supply of workers did not eliminate the armory's manufacturing problems.

While sufficient manpower is essential to any industrial undertaking, the capacity to produce depends on how labor, once acquired, is utilized. Human resources remain relatively inert until they are organized according to some preconceived plan and actually integrated with other factors of production. This is especially true when large quantities of an item demanding some degree of uniformity—such as a firearm—are the object of manufacture. Naturally, the larger the undertaking, the greater the need for systematic organization. Early factory masters soon grasped the importance of developing managerial strategies. Planning, however rudimentary, provided patterns of action that affected every phase of the productive process. At Harpers Ferry the earliest entrepreneurial strategies placed more emphasis on manual than mechanical skills. This as much as anything else would condition attitudes toward work discipline and technological innovation at the government works throughout the antebellum period.

While a few gunsmiths at Harpers Ferry may have fashioned some arms in their entirety between 1799 and 1801, by the time of full-scale operations in 1802, Perkin had adopted a pattern of production closely resembling traditional European craft practices. Under this arrangement individual artisans made a particular part of each gun, such as the lock, stock, mounting, or barrel. Since total production depended on the number of assemblies completed within a given period of time, one of the principal duties of the master armorer was to coordinate output by determining work assignments and allocating resources so that an equal number of parts would be produced simultaneously. Accordingly the physical plant was divided into a forging shop equipped with a water-driven tilt hammer for preparing barrel skelps and three or four forges for welding gun barrels, a smith's shop with ten fires for forging small components on the first floor and a dozen or so benches for lock filers on the second, and a finishing shop provided with water-powered boring, grinding, and polishing machinery on the first floor and twelve to sixteen benches for stocking and assembling arms on the second level.

In 1807 the manufacture of rifles and pistols involved six separate branches of labor: barrel making, lock forging, lock filing, brazing, stocking, and finishing. The completion of each limb required not only different skills, but also special tools for each operation. As artisans completed their tasks, they submitted their work to the master armorer for inspection. He, in turn, sent the parts to another shop, where a "finisher" filed and fitted them, assembling the completed weapon. In other words, each stand turned over to the arsenal storekeeper represented a composite product, the work of several different hands. Yet, despite the rudimentary division of labor involved in the manufacturing process, each gun remained essentially a handcrafted product.[17]

As a public institution whose fortunes necessarily fluctuated with the course of political events, Harpers Ferry could not immunize itself against external forces of change, no matter how much the members of the armory's craft-conscious labor force desired to retain individualized methods of gun making. This became particularly evident in 1808, when, as noted above, the War Department announced that the factory would be tooled to produce at least

17. Stubblefield to Dearborn, September 8, October (undated), 1807, Letters Received, OSW; Stubblefield to Bomford, July 18, 1829, Letters Received, OCO. When visiting the armory in the summer of 1808, John Caldwell of New York observed that only "the weighty part of the business is conducted by the aid of machinery turned by water." Since at the time one usually distinguished between heavy "machinery" and lighter "machines," Caldwell's reference is probably to the use of tilt hammers and grinding wheels rather than to "machines" for shaping small components. This is corroborated by the fact that the "weighty part" of gun making consisted primarily of forging and grinding gun barrels. See John Caldwell, *A Tour through Part of Virginia in the Summer of 1808*, ed. William M. E. Rachal (Richmond, Va.: The Dietz Press, 1951), p. 11.

15,000 muskets annually. Although Stubblefield expected an enlargement of the physical plant, he had no idea that the magnitude of manufacturing operations would be so great. The armory could be equipped to yield 5,000 or 6,000 arms a year without much trouble, but 15,000 presented the superintendent with a severe challenge. Initially he had planned to leave Perkin's six-phase production system intact and simply increase output by adding more armorers to the roster. The success of this scheme, however, depended on having a large number of highly skilled workers. With the quota tripled and the competition for armorers intensified throughout the country by the impending war crisis, Stubblefield found it impossible to secure the needed personnel. He therefore settled upon a strategy which obviated the need for so many highly trained artisans in the productive process. As he explained the situation twenty years later.

I determined to adopt a new plan of manufacturing the arms for the United States, and in the Spring of 1809 commenced making tools and machinery for the purpose of distributing the component parts of the guns so as to make the work more simple and easy. In June, 1810, we got our tools and machinery ready for making arms; and it is upon this uniform plan that they are now made throughout the United States. . . . By this division of labor, a great deal of expense and trouble are saved, a great amount of tools is saved, and the work can be executed with infinitely more ease, more rapidly, as well as more perfectly and uniformly; and, moreover, a hand can be taught, in one-tenth part of the time, to be a good workman when he has but one component part to work upon.[18]

When Stubblefield provided this information in 1829, he also claimed to have invented the division of labor in arms making and petitioned the government to compensate him for first having introduced the system at Harpers Ferry. To this the chief of ordnance curtly replied that the division of labor was a concept, not an invention, and summarily dismissed Stubblefield's claim as one for which no individual could rightly be given credit. "Under such circumstances," he observed, "a great division of labor would seem

18. Stubblefield to Bomford, July 11, 1829, OCO. Also see Stubblefield to Dearborn, May 21, 1808, Letters Received, OSW; Dearborn to Stubblefield, May 25, 1808, Misc. Letters Sent, OSW.

to be a matter of course, that could not well be avoided; as it has been the practice in all manufactories for years past to increase the divisions of labor, as the works and workmen were multiplied."[19]

Inquiry into the origins of the division of labor in the firearms industry reveals that superintendent Benjamin Prescott had introduced the concept at the Springfield armory in 1806 and that Eli Whitney claimed to have used it as early as 1798. Furthermore, existing records establish that Stubblefield made a special trip to the Springfield and Whitney armories in November 1808 for the express purpose "of examining the Works generally and particularly the machinery for facilitating the manufacture of Arms."[20] That he traveled over five hundred miles simply to foster good will and mutual understanding is unlikely, especially during a time of such pressing armory engagements. The journey, in fact, marks the first time a superintendent from Harpers Ferry ever ventured beyond Philadelphia on armory business. In this sense, Stubblefield's visit to the Connecticut Valley was a significant occasion, one that would be repeated many times after 1815. Inasmuch as the trip coincided with his attempt to find an organizational scheme to compensate for the scarcity of skilled labor at Harpers Ferry, there is little question that Stubblefield brought back the "new plan" for manufacturing muskets from New England and, according to his own testimony, introduced it at the armory in the spring of 1809.

In pioneering the division of labor and the concomitant process of piece-rate accounting between 1806 and 1821, the Springfield armory doubtlessly made a major contribution to the development of modern industrial management. Yet, it is to Stubblefield's credit that he recognized the value of these concepts, introduced them at Harpers Ferry, and for a short time outdid his contemporaries from the North in carrying them into effect. The practice of piecework payments was well established at Harpers Ferry by 1816. In contrast with the traditional day-rate method, piece-rate accounting

19. Bomford to Stubblefield, September 16, 1829, Letters Sent, OCO.
20. Dearborn to Prescott, Whitney, November 9, 1808, Misc. Letters Sent, OSW. See also "Monthly Work Returns of Civilian Employees," 1806, SAR; Stephen V. Benet, ed., *A Collection of Annual Reports and Other Important Papers relating to the Ordnance Department*, 3 vols. (Washington: GPO, 1878), 1:15; Deyrup, p. 91.

recorded the production of each armorer by job and measured output in terms of individual operations performed on single components rather than on complete assemblies. Springfield, interestingly enough, did not adopt a complete piece-rate system until 1818, though the concept had prevailed there since 1806.

How much Stubblefield had divided his labor force by 1810 is not known. Because his system was patterned after the one at Springfield, the number probably did not exceed the twenty occupations listed on the New England establishment's work returns for that year. By October 1816, however, Stubblefield had succeeded in dividing musket work into fifty-five different occupations—a remarkable feat considering that a year earlier Springfield had only thirty-four occupational specialties listed on its rolls. The greatest division of labor occurred in the manufacture of lock mechanisms. In 1807 two artisans—one at forging, the other at filing—made the entire lock assembly. By 1816 lock making had been divided into twenty-one separate occupations: seven in forging—four of which had full-time assistants—eleven in filing, and three in finishing. Similarly, work on the mounting and barrel had been divided into eighteen and nine occupations respectively. The only part of the musket unaffected by the division of labor was the stock. As in 1807, twenty-three skilled craftsmen continued to fashion gun stocks complete from the rough blank of walnut to the last coat of oil.[21]

Probably because the labor scarcity problem abated after the War of 1812, Stubblefield became somewhat complacent about his initial successes and saw little need for further divisions of work at the factory. After moving his family to a country plantation in 1815, he assumed the role of an absentee superintendent and more

21. Payrolls and accounts, Harpers Ferry armory, 1816, GAO; Deyrup, pp. 91–92, 108; Whiting to Eustis, March 1810, OSW. For the sake of consistency I have followed the same guidelines used by Deyrup in counting the number of occupational specialties among production workers at Harpers Ferry. Accordingly the term "production worker" excludes the following personnel at the armory: the superintendent, master armorer, paymaster, inspectors and foremen, clerks, machinists, jobbers, carpenters, arsenal workers, shop sweepers, water carriers, watchmen, common laborers—in short, anyone who was employed by the day or month.

or less delegated supervision of the armory to his brother-in-law, master armorer Armistead Beckham. During the next fourteen years occupational specialization revealed few signs of growth at Harpers Ferry. Springfield, on the other hand, made rapid strides in the field under the vigorous leadership of its new superintendent, a former protégé of Whitney, Colonel Roswell Lee.[22] Whereas the number of occupational specialties at Springfield jumped from thirty-four in 1815 to eighty-six in 1820, those at the Ferry increased only from fifty-five to sixty. Five years later, in 1825, Lee had further divided labor at Springfield into one hundred occupations while Beckham and Stubblefield barely augmented the number to sixty-four. At the time of Stubblefield's resignation as superintendent in 1829, the number of occupational specialties at Harpers Ferry had risen 25 percent since 1816. This figure hardly compares with the amazing 188 percent growth recorded at Springfield during the same period. Thus, even if the question of mechanical innovation is completely disregarded, in terms of occupational specialization, organizational sophistication, and managerial vigor, Harpers Ferry clearly had fallen far behind her sister armory in Massachusetts by the 1820s.[23]

In this context one final observation must be made about the connection between the division of labor and the use of special-purpose machinery at Harpers Ferry. One writer believes that "the proportion of piece-work to time work in small arms manufacture bears a direct relation to the development of mechanization and factory organization." Using the Springfield armory as a case in point, she maintains that "the period of most rapid increase [in occupational specialization], between 1815 and 1820, occurred precisely at the time in which many processes were being mecha-

22. A remarkable manager, Lee took charge of the Springfield armory on June 2, 1815. He held the office of superintendent until his death on August 24, 1833.

23. Payrolls and accounts, Harpers Ferry armory, 1816–1829, GAO; Deyrup, p. 240. The increase from fifty-five occupations in 1816 to sixty in 1820 at Harpers Ferry was due to the division of labor brought about by the introduction of Sylvester Nash's barrel-turning lathe in 1817 and Thomas Blanchard's famous stock-turning machine in 1819. The use of Blanchard's invention added to what had previously been a one-man operation four new occupations: rough turning (by machine), letting in the lock (by hand), banding and butting the stock (by hand), and filing and mounting the stock (by hand).

nized for the first time."[24] While this no doubt holds true for Springfield, it is not so for Harpers Ferry. There the most pronounced growth in occupational specialization occurred between 1811 and 1816, a period of few important mechanical innovations. In fact, the techniques used at the Ferry in 1816 differed very little from those of 1807. For the most part, pieceworkers still relied on the dexterous use of hand tools to accomplish their tasks.

Essentially the division of labor at Harpers Ferry represented an attempt by hard-pressed factory managers to compensate for the dearth of skilled labor by simplifying production procedures without radically altering mechanical techniques. Although some new machinery was installed during the expansion between 1808 and 1810, no appreciable relationship can be established between occupational specialization and mechanization before 1816. In short, specialization occurred prior to and independent of mechanical change. Perhaps one reason why the armory failed to effect a division of labor similar to that at Springfield after the War of 1812 was that the system of production had reached a point where it could proceed no further without the introduction of more extensive machine processes. So long as craft-trained artisans occupied key decision-making positions at the factory, the likelihood of substituting machinery for traditional hand methods was remote. Neither Stubblefield nor his principal assistants were prepared to cope with such broad, sweeping technological changes.

24. Deyrup, pp. 92, 108.

CHAPTER 4

Early Manufacturing Techniques, 1816

In 1816 the facilities at Harpers Ferry differed noticeably from those of eight years before. Under James Stubblefield and Samuel Annin the works had more than doubled in size, the labor force had been reorganized and expanded to 214 men, and, though output had not reached the level originally contemplated in 1808, the armory was averaging nearly 10,000 arms a year. Yet, as important as these accomplishments were, they seemed to fade in significance when compared with developments at Springfield. Since 1799 federal officials had commonly acknowledged Harpers Ferry as the "Mother Arsenal," the principal government manufacturing center in the United States. By 1816 Springfield's Roswell Lee was openly challenging this view, arguing that his factory, though smaller in size, was not only outproducing the Virginia installation but doing so at considerably less cost per musket. Eager for the prestige of public recognition, he pressed authorities to "decide at which place the arms have the preference in point of workmanship & which is the most eligible stand for A GRAND NATIONAL ARMORY." In Lee's mind the answer was perfectly clear. While officers at Harpers Ferry seemed content to mark time and bask in the glow of previous achievements, their contemporaries at Springfield were experimenting with new techniques and seizing every opportunity to make their factory one of the most progressive manufacturing establishments in the United States.[1]

To appreciate these events and gain an adequate perspective for

1. Lee to Colonel Decius Wadsworth, December 24, 1816, Letters Received, OCO.

viewing subsequent mechanical developments at Harpers Ferry, we should review the actual technology of arms manufacture as it existed in 1816. This particular year is chosen because, as we shall see, it marked a critical juncture in the armory's operational history—a time when important administrative changes would take place at both local and national levels, a time when a family clique would gain control of the two highest offices at the armory, and, finally, a time when New England workmen and machinery would begin to appear at Harpers Ferry in significant numbers.

In 1816 musket manufacture was divided into four main categories at Harpers Ferry: lock, mounting, stock, and barrel. Undoubtedly the lock, or firing mechanism, represented the most complicated part of the musket. Because a number of fragile components made up the mechanism, it was easily damaged and frequently required repair. Yet, except for fragmentary references found in Ordnance Office correspondence, virtually no records exist on the technology of lock making during the early decades of the nineteenth century. This gap exists, it has been said, because most private gunsmiths did not make their own locks but purchased them ready-made from foreign suppliers.[2] Perhaps a more striking reason is that the process of lock manufacture had changed so little over the years that contemporary observers saw nothing particularly new to comment upon and, hence, either remained silent or limited their remarks to a few brief sentences. Whatever the case, the scarcity of descriptive information about lock making necessitates using diverse sources—including physical artifacts—and making indirect inferences about the art as it stood in 1816.

The lock plate represented the key component around which the lock mechanism was constructed (Figure 2). In 1791 several eminent French scientists described its function "as a species of plan . . . on which are mounted by screws, all those pieces which, by their union, really form that which we call a lock."[3] Its manufacture at Harpers Ferry comprised six basic operations. The first

2. Kauffman, *Pennsylvania-Kentucky Rifle*, pp. 165–166.
3. William F. Durfee, "The First Systematic Attempt at Interchangeability in Firearms," *Cassier's Magazine* 5 (1893–1894):475.

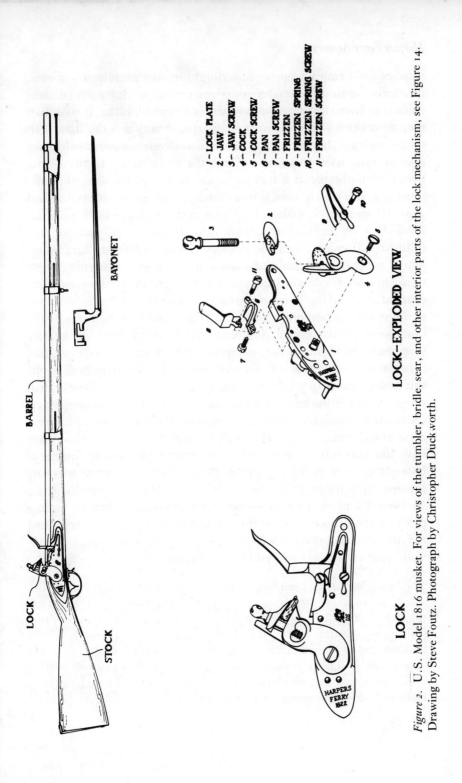

LOCK–EXPLODED VIEW

1 – LOCK PLATE
2 – JAW
3 – JAW SCREW
4 – COCK
5 – COCK SCREW
6 – PAN
7 – PAN SCREW
8 – FRIZZEN
9 – FRIZZEN SPRING
10 – FRIZZEN SPRING SCREW
11 – FRIZZEN SCREW

LOCK

HARPERS
FERRY
1822

BARREL

BAYONET

LOCK

STOCK

Figure 2. U.S. Model 1816 musket. For views of the tumbler, bridle, sear, and other interior parts of the lock mechanism, see Figure 14. Drawing by Christopher Duckworth. Photograph by Steve Foutz.

consisted of bringing a piece of wrought iron to a red heat and then manually forging it between swedges or dies with hardened steel faces that bore the impression of the oblong lock plate. It took two men to do this job—one to hold the piece between the dies with tongs and another to strike it with a sledge hammer. Because several heats were required to forge the piece nearly to shape, the iron became brittle and had to be taken to an annealing furnace, reheated, and slowly cooled in a cast-iron box filled with powdered charcoal to restore malleability. To remove the scale oxide that formed on the surface during annealing, the part was "pickled" in a dilute solution of sulphuric acid and sent to the grinding shop where a workman removed rough spots to give it a flat, more uniform surface. With the completion of grinding, the lock plate entered a second or "finishing" phase of manufacture.

The next operation consisted of drilling and tapping holes in the workpiece for the pins and screws that held the tumbler, bridle, sear, and other interior lock components in place. To facilitate this process, the lock plate was securely fixed in a jig containing properly spaced holes to guide the cutting tool of a vertical-spindle drill press. Although no contemporary description of this particular machine can be found, it probably resembled the water-powered drills used at Springfield and other private armories at the time. More than likely it was made from drawings and patterns Stubblefield brought back from Springfield in November 1808. In any event, the use of upright drill presses and jigs was not essentially an American achievement. Both devices were well known in France prior to 1789 and even described in some detail in a 1791 report to the Royal Academy of Sciences assessing the mechanical contributions of an armorer named Honoré Blanc:

The lock-plate having been forged with two trapezoidal projections on the side of the metal . . . which were made by the same blow. . . . They attach the lock-plate solidly to a die of sufficient thickness for directing the drills truly perpendicular. In this die all the holes are drilled exactly in the position and at the respective distances from each other that they are required to be. The workmen commence by marking the new plate with a drill. This done, they finish the holes, slightly conical. These holes they make cylindrical by means of a tool which they call a "false drill," used with a conductor, in order that the holes shall be exactly perpendicular to both surfaces. Finally, with a finishing bit they give them their finished size, and afterward "tap" such as are to be screw-threaded.[4]

Except for the extreme care taken to insure the uniformity of lock plates, Blanc's basic procedures differed very little from those used at Harpers Ferry in 1816.[5]

Once the lock plate had been drilled and tapped, a workman trimmed off the metal projections that had centered it in the drill jig. With the newly drilled holes serving as bearing points, the piece was next secured between patterns of tempered steel known as a filing jig, the edges of which served to guide the armorer in filing down the part to its proper dimensions. Hand filing constituted the last important operation on the lock plate before it went to the finisher's shop to be fitted with other lock components, case hardened, and polished.[6]

The most important component of the inner lock mechanism was the part that transmitted the power of the main spring to the ham-

4. Durfee, pp. 475–476; Payrolls and accounts, Harpers Ferry armory, 1816, Second Auditor's Accounts, GAO; Colonel John Whiting to William Eustis, January 13, March (undated), 1810, Letters Received, OSW; Eli Whitney to Wadsworth, December 29, 1818, Correspondence and Reports Relating to Experiments, OCO; Stubblefield to Colonel George Bomford, July 11, 1829, Letters Received, OCO. For visual representations of the tumbler, bridle, sear, and other interior components of the lock mechanism, see Figure 14. While this is an exploded view of a Model 1842 percussion musket, the basic shapes of the parts do not differ significantly from those manufactured for flintlock muskets in 1816.

5. That this is not an unwarranted assumption, see "Notes of a Tour of Inspection Commencing 10 December 1822, Armory Harpers Ferry, James Stubblefield Superintendent," Inspection Reports, OIG. This report makes explicit reference to the *Aide-Mémoire de Gassendi* (1809) which contains detailed descriptions of the methods used in French armories. Personnel at the federal armories used Gassendi's memoir as a guide in making tools and equipment. How many other French tracts on the manufacture of firearms were available to American armorers during the early nineteenth century is not known, but they did exist and were frequently consulted thus underscoring the importance of technological transfers.

6. Whiting to Eustis, March 1810, OSW; "Notes of a Tour of Inspection," Harpers Ferry armory, December 1822, OIG; Stubblefield to Bomford, July 11, 1829, OCO. Although Selah North and Eli Whitney are often credited with the invention of filing jigs, Honoré Blanc was using then at the Vincennes armory in France in 1786. See Edwin A. Battison, "Eli Whitney and the Milling Machine," *Smithsonian Journal of History* 1 (1966):14.

mer when released by the trigger and sear. Called the tumbler, its manufacture involved four fundamental stages. Having been first die forged and annealed in a manner similar to the lock plate, a workman milled the face and pivot on both sides of the tumbler. This task completed, the piece passed to another armorer who drilled and tapped a hole in the larger end of the pivot or arbor for the screw that would eventually fasten it to the cock. In 1816 the same man, interestingly enough, who drilled the lock plate also drilled the tumbler, using the same machine but a different set of bits. After the pivot had been drilled, the next step consisted of filing the notches and the square of the tumbler in a jig. This was an extremely critical operation. Special precaution had to be taken that the sides of the square portion of the pivot be correctly filed so the square of the cock would fit it and still be in a perpendicular position when the sear engaged the notch in the tumbler at half-cock. Otherwise the lock mechanism, when assembled, would not function properly. Once this step had been accomplished, the tumbler entered the finishing shop to be hand fitted, case hardened, and polished.

The other parts of the lock mechanism underwent virtually the same processes as the lock plate and tumbler. For example, the bridle—the piece that held the tumbler and sear in place—went through four operations, all of which nearly paralleled those performed on the tumbler. In a similar manner, the lock pins, side pins, and tumbler screw were forged, milled, slit, and threaded by hand. The sear—or part regulating the movement of the tumbler—required only three operations: forging, drilling, and filing. Finally, the cock or hammer was forged, ground, punched between dies in a jumper, drilled, and filed down in a jig. Including the main and sear springs, these components made up the complete lock assembly as it left the finishing shop.[7]

The origins of the process used at Harpers Ferry in 1816 for milling tumblers and other components offers valuable insight into the development of mechanized lock-making techniques during the

---

7. Payrolls and accounts, Harpers Ferry armory, 1816, GAO; Durfee, p. 476. For a concise discussion of screw-making techniques as practiced around 1820, see Battison, pp. 21–23.

early nineteenth century. Recent research indicates that workmen carried out these operations by means of a machine equipped with a hollow mill cutter. One writer suggests that the use of the term "milling" probably originated from the resemblence between the shape of the cutter and the face of millstones commonly used for grinding grain. The same writer also adds a description:

Such a mill . . . has its cutting teeth around the central hole at a right angle to its axis. The central hole would then . . . fit over the tumbler. As the hollow cutter advances onto the piece to be worked, it is supported by that part of the workpiece which has already been finished by the cutter; the cutter is thus supported and guided without the need of a highly precise machine to furnish absolute and continuous guidance. For this reason hollow milling is not a true milling operation.[8]

Although European and American gunsmiths traditionally used hollow mills secured in hand vises to turn screw blanks, the Frenchman Blanc evidently mechanized the process around 1786 and extended it to cutting the faces of tumblers and hammers. In the United States Eli Whitney claimed to have used a machine for milling screws sometime before the completion of his first government contract in January 1809. In fact, he probably gained direct inspiration from Blanc's work. Whatever the case, the first definite use of a hollow milling machine appears not at Whitney's factory but at the Springfield armory in December 1809. Writing to the secretary of war, a government inspector reported that "a Machine just put into operation by Water" at Springfield could slit and cut wood screws and mill tumblers, side screws, and trigger plates "at half the expense" they formally cost. Soon after Stubblefield introduced a similar machine at Harpers Ferry and, by 1816, had at least four in operation for milling tumblers, tumbler screws, lock pins, side pins, breech pins, and wood screws. Again, we have no positive evidence that Stubblefield built these machines from Springfield patterns, but, considering their functional similarity and the time of their introduction, such a conclusion seems warranted.[9]

8. Battison, pp. 16, 26.
9. Battison, pp. 19, 21, 23–26; Durfee, p. 476; Whiting to Eustis, January 13, 1810, OSW; Payrolls and accounts, Harpers Ferry armory, 1816, GAO.

With Roswell Lee's appointment as superintendent in April 1815, the Springfield armory began with renewed vigor to expand the milling of iron. Between 1815 and 1819 Lee supervised the construction of four improved machines for milling and drilling components "perpendicular" in a manner similar to a drill press, built additional machinery for milling and slitting pins, and introduced a new process known as "clamp milling" to form the heads of cock pins and ramrods between dies. Nothing indicates that similar developments occurred at Harpers Ferry during these years. Although Stubblefield did install a new machine for milling and drilling trigger plates in 1819, the armory had not yet introduced clamp-milling techniques or extended the mechanization of lock making on a scale equal to that at Springfield. As in 1810, the machining of lock components at Harpers Ferry in 1816 remained limited essentially to the borrowed techniques of drilling, hollow milling, and slitting.[10]

In marked contrast to these procedures, labor on the gun stock was neither divided nor mechanized in 1816. Except for the techniques of Simeon North, a Connecticut contractor who had introduced a new mode of stocking arms "by the help of water machinery" at his factory in Middletown, the making of gun stocks remained fundamentally a one-man, hand-wrought operation differing little from traditional European practices followed since the seventeenth century.[11]

Furnished with a lock, barrel, and set of mounting by the master armorer, a workman literally fashioned the stock to fit these limbs. Other than several hardwood patterns for tracing and centering the form of the stock on a rough-sawed walnut blank, the artisan had only the finished iron parts to serve as templates. No special-pur-

10. *American State Papers: Military Affairs*, 2:551; Ethan A. Clary, "A Description of the United States Armory, Springfield [1817]," in Milton Bradley, *History of the United States Armory, Springfield, Massachusetts* (Springfield: Milton Bradley, 1865), n.p.; Payrolls and accounts, Harpers Ferry armory, 1819, GAO; Lee to Wadsworth, December 24, 1816, OCO. Because the procedures for making the mountings of muskets (trigger guards, bands, heel plates) were nearly identical to those for lock production, they need not be discussed here. Battison discusses the origins of clamp milling, see p. 26.

11. Lee and Stubblefield to Wadsworth, March 20, 1816, Letters Sent, SAR.

pose gauges were used. Accordingly the first operation consisted of reducing the rough stock to its proper dimensions with hand planes. Once he accomplished this, the workman drew a center line along the top edge of the stock for the barrel groove which he then cut out using chisels, gouges, and planes. Usually several fittings were required to remove rough spots and to insure that the barrel bedded securely in the groove. The next procedure entailed centering the lock plate on the stock and recessing the area beneath it with chisels, gouges, and gimlets to accommodate the inner workings of the lock mechanism. The workman also made the mortice for the trigger and trigger plate in a similar manner. At the same time, he took special precautions to insure that the sear and trigger played freely in their positions without rubbing against the recessed area of the stock. He then fitted the butt plate, trigger guard, bands, and band springs to the stock. Finally, he worked the stock down to its final dimensions with drawer knives, floats, and sandpaper. Including the holes drilled for the sear, tang screw, and side pins, all labor on the gun stock was performed by hand in 1816.

As the difficulty of such work required a high degree of skill, all twenty-three stockers employed at Harpers Ferry had either learned the trade in Pennsylvania gun shops or served apprenticeships to someone who had. Productivity varied according to the ability and experience of each artisan. In October 1816, for example, George Cooper, Caspar Creamer, Daniel Creamer, John Ganett, and Michael Gumpf each stocked over forty-five muskets. At the same time Martin Breitenbaugh, Peter Hoffman, Christian Kreps, and Conrad Yaeger stocked over thirty-five arms apiece. Several younger workmen such as Charles Cameron, Charles Lancaster, and James Sargent offset these high production figures by finishing less than fifteen stocks in a month's time. As a group, the stockers at Harpers Ferry averaged twenty-eight musket equivalents per man in October 1816 for which they received piece-rate wages of $1.40 for muskets and $1.90 for rifles.[12]

12. Whiting to Eustis, January 13, March (undated) 1810, OSW; Stubblefield to Bomford, November 9, 1821, Letters Received, OCO; *American State Papers: Military Affairs*, 2:544; Payrolls and accounts, Harpers Ferry armory, 1816, GAO. Again, the designation "musket equivalents" is used because some workmen stocked rifles during the month of October 1816.

Just as individual artisans influenced production, they affected the finished stock itself. Many variations in workmanship appeared reflecting "the whims, the styles, and even the carelessness" of the workmen.[13] As a result, the stocks made at Harpers Ferry may have been similar—even uniform to some extent—but they were far from being interchangeable.

More descriptive information exists on the manufacture of gun barrels than on any other part of the musket. Writers have pointed out that most technical treatises on arms making since the seventeenth century devote more space to the barrel simply because it represents the essential part around which a gun is constructed. No one can seriously question its functional significance since a firearm is inconceivable without a tube or barrel to guide a projectile once discharged toward its target. Perhaps a more cogent reason for the abundant information on the barrel is that its production was not only the most expensive but also the most physically demanding part of firearms manufacture. Certainly Harpers Ferry's experience serves to corroborate this point. During the second decade of the nineteenth century nearly all correspondence about technical problems dealt with the manufacture of gun barrels. The armory continually had difficulties producing barrels that would stand the proof of inspection. Losses were so high, in fact, that they exceeded the total of all other components rejected at various stages of production.[14]

In 1816 the process of barrel making encompassed fourteen separate jobs. Except for a greater division of labor and the more extensive application of water power to machines for boring and rifling, the procedures followed at Harpers Ferry closely paralleled those used throughout the small shop industry of Pennsylvania. The first step involved the preparation of a ruler of metal, called a "skelp," at the forging shop. Taking a piece of bar iron, a forger carefully measured 10½ pounds of the material in a boxlike water gauge and cut it off with shears moved by water power. He then heated the iron in a coal fire and, with the help of an assistant, forged it under a sixty-five-pound tilt hammer. Two heats were needed to finish

13. Brown, *Guns of Harpers Ferry*, p. 14.
14. General John E. Wool to John H. Eaton, May 28, 1829, Inspection Reports, OIG; "Proceedings of a Board of Officers," February 1842, OCO.

the skelp, which measured three inches wide, half an inch thick, and about forty-four inches long. Upon completion, if an inspector found it free of cross cracks and other imperfections, the piece passed to the hands of a welder to be converted into a musket barrel.[15]

Although at the Springfield armory water-driven triphammers were adopted for welding barrels in 1815, at Harpers Ferry workmen continued to weld them solely by hand. Essentially the welding process consisted of two operations. First, the barrel welder put the skelp into a bituminous-coal-fired forge and brought it to a cherry red heat. Then, with the aid of two assistants, he proceeded to fold it around a mandrel, or bick iron, beginning at the muzzle and moving toward the breech until the two edges met. Unlike the practice usually followed at other armories, however, he did not lap the edges one over the other but left them square. It required two heats to close the skelp: during the first heat the welder closed about two-thirds of its length, during the second heat, the remainder.

The second and more tedious operation consisted of actually welding the barrel skelp with hand-held hammers and shaping it over a grooved anvil. The forger began his weld at the middle of the barrel and worked toward the breech. Because the metal cooled rapidly, only a small section could be forged at a time. Consequently the skelp had to be reheated four times to reach the breech end of the barrel and two more times to weld and swedge the breech itself. This operation completed, the welder returned to the middle section of the barrel and began welding and swedging it toward the muzzle. In all it took ten heats, about an hour and a half, and a three-man team to complete a musket barrel.[16]

15. Payrolls and accounts, Harpers Ferry armory, 1816, GAO; "Notes of a Tour of Inspection," Harpers Ferry armory, December 1822, OIG. Unless otherwise cited, all references to the process of barrel making are based on these documents. Cf. Kauffman, *Pennsylvania-Kentucky Rifle*, pp. 160–165; John G. W. Dillin, *The Kentucky Rifle* (4th ed., York, Pa.: Trimmer Printing, 1959), pp. 30–31, 43–45.

16. In 1816 eight barrel welders and sixteen assistants were employed at Harpers Ferry. The welder received 46 cents and his strikers 20 cents for each barrel completed. The most productive forger, Soloman Eador, welded 150 musket barrels in October 1816. He, in turn, was followed by Michael Crowl, Henry Best, and William Miller who each welded 138 barrels during the same period. Total output for October was 861 barrels.

After a total of fourteen heats during the forging process, the barrel became hard and brittle. To restore malleability, it had to be annealed in a manner similar to that described for the lock. After the barrel had been slowly cooled for about forty-eight hours, it was removed from the annealing furnace, straightened by hand, and sent to the machine shop. Here the first operation involved reaming the barrel to clear the bore of any lumps or irregularities that may have remained from welding and to enlarge the interior enough to admit an auger for rough boring. In 1816 one man performed this task using a single-spindle machine powered by water. From all available accounts, it probably resembled common boring machines used in Pennsylvania rifle shops during the early nineteenth century.[17] The device had a wooden bed and a sliding carriage to which the barrel was fastened and pushed by hand against the rotary motion of the reaming bit. The reamer consisted of a cylindrical rod, one end of which screwed to the spindle while the other end held a cutter shaped like the square lip of a whistle. Because the reamer measured little more than half the length of the barrel, the latter had to be removed from the machine, turned around endwise, and reattached to the carriage to finish the operation.

Upon completing his task, the workman submitted the barrel to another armorer to be rough bored. Using a machine similar in construction to those at Springfield, the operator mounted the barrel on a carriage moved by a hand-turned crank connected to a rack-and-pinion feed. But instead of drawing the cutting bit through the barrel as practiced at Springfield, he moved the work against the cutter with the same motion as reaming.[18] During this operation

17. Machines of this type are illustrated by Kauffman, *Pennsylvania-Kentucky Rifle*, p. 162; and by Dillin, plate 45.

18. Jenks & Wilkinson to Lee, February 13, 1818, Letters Received, SAR; Lee to Bomford, September 15, 1821, Correspondence Relating to Inventions, OCO. By 1821 the Springfield armory had adopted an improved boring machine which moved the carriage by means of two parallel screws. This device is illustrated in Daniel Pettibone's patent for boring gun barrels (see Figure 3) dated February 12, 1814, "Name-and-Date" Index to Restored Patent Drawings, No. 2064, RPO. In Lee's opinion, the method of advancing the barrel against the cutting tool could not compare with the steadiness and regularity produced by drawing the tool backwards through the barrel. Neither this machine nor the method of feeding the work to the auger was being used at Harpers Ferry in 1822.

three augers of the same diameter but of increasing length were successively passed through the barrel (Figure 3). Commonly known as a "screw auger," its invention had been claimed by numerous individuals, including the first machinist at Harpers Ferry, James Greer. While Greer maintained that he had developed the auger in 1797 and introduced the process at Robert McCormick's Globe Mills armory in Philadelphia in 1798, William Holmes of the Springfield armory, Daniel Pettibone of Philadelphia, and Lemuel Pomeroy of Pittsfield, Massachusetts, also entered similar claims. Others asserted that the auger had been used in the Pennsylvania rifle industry as early as 1796 and in England even earlier. Since the screw auger appeared almost simultaneously in several arms factories and since a series of acrimonious patent litigations failed to establish any definitive priority, the tool appears to have been the creation of many inventive minds.[19]

The auger operated on the same principal as the older nut boring bit. As one armorer observed, "They both cut & delivered the cuttings the same," the only difference being in the method of making the tools.[20] Instead of forging a round piece of iron and filing out the spiral grooves as done with the nut bit, a workman forged the screw auger flat then "twisted it in the manner of a carpenter's auger" thereby forming the cavities for the grooves. Once the boring tool had been attached to the spindle and the machine set in motion, the spiral cutter advanced through the barrel taking off metal at the point of initial contact and discharging the shavings through the grooves at the other end of the bit. During each cut a steady stream of water passed through the bore to cool and lubricate the barrel and bit.[21] In 1816 three men worked at rough boring gun barrels. Although productivity varied, Nathan Benton, John Sickafuse, and Samuel Sickafuse usually averaged

19. Greer to Bomford, March 7, 1818, to Wadsworth, September 24, 1818, Letters Received, OCO; Greer to members of the U.S. Senate and House of Representatives, March 2, 1820, Correspondence Relating to Inventions, OCO; Wadsworth to John C. Calhoun, February 19, March 18, 1820, Letters Sent to the Secretary of War, OCO; Stubblefield and Lee to Bomford, January 20, 1825, Letters Sent, SAR. See also Deyrup, *Arms Makers*, p. 94.

20. M. T. Wickham & Co. to Lee, August 24, 1824, Letters Received, SAR.

21. Lee to Bomford, September 15, 1821, SAR; Stubblefield to Bomford, February 2, 1827, Letters Received, OCO.

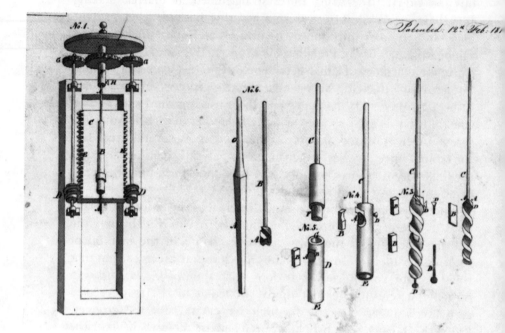

Figure 3. Machine for boring gun barrels (*left*) and various augers (*right*) as depicted in Dan Pettibone's patent of February 12, 1814. Photograph from RPO, National Archives.

one thousand a month. For this work they received 9½ cents for each barrel completed.

Smooth boring followed rough boring. Prior to commencing this task, however, the smooth borer straightened the barrel, making sure that the breech portion aligned with the axis of the bore. Other than requiring two bit changes instead of three, the procedure for smooth boring closely paralleled that of rough boring. Again, one man performed the operation using a machine with a hand-cranked rack-and-pinion feed and a water-powered spindle to propel the rectangularly shaped cutting bit. Nevertheless, smooth boring proved to be a time-consuming process. As each bit had to be passed through the barrel several times, five men and five machines working full time were needed to keep pace with the output of three rough borers in 1816.

When the barrel had been smooth bored, it returned to the man who had originally reamed it to be chambered for a breech plug. Evidently with the same machine, he used a two-inch reaming bit to make the chamber concentric with the bore. Following this operation the barrel was taken to a workbench and secured upright in a "hold fast" vise; then the chamber was tapped for the threaded plug. Two conically shaped taps of increasing diameter were successively screwed into the chamber by means of a hand-turned "scior." They, in turn, were followed by a completely cylindrical tap having the same diameter as the breech plug. When this tap came to rest in the chamber, the female thread had been formed. Only drilling a hole for the vent and filing off the bur the taps had raised on the extremity of the breech remained to complete the job.

Up to this point all operations on the barrel primarily aimed at finishing the inner bore. Consequently hammer dents and other irregularities that appeared on the exterior part of the barrel had to be removed by reducing the surface to a rounded, more uniform size. For this purpose the workpiece entered the grinding shop where a worker fitted it with a steel mandrel which extended lengthwise through the bore and screwed into the tapped breech chamber. He then fixed the ends of the mandrel to the center points of a swinging iron frame and turned the barrel against a revolving grindstone with a hand-operated crank. This process had two phases. First, the worker ground both the round and oval sections of the barrel on a coarse grit "yellow" stone measuring five to six feet in diameter and fourteen to sixteen inches in width. Finishing this operation, he moved to another wheel and ground the piece to size on a hard but less coarse "blue" stone of the same dimensions. As two, sometimes three pounds of metal had to be removed from the barrel, grinding proved to be a very slow and tedious process. Moreover, workers considered it "a deadly business." The bursting of grindstones claimed the life of at least one armorer at Harpers Ferry, while similar accidents left several others either without limbs or permanently blinded by flying particles. Still others succumbed to lung disease contracted by breathing the metal shavings and the dust-contaminated air that permeated the grinding shop. Likewise, the dampness produced by water-drenched grinding machinery constituted another serious health menace. Even if a grinder survived these hazards, the occupation left its mark because he could easily be identified by his bloodless fingers and granitelike hands. "When he left the work," one contemporary remarked, "he never recovered from its effects."[22]

A crucial test followed grinding. To determine the strength of the weld, a proof master screwed the breech plug into the chamber, charged the barrel with powder and ball, and ignited it. If the barrel stood two successive charges without bursting, it received a proof mark and was sent to the boring shop to be straightened for the last time. In 1816 the Pennsylvania-trained armorer George Nunnemacker performed this job by passing a taut silk thread through the barrel. If the thread touched the inner wall of the bore from one end to the other, nothing further had to be done. But if it cast a shadow, the straightener marked the spot, placed the barrel on an anvil, and tapped it lightly with a copper hammer. He re-

22. Charles H. Fitch, "The Rise of a Mechanical Ideal," *Magazine of American History* 11 (1884):518. Other pertinent references are Benjamin Moor to Major John Symington, March 27, 1845, Letters Received, OCO; Petition of Benjamin Moor, 32A-H13.1, ca. 1851, HR; Jacob Abbott, "The Armory at Springfield," *Harper's New Monthly Magazine* 5 (July 1852):452–453; Barry, *Strange Story of Harper's Ferry*, pp. 20–21.

peated this process until the thread touched the bore throughout its length.[23]

Once straightened, the barrel passed to the hands of another armorer to be finish bored and polished. Following this operation, the same individual who had tapped the chamber screwed the breech plug into position and filed it down in a jig. Next, the exterior surface of the barrel was draw-filed and carefully finished to remove any marks left by the grindstone or straightener's hammer. Finally, a workman drilled and countersank a hole for the tang screw. With polishing, the barrel was ready to be sent to the finisher's shop for final assembly.

While the arms manufactured at Harpers Ferry compared favorably in quality with those made by private contractors, the weapons produced at Springfield were generally preferred by military authorities. It was acknowledged that the Potomac armory excelled at making highly finished pattern and presentation pieces, but it could not equal Springfield's record for consistently producing a sound, reliable, and—after 1815—more uniform product. On many occasions Stubblefield was criticized for turning out muskets so defective in workmanship that they could not be repaired at outlying arsenals without great expense and inconvenience. In particular the War Department noted defects in barrels, ramrods, bayonets, and lock mechanisms. On one occasion the secretary of war complained that a parcel of muskets sent to Georgia did not even have vents drilled in the barrels and, therefore, could not have been proven. "You may rest assured," he told Stubblefield, "that the information is too well founded; and I am bound by every sense of duty to enjoin upon you a more rigorous Inspection."[24] Such reports obviously hurt the armory's reputation and constituted one

23. Joseph Hoffman to Captain William Wade, January 10, 1825, Letters Received, OCO; Charles H. Fitch, "Report on the Manufactures of Interchangeable Mechanism," *Tenth Census of the United States (1880): Manufactures*, 22 vols. (Washington: GPO, 1883), 2:626.

24. Secretary of War to Stubblefield, September 7, 1812, Letters Sent Relating to Military Affairs, OSW.

reason why Lee claimed the title of the "GRAND NATIONAL ARMORY" for Springfield.

Since no specific regulations had been established by the War Department for the inspection of finished firearms prior to 1822, criteria varied not only from one armory to another but within different branches of the same armory. This led to chaotically different standards and numerous variations among muskets purportedly of the same model. In 1816 the procedures followed at Harpers Ferry were primitive, though no more so than at many other establishments. For the most part the inspection process consisted of qualitative comparisons made with a pattern arm and its parts. That is, an inspector discovered defects in work mainly by eye rather than by instrument. The basic criteria for inspecting the lock in 1816, for example, consisted of taking the mechanism apart and examining its parts to make sure they had been made in a workmanlike manner. If the inspector found the components properly shaped, filed, and fitted, he reassembled the piece and tried its action to determine whether it functioned smoothly and gave "good fire." If the lock met these requirements, it received the inspector's stamp of approval. Other than calipers which served to check dimensions against patterns, no gauges were used in the inspection of gun barrels, stocks, and other components.[25] The subjective nature of this process made it exceedingly difficult to maintain any uniformity between muskets made at Harpers Ferry and those produced elsewhere.[26]

Even though Stubblefield had introduced the division of labor and generally kept abreast of contemporary mechanical improvements during the early years of his superintendency, his ardor for upgrading productive techniques cooled appreciably after the War

25. These calipers or "spring dividers" were purchased from European importers. More than likely gauge plugs were used to verify the bores of gun barrels, although no evidence of their use can be found at Harpers Ferry before 1821. See Whiting to Eustis, March 1810, OSW; Eustis to Annin, September 28, 1812, Letters Sent Relating to Military Affairs, OSW; Bomford to Lee, September 21, 1821, Letters Sent, OCO; Stubblefield to Bomford, November 9, 1821, Letters Received, OCO.

26. Lee to Stubblefield, August 6, 1816, November 20, 1817, to Bomford, September 11, 1821, Letters Sent, SAR; Bomford to Lee, August 7, 1821, Letters Received, SAR.

of 1812. Owing in part to complacency, failing health, and more pressing private business interests, he no longer seemed to provide the entrepreneurial leadership the armory desperately needed. This proved most unfortunate, for the decade after 1816 marked an important transitional period in the development of novel mechanical concepts and their application to the manufacture of firearms. It would be incorrect to say Stubblefield completely ignored these changes, but he certainly did not take full advantage of them. As a result, Harpers Ferry fell far behind her sister armory to the north in almost every phase of production during the late 1810s and 1820s. Indeed, as we shall see, the most significant mechanical improvements at the armory during these years were due not so much to Stubblefield's initiative as to the constant prodding of outsiders, particularly Colonel George Bomford of the Ordnance Department and Roswell Lee of Springfield. Even then proposed changes occurred slowly or, in some cases, not at all.

CHAPTER 5

Cooperation between the
Armories, 1815–1829

S ince the diffusion of new ideas and processes depended upon the flow of information from one factory to another, geographic factors had an important bearing on the timing and extent of technological development in early nineteenth-century America. Prior to the appearance of a bona fide machine-tool industry in the 1840s, the quasi-educational phenomenon one writer describes as "technological convergence" already existed among New England arms makers, though on a more limited shop-to-shop basis.[1] Springfield, situated in a region that abounded with foundries, machine shops, and mills of all sorts, provides an instructive case in point. Under Roswell Lee the national armory adopted a Worcester firm's method of welding gun barrels with triphammers, purchased castings for engine lathes from David Wilkinson of Pawtucket, Rhode Island, lent tools and machine patterns to private business companies, and readily shared information with Eli Whitney, Lemuel Pomeroy, and many other arms contractors. These as well as other exchanges would have been far less frequent and, in some cases, even impossible without the close proximity of Yankee manufacturers to one another. Easy accessibility not only cultivated common interests but also aided the transmission, adoption, and further elaboration of labor-saving improvements. By accelerating technological change and advancing mechanical know-how, these

1. Nathan Rosenberg, "Technological Change in the Machine Tool Industry, 1840–1910," *Journal of Economic History* 23 (1963):423–430.

factors gave Springfield and other Connecticut River towns advantages that more isolated communities rarely enjoyed.[2]

Harpers Ferry, lodged "among the rocks and mountains" of an agrarian hinterland, sorely lacked such an essential prerequisite for collaboration and convergence. Except for two iron works upriver and the Foxall-Mason foundry in Georgetown, there were no places where workmen could conveniently go to compare notes or to see what was being done to upgrade metalworking techniques of production. This remoteness from other machine-using establishments—especially other armories—reinforced provincial attitudes and severely constrained the Potomac work's progress as a center of innovation.

As early as 1810 Colonel John Whiting had detected this problem and had written a report emphasizing the importance of a regularized cooperative program "in order that any improvements in either of the Armories at Harper's Ferry and Springfield, [might] be completely possessed by both."[3] Secretary of War William Eustis, however, failed to note Whiting's proposal and subsequently relegated the report to oblivion in the Department's massive correspondence files. Thus communications between Harpers Ferry and Springfield remained infrequent up through the War of 1812.

With the end of hostilities came an important change that helped remedy the situation. Under the original legislation of 1794 the War Department was to exercise direct control over all armory affairs, but, owing to a chronic shortage of staff, it had allowed local officials a relatively free hand in determining their own policies. Early in 1815, Secretary of War James Monroe, well aware that this makeshift organization had many shortcomings, moved to readjust the existing framework of authority. Intent upon tightening administrative procedures and returning power to the national

2. This is well documented by Deyrup, *Arms Makers*, p. 67; Gene S. Cesari, "American Arms-Making Machine Tool Development" (Ph.D. diss., University of Pennsylvania, 1970), chap. 5; Michael H. Frisch, *Town into City: Springfield, Massachusetts, and the Meaning of Community, 1840–1880* (Cambridge, Mass.: Harvard University Press, 1972), pp. 15–22.

3. Whiting to Eustis, January 13, 1810, OSW. Also see Lee to Stubblefield, June 11, 1816, Letters Sent, SAR.

level, he successfully obtained Congressional legislation placing both national armories under the immediate jurisdiction of the Ordnance Department. The reorganization took effect on February 8, 1815.[4]

Originally established in 1812 as an agency for the inspection and distribution of military supplies, the Ordnance Department had distinguished itself during the ensuing war years under the able leadership of Colonel Decius Wadsworth. An experienced administrator and a stickler for discipline, Wadsworth took his job seriously—too seriously, in fact, for many at Harpers Ferry—and within three years had formed a small but tightly knit organization centered in Washington. His motto, "*Uniformity, Simplicity and Solidarity,*" was a guiding principle in the formulation of departmental policy, and he expected all who served under him to live up to its precepts or suffer the consequences—usually transferal to less desirable duty. Accordingly he surrounded himself with a group of bright young engineering officers, the most able of whom was his chief assistant and successor, Lieutenant Colonel George Bomford. To Bomford, a West Pointer, fell the responsibility of promoting greater systematization and efficiency at the national armories.[5]

After relieving the secretary of war in 1815 of duties as the sole arbiter of armory affairs, Wadsworth and Bomford drew up an agenda for Harpers Ferry and Springfield that went far beyond administrative reform. Thousands of arms had been damaged and rendered virtually useless during the recent war with England, and so the immediate concern was the production of cheaper, more uniform weapons that could be repaired in the field by substituting new parts for broken ones. The underlying principle was not new. It had been successfully advanced at several French armories in the 1780s, publicized in the United States by Thomas Jefferson and Eli Whitney, and partially applied to the manufacture of pistols by Simeon North of Middletown, Connecticut, as early as 1813.

4. U.S. *Statutes at Large*, 3:204.
5. Wadsworth to Secretary of War, August 8, 1812, Letters Sent to the Secretary of War, OCO. This and several succeeding paragraphs follow closely the author's "From Craftsman to Mechanic," in *Technological Innovation and the Decorative Arts*, eds. Ian M. G. Quimby and Polly A. Early (Charlottesville: University Press of Virginia, 1974), pp. 112–115.

Wadsworth and Bomford were well aware of these developments and appreciated their potential for eliminating a costly maintenance and repair problem that had plagued military authorities since the days of the Revolution. Intrigued, even infatuated with the possibility of making arms with interchangeable parts, both men became zealous advocates of the "uniformity system" and relentlessly pursued the idea of introducing it at the national armories. To do so, however, some means had to be found to eliminate the countless variations that inevitably crept into guns made primarily by hand.

The first indication that the Ordnance Department intended to take action came in June 1815, when Colonel Wadsworth called a special meeting at New Haven, Connecticut, to discuss establishing more rigorous standards for the manufacture of military muskets. In addition to the chief of ordnance, the participants included Superintendent Roswell Lee of Springfield, James Stubblefield of Harpers Ferry, and Wadsworth's long-time friend and confidant, Eli Whitney. After several days of deliberation they agreed upon a new model musket; standardized pattern pieces, however, should first be prepared and tested at the national armories. If the experiment proved successful, the program would then be extended to arms made by private contractors.[6]

Although the uniformity idea looked promising on paper, neither acceptance nor success came easily. To some the establishment of industry-wide standards seemed fatuous and beyond the bounds of administrative control. Since manufacturing methods and inspection procedures differed markedly from one armory to another, too many variables had to be reckoned with. One member of the military voiced a common sentiment when he cautioned superiors, "It is the easiest thing in the world to change the pattern at Washington and to make, in imagination, thousands of arms upon the new

---

6. Wadsworth to Bomford, May 15, June 13, 1815, Correspondence Relating to Inventions, OCO. Although the Ordnance Department envisioned the manufacture of completely interchangeable firearms, in 1815 the concept of uniformity meant little more than a degree of similitude between finished components which would allow field armorers to repair damaged muskets from salvaged parts with a minimal amount of filing and fitting. Until the mid-1840s common military muskets were serially produced with numbered components in order to distinguish one musket assembly from others in the same batch.

pattern; but it is far otherwise in the practice."[7] Moreover, no matter how perfect the pattern, constant usage wore down its original dimensions causing both it and the arms being manufactured to become increasingly destandardized. These and other technical problems led skeptics to conclude that the new system, with all its inherent weaknesses, would also be prohibitively expensive. Uniformity would require not only the painstaking preparation of many patterns, but also the construction of precision gauges and special-purpose machinery. Moreover, once built, these implements would have to be integrated with other units, constantly checked and rechecked for accuracy, and coordinated with the total production process.

To introduce such an intricate system at one factory was difficult enough; to do so for two widely separated establishments presented mind-boggling problems. Even the dynamic and resourceful Lee remained wary about finding any quick technological solutions. Having frequently consulted with Stubblefield while preparing the new Model 1816 musket (see above Figure 2), he was painfully aware that many discrepancies existed between the patterns used at Harpers Ferry and at Springfield. He also knew that frequent adjustments of the model at both armories had failed to "remedy the evil." After more than two years of assiduous labor neither Lee nor Stubblefield had much to show for their efforts. Colonel Bomford continually complained that "a total disagreement" existed between the muskets made at the two armories, and Lee was in a dispirited mood by November 1817. "It is difficult . . . to *please every body*," he reminded the chief of ordnance. "*Faults* will *realy* exist & *many imaginary ones will be pointed out.* . . . It must consequently take some time to bring about a uniformity of the component parts of the Musket at both Establishments."[8]

Lee's continued pleas for "time, patience & perseverence" failed to make much of an impression at the Ordnance Department. Capti-

7. *American State Papers: Military Affairs*, 2:553.
8. Lee to Senior Officer of Ordnance, November 20, 1817, Letters Received, OCO. Also see Lee to Stubblefield, August 6, 1816, November 20, 1817, Letters Sent, SAR; Stubblefield to Lee, December 4, 1817, Letters Received, SAR; Wadsworth to Lee, November 23, 1816, Bomford to Lee, November 14, 1817, Letters Sent, OCO.

vated by the engineering ideal of uniformity and convinced of its urgency, Bomford wanted results, not alibis, and on more than one occasion made his feelings pointedly clear to the Springfield superintendent. Yet, despite their differences, both men realized that the two armories could coordinate their efforts only through cooperation. Otherwise all would be lost. They also knew that Harpers Ferry stood in dire need of technical assistance. Owing to its isolated position and strong craft orientation, the armory's threshold for developing and digesting the new technology lay far below that at Springfield. Stubblefield admittedly knew very little about precision methods and seemed "at a loss" to understand exactly what the Ordnance Department wanted. Likewise, James Greer, the sole machinist at the Potomac works since 1801, had grown old and, at least in terms of know-how, had lost touch with the times. Given these circumstances, the principal responsibility for preparing pattern pieces and inspection devices devolved upon Springfield's Lee.[9]

While accurate patterns were essential for achieving uniform dimensions, the really novel aspect of Lee's work involved the use of inspection gauges. Urged by the Ordnance Department, he began experimenting around 1817 with a method of gauging components both during and after the process of manufacture. Although the idea needed to be perfected, it had reached a point of sophistication far in advance of other armories by 1819. As Major James Dalliba, no great enthusiast of the system himself, reported in November of that year:

In order to attain this grand object of uniformity of parts, the only method which can accomplish it has been adopted at Springfield, . . . viz: making each part to fit a standard gauge. The master armorer has a set of patterns and gauges. The foremen of shops and branches and inspectors have each a set for the parts formed in their respective shops; and each workman has those that are required for the particular part at which he is at work. These are all made to correspond with the original set, and are tried by them occasionally, in order to discover any variations that may have taken place in using. They are made of hardened steel. The workman makes

---

9. Lee to Bomford, September 11, 1821, Letters Sent, SAR; Stubblefield to Lee, October 12, 1821, Letters Received, SAR.

every similar piece to fit the same gauge, and, consequently, every similar piece must be nearly of the same size and form. If this method is continued, and the closest attention paid to it by the master workmen, inspectors, workmen, and superintendent, the desired object will finally be attained.[10]

Nearly four more years elapsed before similar instruments were introduced at Harpers Ferry. Consisting of eight, perhaps as many as eleven "go-no go" steel gauges, they were undoubtedly primitive by modern standards. Nonetheless, as later inspection reports indicate, their eventual adoption in 1823 marked a significant step toward the later development of closely finished interchangeable parts.[11]

While promoting its cherished goal of uniformity, the Ordnance Department repeatedly stressed the importance of maintaining strong administrative ties and close technical collaboration between the national armories. To his credit, Lee needed little prodding in this respect. Eagerly seizing the opportunity to improve inter-armory communications, he made a point to become acquainted with Stubblefield and by the end of 1815 had established a continuing dialogue with the Harpers Ferry superintendent. The two men regularly exchanged information relating to annual inventories, accounting methods, wage rates, and general shop procedures. They also assisted each other in arranging for the purchase and inspection of raw materials. In return for rough-sawed gun stocks, Lee provided Stubblefield with Nova Scotia grindstones and sperm oil. On other occasions he sent samples of bar iron to Harpers Ferry and, at

10. *American State Papers: Military Affairs*, 2:544.

11. As much as this gauging method improved the overall quality of muskets made at the armories, serious lapses did occur. In October 1827, for example, Lee notified Stubblefield, "the Arms made at your Armory do not agree with the Arms made here. . . . The variation between our Arms seems to be greater now than at any time previous, which circumstances I very much regret. I will not say who is most in fault, but I believe there is some at both places. I hope we shall be able to make improvements on this point. If not, we shall surely be censured. Let us exert ourselves to the utmost." See Lee to Stubblefield, December 10, 1822, October 31, 1827, to Bomford, November 23, 1822, June 6, 1826, Letters Sent, SAR; Stubblefield to Lee, October 12, 1821, Letters Received, SAR; Lee to Bomford, December 18, 1821, Correspondence Relating to Inventions, OCO; Bomford to Lee, September 21, 1821, Letters Sent, OCO.

Colonel Bomford's request, even arranged for the purchase and shipment of a large quantity of privately made musket barrels and bayonets to the Virginia factory in 1826 and 1827. With the completion of a rolling mill at Springfield in 1828, Lee also supplied Harpers Ferry with Salisbury iron, drawn to various sizes and shapes for the manufacture of lock components, barrel bands, and other musket parts. This service lasted until 1854 when the Ferry finally constructed a rolling mill of its own based on designs and machinery supplied by Ralph Crooker of Boston.[12]

During his eighteen-year tenure as superintendent at Springfield, Lee remained a zealous advocate of cooperation between the armories. Through him important liaisons were established between Stubblefield and prominent New England manufacturers. Cognizant that the general welfare of the two national armories went hand in hand, he repeatedly reminded Stubblefield that maintaining constant communications was important. On more than one occasion he became impatient and chided his friend for neglecting the collaborative effort. A typical instance occurred in April 1823 when, having previously arranged to meet Stubblefield in Washington and accompany him to Harpers Ferry, Lee waited in vain for the superintendent's arrival. Frustrated and perplexed, he informed the Virginian: "It is with regret as well as mortification that I have to leave this place without seeing you, but I have waited here eight days & not hearing of or from you, I dare not proceed to your place for fear of missing you entirely & having waited thus long, I cannot now spare the time. One minutes writing would have saved all this, but so it is, I must leave here this day, & it is realy to be regreted that I could not have had the pleasure of seeing or at least hearing from you. I certainly would not have neglected you or your communication in this way."[13] This as well as other letters indicate that, despite the close personal relationship, Lee had difficulties with his comrade from Harpers Ferry. While Stubblefield applauded cooperative measures, he seemed to lack

12. Lee-Stubblefield correspondence, 1815–1829, Letters Sent-Letters Received, SAR; Ralph Crooker to J. S. Whitney, March 17, 1856, Letters Received, SAR; Colonel Benjamin Huger to Colonel Henry K. Craig, February 25, 1852, Letters Received, OCO.

13. Lee to Stubblefield, April 6, 1823, Letters Sent, SAR.

firmness and determination in carrying them out. Indeed, had it not been for Lee's persistence, it is doubtful whether fruitful communications between the two armories would have lasted for long. Certainly the best testimony to Lee's perseverance in this respect is supplied by the Springfield records themselves. His letter books provide a mine of information on Harpers Ferry up through his death in 1833. After that date armory managers continued to cooperate with one another, but the frequency and content of their correspondence declined appreciably.

One of the most lasting effects of Lee's and Stubblefield's association involved the transfer of both men and machinery. As so often was the case, Springfield, which possessed a rich reservoir of mechanical expertise, did the lending and Harpers Ferry the borrowing. Between 1816 and 1829 only three armorers left the Potomac works for Springfield. During the same period Lee sent at least twenty-five workmen to the Ferry. Only a few of these New Englanders, however, established permanent ties in Virginia. Most of them were conventional armorers who, like Elijah Allen and Daniel Stevens, stayed at the factory only long enough to accumulate a grubstake before moving westward to the Ohio country. Others such as Oliver Allen soon became homesick and returned to Springfield. Still others, for reasons ranging from disgust at their poor treatment to fear of the unhealthy climate, left Harpers Ferry as soon as the first opportunity arose for more gainful employment elsewhere. In addition, Lee loaned machinists, millwrights, and pattern makers to Stubblefield on a temporary basis. Included among them were Sylvester Nash, David Parsons, and Leonard Cutler. As carriers of fresh technical information, these men played an increasingly important role in updating production methods at the Musket Factory.[14]

Just as precision instruments and stricter inspections accompanied the introduction of the uniformity system, so did special-purpose machinery. Both Bomford and Lee recognized that, no matter how refined manual skills became, musket parts fashioned by hand simply could not compare in accuracy and consistency

14. Lee-Stubblefield correspondence, 1816–1828, Letters Sent-Letters Received, SAR.

with those finished by machines. In this opinion they were joined by such private contractors as Simeon North, Nathan Starr, and Asa Waters who, upon being presented with the Army's demand for higher quality weapons and the promise of continued government patronage, expressed a willingness to mechanize their operations.[15] Stirred by the motives of patriotism and profit, an unprecedented spirit of cooperation, and the invaluable assistance of a new generation of forward-looking mechanics, these men promoted a number of boldly original mechanical innovations between 1816 and 1830 which laid the foundation for what eventually became known abroad as the "American System" of manufacturing. The role Harpers Ferry played in these developments is most interesting. Though a bastion of craft tradition, the armory, due to its proximity to Washington, ironically emerged as an important testing ground for many novel mechanical ideas.

The lock mechanism represented the most complicated part of the military musket, but when arms makers and government authorities spoke of the uniformity system between 1816 and 1822, they seemed more concerned with its application to gun barrels. Several reasons underlie this emphasis. On the one hand, the use of filing jigs, hollow milling, and drilling machines had become sufficiently extended by the end of the War of 1812 to insure some uniformity in the manufacture of lock components without necessitating radical mechanical innovations. The situation, however, was quite different with respect to the barrel. Besides normally being the most expensive limb of the musket to make in terms of labor and raw materials, high losses sustained during boring and proving further increased its production cost. This problem was partially corrected through the introduction of triphammers for welding skelps. While this process had little to do with improving standards of uniformity, a closer look at it helps illustrate the indifference toward mechanization that, even with cooperation between the armories, permeated attitudes at Harpers Ferry.

15. Wadsworth to John C. Calhoun, October 13, 1818, in *The Papers of John C. Calhoun*, ed. William E. Hemphill (Columbia: University of South Carolina Press, 1967), 3:201–204; Nathan Starr to John H. Eaton, July 10, 1829, Correspondence Relating to Inventions, OCO; Deyrup, pp. 55, 89–99; George A. Stockwell, *The History of Worcester County, Massachusetts* (Boston: C. F. Jewett, 1879), pp. 111–113.

Although groove-faced triphammers had long been used in New England to forge bar iron into particular shapes, Asa Waters' adaptation of the process at his factory in Sutton, Massachusetts, around 1808 proved to be an important innovation for the arms industry. The heavy regulated blows of the water-powered hammer produced a much sounder seam in about half the time needed to weld a barrel by hand and, because only one assistant was required instead of the usual two in hand welding, the triphammer method cut the cost of labor by nearly twenty cents a barrel. Combined with stock saved, total expenditures per barrel were reduced by thirty cents. Apprised of this fact, other arms makers soon began to copy Waters' time- and labor-saving invention. Within a year, for example, both Stephen Jenks of Pawtucket, Rhode Island, and the Leonard brothers of Canton, Massachusetts, had adopted the technique.[16] In 1815, having been in office less than three months, Lee installed a triphammer at Springfield on a trial basis. The experiment proved so successful that he constructed three more units in 1816 and in December reported to his former employer, Eli Whitney: "I have put in motion four trip hammers for welding barrels (three are attached to one wheel)—they operate extremely well . . . . The [tub] wheel is kept in motion & the hammers are stopt at pleasure by the help of a Spring. —They can be made to give 400 blows in a minute—the wheel is made on Tyler's plan with perpendicular shafts."[17]

16. In a memorial to the War Department dated October 10, 1817, Waters claimed that he had "invented a new and useful improvement in the manufacture of Fire-Arms by means of semi-cylindrical Steel Dies, infixed in the anvil and hammer of the common trip-hammer for the purpose of closing and welding gunbarrels in a much more expeditious and in a much less expensive and laborious manner than was ever before practiced." While Lee testified that Waters was the first to use the process in New England, he noted, "I understand that some triphammers have been in opperation several years for welding Rifle and Musket barrels in Lancaster & Reading in Pennsylvania; but whether the time those hammers were first used is such as to supercede Mr. Water's claim I am not able at present to determine." See Waters to Calhoun, October 10, 1817, to Bomford, November 25, 1817, Lee to Senior Ordnance Officer, July 7, August 6, 1818, Correspondence Relating to Inventions, OCO; Lee to Junior Ordnance Officer, April 20, 1816, Statement of Charles S. Leonard, July 30, 1818, Edward Lucas, Jr., to Colonel George Talcott, March 19, 1840, Letters Received, OCO; Jonathan Leonard to Lee, June 26, 1818, Letters Received, SAR; Deyrup, pp. 94–95.

17. Lee to Whitney, December 30, 1816, to Stubblefield, January 16, 1817, Letters Sent, SAR. By 1818 Lee had ten hammers in operation.

Ten months earlier, Stubblefield had visited Springfield and observed Lee's triphammer in operation. Apparently impressed with its performance, he announced plans for building four hammers at Harpers Ferry and asked Lee to accompany him to Virginia to assist in the preparations. In 1817 Stubblefield even procured the services of the Springfield mechanic David Parsons to help with the project. For some unexplained reason, however, the machines were never actually constructed. In 1819 and again in 1828 the superintendent expressed his "immediate" intention of erecting triphammers for welding barrels. On both occasions, nothing materialized. As a result, the manner of welding gun barrels remained exactly the same at Harpers Ferry in 1829 as it had in 1807—a handwrought operation.[18]

Stubblefield's failure to adopt such a fundamental technique doubtlessly cost the government a great deal of money. Whereas the loss of gun barrels welded by triphammers at Springfield averaged around 10 percent between 1815 and 1830, those welded by hand at Harpers Ferry exceeded 25. For example, 82,520 musket barrels were welded at the Ferry between 1823 and 1829. Of this number 10,315 failed during boring and 11,117 burst in proof, making a total loss of 21,462 barrels which cost the United States at least $42,470. Directed by the secretary of war to investigate the matter in May 1829, a board of inquiry chaired by Inspector General John E. Wool concluded:

From all the information obtained on this subject the board is of the opinion, that the loss is unusual & extraordinary. It is, however, stated that in consequence of welding by hand, which has been the practice at Harpers Ferry the loss is much greater than with the trip Hammer. The loss at Springfield as stated by Col. Lee, with the trip Hammer is about from 8 to 12 per cent, and from 12 to 15 when welded by hand. It is therefore inferred that the loss at Harpers Ferry, ought to be ascribed, either to bad metal or great neglect in welding.[19]

18. Lee to Simeon North, February 21, 29, 1816, to Stubblefield, April 14, June 11, 1816, January 16, 1817, Letters Sent, SAR; Stubblefield to Lee, January 4, 1817, February 27, 1819, Letters Received, SAR; Stubblefield to Bomford, October 28, 1817, Correspondence Relating to Inventions, OCO; Stubblefield to Wadsworth, August 31, 1819, to Bomford, January 7, 1828, Letters Received, OCO; Payrolls and accounts, Harpers Ferry armory, 1817, GAO.

19. Wool to Eaton, May 26, 28, 1829, Inspection Reports, OIG.

Naturally the barrel forgers at Harpers Ferry attempted to exonerate themselves by contending they had been supplied with faulty raw materials. Undoubtedly part of the problem did lie with poor iron palmed off on the armory by unscrupulous mongers, particularly Peter Shoenberger of Huntingdon, Pennsylvania. Once Stubblefield's successors began purchasing barrel iron in 1830 from the same firms that supplied Springfield, losses did diminish. Nevertheless, as master armorer Benjamin Moor later affirmed, losses on hand-welded barrels sometimes exceeded 40 percent during the 1830s. Even assuming the number of condemnations to have been no greater than at Springfield, the prohibitive cost of handwork should have dictated against continuing the method at Harpers Ferry. Superintendent George Rust, Jr.—himself no mechanic—clearly recognized this fact when in 1832 he informed Colonel Bomford:

It appears . . . the total cost of the *labor* in manufacturing a musket at the Harpers Ferry Armory is $6.40 and at the Springfield Armory $5.54½ making a difference of nearly 15½ pr. ct. in favor of the latter.

The cause of this difference is attributable, in a great degree, to the superiority of the machinery at the Springfield Armory—the welding of barrels by water power, for instance, diminishing the cost about 22½¢ per barrel.[20]

Though Rust and Moor began the construction of a new barrel-forging shop in 1834, triphammers did not completely replace hand-welding methods at the Potomac works until 1840, nearly twenty-five years after the process had first been introduced at Springfield.

Whether gun barrels were welded by triphammers at Springfield or by hand at Harpers Ferry, a particularly bothersome technical problem still remained in finishing the outside surface. The traditional method of grinding barrels down to their final dimensions had many undesirable features. Basically a rule-of-thumb operation, it often left weak spots where the bore ran eccentric to the

20. Rust to Bomford, March 3, 1832, Letters Received, OCO. Also see Moor to Bomford, June 19, 1835, OCO; "Proceedings of a Board of Officers," February 1842, Inspections of Arsenals and Depots, OCO. Deyrup, pp. 72–79, discusses the Salisbury, Connecticut, iron makers who supplied Harpers Ferry after 1829.

outer circumference. This not only prevented uniform dimensions but also resulted in further losses at the proof house. Grinding, as indicated, was also very slow, tedious, and hazardous work. Faced with these conditions, masters as well as workmen naturally sought a more effective and less dangerous means of finishing musket barrels.

Several attempts had been made since the 1790s to substitute engine lathes for grindstones. The first known instance was at the Springfield armory in 1799 when superintendent David Ames reported that he was building a turning lathe and had "grate hopes it will answer a very valuable purpose."[21] Although the effort proved unsuccessful, other arms makers intermittently experimented with barrel-turning machines. One of the most prominent was New Haven's Eli Whitney who, years later, claimed to have invented such a device around 1808 but refrained from introducing the improvement for fear "that some person would contract to make Barrels & not only take advantage of my invention but intise away the workmen whom I had instructed in the use of the Machine before I could be half compensated for the expense of making it."[22] Whether or not Whitney actually built the machine will probably never be known. In any case, the first sustained effort to apply the lathe principle to turning the exterior surface of gun barrels came after the War of 1812.

The simultaneous appearance of at least five different barrel-turning machines between 1816 and 1819 provides a valuable commentary on technological innovation as the offspring of production needs, accumulated experience, and cooperative effort. The earliest of these was a lathe constructed by Simeon North in 1816. Soon afterwards the mechanic Sylvester Nash devised a similar machine at Springfield. Evidently the Ordnance Department thought the invention had merit because Bomford instructed Lee to send Nash to Harpers Ferry for the purpose of perfecting his model and assisting Stubblefield with the various technical problems of uniform production. When Nash arrived at the armory in

21. Ames to Samuel Hodgdon, October 10, 1799, Hodgdon-Pickering Papers, AGO.
22. Whitney to Lee, January 3, 1818, Letters Received, SAR.

November, Stubblefield appointed him as a machinist at forty-five dollars a month, a salary five dollars higher than the regular machinist, James Greer, received.[23]

By February 1817, Nash had constructed a full-scale version of his turning lathe but experienced difficulty in getting it to operate correctly. Forced to make several alterations, he disassembled the machine and continued to remedy defects during the next eight months. Evidently he added a number of important changes to the prototype, for armory purchase vouchers indicate that a nearby foundry cast a new bedplate, two slides, a head stock, and a tail stock for the lathe in November. Stubblefield seemed pleased with the result. Without mentioning Nash's name, he informed Lee on December 4: "I have nearly completed our Barrel turning Machine, and find it will answer a Grand purpose, and if I meet you in Phila. I will take a barrel with me for your Inspection."[24]

Three months' trial did not alter Stubblefield's opinion. Convinced that the machine would serve a "valuable purpose," he asked Bomford's permission on February 28, 1818, "to erect two others" at the armory. We do not know whether Bomford granted this request. We do know, however, that during the summer the Ordnance Department sent Charles Artzt, a German mechanic who had recently designed another barrel-turning machine, to Harpers Ferry to conduct similar experiments. Meanwhile, Nash had gone to Washington, made application at the Patent Office, and received letters patent for his invention on April 11, 1818. After selling the rights to use his improvement at Harpers Ferry for $500, the New Englander lost little time in packing his bags and returning to Springfield. Despite his short stay at Harpers Ferry, Nash's presence signaled the beginning of a slow but steadily growing reliance on Yankee know-how at the government works.[25]

23. Lee to Stubblefield, April 10, 1817, Letters Sent, SAR; Nash to Lee, April 28, 1818, Letters Received, SAR; Lee to Captain John Morton, December 6, 1817, Edward Savage to Secretary of War, April 20, 1852, Letters Received, OCO; Payrolls and accounts, Harpers Ferry armory, 1816–1817, GAO.
24. Stubblefield to Lee, February 28, December 4, 1817, Letters Received, SAR; Payrolls and accounts, Harpers Ferry armory, 1817, GAO.
25. Stubblefield to Bomford, February 28, 1818, Artzt to Wadsworth, September 8, 1818, Letters Received, OCO; Nash to Lee, April 28, 1818, Stubblefield to

A solid and rather complicated piece of machinery, Nash's lathe consisted of "a common turning Engine" moved by water power and equipped with a slide rest and a rack-and-pinion feed (Figure 4). In view of his background, Nash probably based this design on earlier Springfield armory patterns. In any case, he attributed no novelty to these features. Rather, he stated, "the principal parts which I claim as my invention in this machine are the steadying supporters, and the wrought iron plate so shaped upon the top as to give the carriage the necessary motion for giving the barrel the true and regular taper or scolop."[26]

The purpose of the supporters or props was to prevent the barrel from trembling and springing out of position once the cutting tool engaged its surface. Used in pairs and fixed by hinges to both sides of the cast-iron bed, these arched supporters could be opened or closed at will. When in a closed position, two semi-circular brass dies, one on each support, fitted around the barrel by means of an iron clamp secured at each end by a thumbscrew. In the center of the clamps, and over the brass dies, "an iron stirrup furnished with a thumb screw" served "to keep the top of the supporters fast and steady in their proper position." "In using this machine," Nash noted, "it is found that while the cutting tool is near the end of the barrel the strength of the materials are sufficient to prevent the barrel from trembling, but as it moves forward and advances to the distance of about 3 or 4 inches, it is found necessary to raise the first pair of supporters." Although the machine had as many as eight pairs of supports, no more than three or four engaged the work-piece at the same time. Accordingly, as the sliding carriage advanced toward the muzzle end of the barrel, the operator raised and lowered the props until the work cycle was completed.

Like most machines of the pre-1840 period, Nash's lathe rested on a sturdy wooden frame. The cast-iron bed measured eight feet

---

Lee, June 5, 1818, Letters Received, SAR; Patent specifications of Sylvester Nash, April 11, 1818, "Name-and-Date" Index to Restored Patent Drawings, No. 2939, RPO.

26. Patent specifications of Sylvester Nash, April 11, 1818, Restored Patents, vol. 4, RPO; "Notes of a Tour of Inspection," Harpers Ferry armory, December 1822, OIG. Unless otherwise indicated, the following paragraphs are based on these two sources.

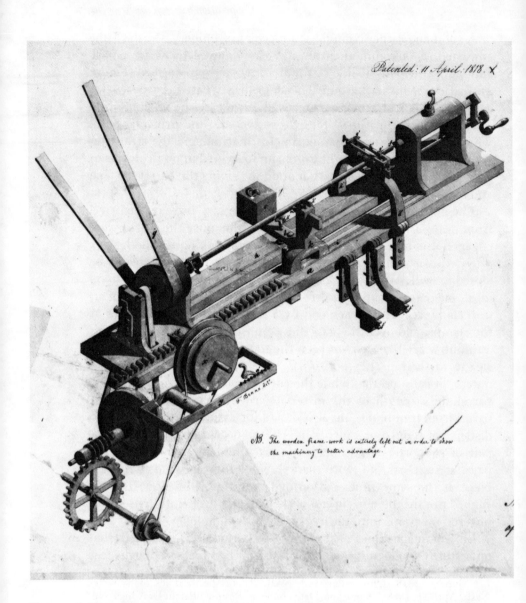

Figure 4. Sylvester Nash's lathe for turning musket barrels, showing the workpiece (*g*), slide rest (*k*), cutting tool (*m*), gauge screw (*n*), guide plate (*h*), supporters (*d'*), and rack-and-pinion feed (*b'*, *c'*). Patented April 11, 1818. Photograph from RPO, National Archives.

long, eighteen inches wide, and two and a half inches thick with an opening in the middle three inches wide and fifty-four inches long. At one end of the bed stood the cast-iron headstock which supported the main pulley and live arbor in its bearing. At the other end lay the tailstock with a screw-adjusted dead center. During the process of turning, the gun barrel was mounted between the bearings of these "poppet heads" by means of a circular steel mandrel which passed lengthwise through the bore and attached to the revolving spindle axis by a collar-and-dog at the muzzle and a center point at the breech. When set in motion, the slide rest carrying the cutting tool moved laterally along the barrel by means of the rack-and-pinion feed. "This rack," Nash explained, "is connected with, and gives motion to the carriage, by means of a connecting rod; this rod has a socket joint at each extremity by which the motion of the carriage is caused, and by turning the gauge screw prevents the racks from being effected."

The second patented principle of Nash's machine concerned the so-called "guide plate" which was positioned horizontally along the bed piece and parallel to the triangular steel-ribbed plate for the sliding carriage. To stabilize the carriage and at the same time "allow it a free motion along the plates," Nash attached weights to the tongue and top of the slide rest. Also, he noted, "Any round shapes can be given to the materials by making the shape required on the wrought iron guide plate." Hence, upon coming in contact with a gauge screw extending downwards from the tongue of the carriage, the tapered guide plate regulated the position of the cutting tool in relation to the workpiece. According to Nash, this action produced "the true and regular taper" of the barrel. Yet, because a planer was not used in building this machine, one wonders just how accurate these guideways may have been. One thing is certain: Nash's lathe was not self-acting. In addition to raising and lowering the steadying supporters, the operator had to "stand constantly by with his caliper or compass in hand trying the exactness of the work and altering from time to time the bearing of his chisel against the barrel."[27]

27. Extracts from Colonel Archer's Report, November 25, 1823, Letters Received, OCO.

While Nash was busy putting the finishing touches on his machine at Harpers Ferry, Roswell Lee received instructions from the Ordnance Department to give "a critical & candid trial" to another barrel lathe designed by Daniel Dana and Anthony Olney of Canton, Massachusetts. Furnished a model by the inventors, Lee and his master armorer, Adonijah Foot, prepared for the experiment in November 1817. Evidently Dana and Olney based their design on a David Wilkinson lathe, for Lee soon ordered a "Roller Engine" from the Pawtucket manufacturer for the purpose of preparing a full-sized unit. Even though the Canton machine differed "but little" in principle from Nash's, Lee nonetheless thought it would likely "answer the purpose" better than any he had seen. Probably the most fundamental difference between the two machines was in the manner the sliding carriage moved on its guideways. Unlike Nash's sensitive gauge-screw and connecting rod attachments, Dana and Olney's screw-driven carriage exhibited greater stability by resting directly on its guideways.[28] As one observer put it, "the two sides of the rail-way on which the frame or carriage . . . moves are parallel, but the superior faces of the railway are carried in a manner similar to that of the exterior line of the barrel." By this arrangement the rounded surface of the barrel could be turned to the proper dimensions without altering the position of the cutting tool.[29]

Upon receipt of Wilkinson's roller engine in mid-January 1818, Lee immediately put his men to work making Dana and Olney's adaptations. By February 24 the lathe had been completed and tried for the first time. Like Nash with his machine, however, Foot experienced difficulties in getting it to turn a good barrel. In a letter dated March 1, the master armorer issued an interesting progress report:

On Thursday . . . [we] turned four Barrels, but they were not very Smooth owing to the Sideways not agreeing one with the other. We

28. Lee to Stubblefield, December 6, 1817, to Whitney, December 12, 1817, to Wilkinson, November 18, 1817, to Morton, December 6, 1817, Letters Sent, SAR; Lee to Wadsworth, July 27, 1818, Correspondence Relating to Inventions, OCO; Robert S. Woodbury, *History of the Lathe* (Cambridge, Mass.: The M.I.T. Press, 1961), p. 92.

29. Extracts from Colonel Archer's Report, November 23, 1823, OCO.

altered them several times to make them agree, & get the shape of the Barrels, but found the Wooden sideways not sufficiently hard to keep the rests steady. We then placed some sheet Iron on them which made them move more steady. We also found it necessary to apply more weight to keep the Box [slide rest] in its place. On Friday We turned several more Barrels, but not so smooth as I wish to see them, the chips appeared to catch between the Rests and the barrel which causes some Rings. We then applied some Leather under the rests but it was soon torn away. but I think the difficulty will all be overcome and the Machine made to answer an excellent purpose. It throws off the Iron with ease, and requires 20 minutes only to turn a Barrel, a good chisel goes through without sharpening and with little grind, produces a straight even Barrel but it will require a Week or two to become acquainted with its bearings and make it go well.

We are now at work fitting the Iron side-ways, which will require a few days to get them right and make some other alterations in the Rest & *endless screw*.[30]

After a month of making cut-and-try adjustments, Lee informed Stubblefield that "our turning Machine opperates very well." Moreover, he added, "I scarcely know which is the best, yours or ours, I think however by adding one or two Standing rests, it [the latter] would rather be entitled to the preference."[31] While Dana and Olney's machine effected a savings of eight cents a barrel over the old method of grinding, Lee thought its greatest advantage lay in the superiority of work produced. "By turning," he informed Colonel Bomford, "we can get the exact size, shape and uniformity. The Barrel being of the same thickness on every side is less liable to fail in proving and is much better on every account, than when the sides are of unequal thickness, as is frequently the case when ground, one Machine will turn from twenty to twenty five Barrels a day and one man will tend two Machines."[32]

Because he believed the machines by Nash and Dana and Olney would be greatly improved by a combination of their principles, Lee recommended purchasing the rights to both inventions. Bom-

30. Foot to Lee, March 1, 1818, Letters Received, SAR.
31. Lee to Stubblefield, April 15, 1818, Letters Sent, SAR.
32. Lee to Bomford, May 27, 1818, Correspondence and Reports Relating to Experiments, OCO.

ford immediately approved. As a result, the barrel-turning lathe ultimately adopted at Springfield in 1818 consisted of a Wilkinson engine equipped with Nash's steadying supporters and Dana and Olney's carriage guides. No similar mechanical synthesis occurred at Harpers Ferry. Even though Nash's lathe required constant adjustment of the cutting tool and a full-time attendant for each machine, Stubblefield seemed perfectly satisfied with the unaltered version and continued to rely on it as late as 1829.

In addition to Nash and Dana and Olney, Asa Waters also designed a machine for turning gun barrels. Considering his proximity to Springfield and his frequent correspondence with Lee, the Millbury contractor was doubtlessly aware of the experimental work at the national armory before he embarked on his own project. By the spring of 1818 he had completed the machine and received letters patent for his improvement the following December.[33] However, as gun barrels became flat and oval shaped near the breach, neither Waters' machine nor others like it could finish the entire length. Once the cylindrical portion of the barrel had been turned, therefore, the breech section had to be hand filed in a jig at the loss of considerable time, uniformity, and money. According to his son, Waters spent "a year or more in fruitless attempts to solve the problem of reducing the butt by machine." Baffled completely and anxious to find some means of eliminating handwork on the barrel, he asked a local mechanic named Thomas Blanchard to help devise some mechanical method to turn the breech. Destined to become one of the most celebrated inventors of the nineteenth century, Blanchard already had two patents to his credit by 1818 but no experience as an arms maker. Yet, as the traditional story goes, upon "being told what was wanted, he glanced his eye over the machine, began a low monotonous whistle, as was his wont through life when in deep study, and ere long suggested an additional, very simple, but wholly original cam motion, which upon being applied relieved the difficulty at once, and proved a perfect success."

33. M. D. Leggett, *Subject Matter Index of Patents for Inventions Issued by the United States Patent Office, from 1790 to 1873*, 3 vols. (Washington: GPO, 1874), 1:644–645.

Other than Asa Holman Waters' statement that a key feature of Blanchard's lathe consisted of a cam motion for tracing a master pattern, no further description or representation of this particular machine can be found. It unquestionably represented an important transitional invention for, as Blanchard was reported to have said, "it was then and there . . . that the idea of his world-renowned machine for working out irregular forms first flashed through his mind."[34] Although the original cam-former invention had been attributed to various people, the principle itself was not new. Nor, for that matter, did Blanchard ever claim it to be. In his revised patent specifications of 1820, he readily acknowledged previous modes of turning irregular surfaces as described in Diderot's *Encyclopédie* and also Marc I. Brunel's famous block-making machinery as depicted in the *Edinburgh Encyclopedia*. In Blanchard's opinion, the mechanical principles embodied in these machines not only had become "common property" but had been so for many years. Consequently he based his claim not on general principles per se, but on their systematic arrangement and specific method of operation. His genius was in designing more versatile and productive "self-directing" machines capable of reproducing irregular shapes with efficiency and dispatch.[35]

Since government inspectors regularly visited Waters' factory, news of Blanchard's improvement reached Springfield very quickly. With typical alacrity, Lee contacted the inventor and invited him to demonstrate his principle at the armory. Impressed with what he saw, Lee immediately signed a contract with Blanchard for the construction of two machines: "one for turning flats or squares & the other for draw grinding Barrels." After their completion toward the end of June there followed a familiar train of events. On July 11 the cooperative-minded Lee wrote Stubblefield of his latest technical coup. In particular he believed the new lathe would "be very useful in turning the butts of our barrels & will

34. Asa Holman Waters, "Thomas Blanchard, The Inventor," *Harper's New Monthly Magazine* 63 (July 1881):255; Waters, *Biographical Sketch of Thomas Blanchard and His Inventions* (Worcester, Mass.: Lucius P. Goddard, 1878), pp. 6–7; Stockwell, p. 112.

35. Patent specifications of Thomas Blanchard, January 20, 1820, Correspondence Relating to Inventions, OCO.

answer well for turning the squares on Rifle Barrels, as it may be constructed to turn eight square." While characterizing Blanchard as "an ingenious man though not a very close workman," Lee notified Stubblefield, "I have advised him to proceed to your works, not doubting you will be willing to encourage him in bringing forward those improvements which I am convinced will be useful to our Establishments." Clearly the Ordnance Department approved and, most likely, had suggested this measure, for Lee concluded:

I have received instructions to correspond with you in relation to all important improvements; that we may have a mutual understanding respecting their utility, that they may be adopted in both Establishments. Should you have his Machines put in opperation I will thank you to write me your opinion of them soon as convenient, & if they meet your approbation, please inform me what amount would be a reasonable compensation. I have said to him that if his machines opperate well, I should be willing to recommend that $2000 be paid him for both Machines for the two Establishments & the expense of building & setting them in motion.[36]

After arriving at Harpers Ferry in mid-July 1818, Blanchard spent the better part of two months preparing a draw-grinding machine and an improved version of his gun barrel lathe. Primary interest and attention, however, focused on the latter. Whereas the lathe originally installed at Springfield turned only the short breech section of the barrel, the one built at Harpers Ferry finished the entire length and, as Blanchard described it, "changes from turning round to turning flat & oval of itself and turnes them so well that the draw grinding will grind them in 8 minutes." Though at first skeptical, Stubblefield seemed pleased with the result. "This machine," he wrote Colonel Wadsworth, "makes the barrel more complete than any thing ever set into operation to my knowledge." Wadsworth thoroughly agreed. After examining a "very fine" specimen brought to Washington by the inventor, he told Stubblefield "if you can finish 20 a day you will be able with the help of another similar engine to finish as many barrels as you may want, and save a

36. Lee to Stubblefield, July 11, 1818, Letters Sent, SAR; Blanchard to Lee, June 2, 1818, Letters Received, SAR.

great expense in grinding stone."[37] For some unexplained reason, however, neither armory permanently adopted Blanchard's lathe for turning the entire barrel. Springfield never even put one in operation on a trial basis. Instead, it continued to use two machines, namely, Dana and Olney's lathe for turning round and Blanchard's earlier "short" lathe for shaping the breech. Similarly, Harpers Ferry reverted to the combined use of the Nash and Blanchard methods around 1821. The practice of employing two machines to turn gun barrels still prevailed at both armories as late as 1854.[38]

When Blanchard left Harpers Ferry in October 1818, he was already pondering the application of the eccentric-turning principle to the manufacture of gunstocks. Back in Massachusetts, he set up a small shop near Millbury and, after several months of incessant labor, came up with a model that "operated so well that a full-sized working machine was decided upon." With the assistance of Asa Waters and several other workmen, Blanchard had the lathe running by February 1819. Awkward looking but amazingly efficient, it represented one of the truly outstanding American contributions to nineteenth-century technology.[39]

The rapidity with which news spread among New England arms makers is well illustrated by the fact that Lee knew about Blanchard's stocking machine several weeks before it had been com-

37. Blanchard to Lee, October 13, 1818, Letters Received, SAR; Stubblefield to Wadsworth, September 30, 1818, Correspondence and Reports Relating to Experiments, OCO; Wadsworth to Stubblefield, October 7, 1818, Letters Sent, OCO.

38. Lee to Stubblefield, July 6, 1819, to Brooke Evans, April 3, 1821, Letters Sent, SAR; Lee to Bomford, August 7, 1822, Correspondence and Reports Relating to Experiments, OCO; Extracts from Colonel Archer's Report, November 25, 1823, Letters Received, OCO; "Notes of a Tour of Inspection," Harpers Ferry armory, December 1822, OIG; Rosenberg, *American System of Manufactures*, pp. 130–131.

39. Waters, "Thomas Blanchard," p. 255. Even though Blanchard erected his first lathe at Water's armory, the Millbury contractor certainly did not install the machine on a permanent basis until the mid-1820s or perhaps even later. The payment of royalty fees seemed to be the main barrier. While Waters wanted to acquire a stocking machine, he complained to Bomford that "Blanchard . . . demands such a price for the right of using it that at the price which I am to receive for the Muskets on my new contract I cannot well afford that extra expence" (Waters to Bomford, April 6, 1824, Letters Received, OCO).

pleted at the Waters armory. Since a Springfield employee named Justin Murphy had inspected a parcel of contract muskets at Millbury during the second week of January 1819, he probably carried word back to Lee. In any event, the vigilant superintendent dashed off a letter of inquiry to the inventor on January 21 asking him to demonstrate his new process at the national armory. Blanchard, an obviously poor speller, replied two weeks later:

Yours of the 21 ultame. come safe to hand—you wished me to wright you respecting macenory—I conclude you meen a machine I have recently invented for turning gun stocks and cuting in the locks and mounting. Doubtless you have heard concerning it But I would inform you that I have got a moddle built for turning stocks and cuting in the locks and mounting. I can cut a lock in by water in one minute and a half, as smooth as can be done by hand. The turning stocks is verry simple in its operation and will completely imatate a stock made in proper shape. I shal bring the moddle to Springfield in the course of three weeks—I shal want your opinion of its utility.[40]

Blanchard carted his machine to Springfield and exhibited it at the armory in March 1819. Fascinated by its operation, Lee recommended an immediate trial to the Ordnance Department. This time, however, Colonel Wadsworth decided that the experiment could be more conveniently conducted nearer Washington. Thus, instead of having the stocking machine set up at Springfield, Lee accordingly sent Blanchard on to Harpers Ferry. On April 17, Stubblefield notified his friend, "Mr. Blanchard is here fitting up his turning lathe" and "as soon as I get it in operation I will inform you how it operates etc."[41]

Completion of the machine lasted most of April and all of May. But once set in motion it silenced those who scoffed at its ungainly appearance. Indeed, its performance exceeded Stubblefield's fondest expectations. "It shews," he wrote Wadsworth, "that Stocks can be turned in a Lathe, and I think it will be a great improvement in the Stocking of Muskets, and will also save expence, and insure uniformity in the stocks." For his part, Blanchard considered the

40. Blanchard to Lee, February 5, 1819, Letters Received, SAR.
41. Stubblefield to Lee, April 17, 1819, Letters Received, SAR; Waters, *Biographical Sketch*, pp. 7–8.

trial a resounding success. In a jubilant mood upon returning to Millbury he informed Lee, "I have got the machine for turning stocks into operation by water at the Ferry, it answers every purpose—it turns a stock in 9 minutes." Furthermore, he noted, "I am now ready to assist in building one for you agreabel to Col. Wadsworth's request."[42]

Although Lee eagerly awaited construction of the machine, a legal dispute between Blanchard and a Millbury brazier named Asa Kenny forced postponement of the project. The plaintiff Kenny contended that Blanchard had copied certain improvements made by him and incorporated them in the stocking machine. Naturally Blanchard denied these charges. Nevertheless, the institution of a lawsuit and the subsequent issuance of a restraining order meant that no patent could be issued nor could Blanchard build any lathes until the Superintendent of Patents, Dr. William Thornton, rendered a decision in the case. Ultimately Lee and three other commissioners appointed by the secretary of war investigated the question and in August 1819 issued a report unfavorable to Kenny. With the case settled, Blanchard secured letters patent on September 16, 1819, and, two days later, placed an order for several brass wheel castings with Ward, Bartholomew & Brainard of Hartford, Connecticut, indicating he had begun work on a machine for the Springfield armory.[43]

In his revised patent specifications of 1820, Blanchard described his improvement as "being an Engine for turning or cutting irregular forms out of wood, iron, brass, or other material or substance, which can be cut by ordinary tools" (Figure 5). Built on a heavy wooden frame, its operation—tracing a master pattern and reproducing it—involved three basic mechanical movements: a sliding

42. Stubblefield to Wadsworth, May 31, 1819, Correspondence and Reports Relating to Experiments, OCO; Blanchard to Lee, June 9, 1819, Letters Received, SAR. On August 14 Stubblefield informed Lee, "I have taken off twenty five cents from the price of our Musket Stocks since I have had the turning Machine erected, & find it to answer very well" (Stubblefield to Lee, August 14, 1819, Letters Received, SAR).

43. Lee to Wadsworth, June 18, 1819, to Kenny, July 9, 1819, to Ward, Bartholomew & Brainard, September 18, 1819, Letters Sent, SAR; Kenny to Lee, June 27, 1819, Blanchard to Lee, July 9, 1819, Letters Received, SAR.

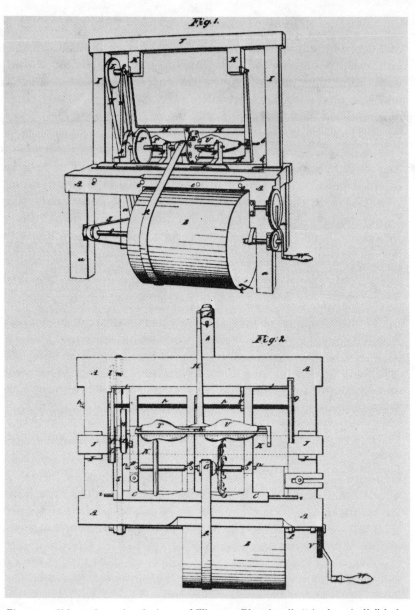

Figure 5. Side and overhead views of Thomas Blanchard's "single spindle" lathe for turning irregular forms. Note the hanging lathe (*H*), sliding carriage (*C*), cutter wheel (*E*), friction wheel (*F*), threaded holdfast (*g*), feed screw (*p*), master pattern (*T*), and workpiece (*V*)—in this instance, a shoe last. Patented September 6, 1819; reissued January 20, 1820. U.S. Department of Commerce, Patent Office. Drawing photographed by Frederick J. Wilson.

carriage which moved along parallel guideways and supported a cutter wheel and dumb tracer in three "poppet heads" or bearings with adjustable center screws; a pincerlike threaded holdfast which, upon engaging the revolving feed screw, gave longitudinal motion to the sliding carriage; and a swinging cast-iron frame suspended from pivots nearly seven feet above the floor which held the master pattern and workpiece in place on the spindle axis by means of a dog, center points, and an adjustable screw. The machine received power from three leather belts attached to a "drum" and two smaller pulleys located on a revolving shaft between the forelegs of the wooden frame. Its operation was summarized as follows:

When the drum is put in motion, the band which passes about it, puts in rapid motion the cutter wheel, while the band which passes from one of its little pullies to the feeding screw pulley, puts this in motion, and gives motion to the sliding carriage from left to right, and the other pulley on the drum axis puts in slow movement the pattern wheel and rough material in a direction opposite to that of the cutter. The friction of the pattern against the friction wheel, by the bearing of the hanging lathe against it, puts this in rotation at the same time that this prevents the swinging lathe from bringing the axis of the rough material in the smallest degree nearer to the cutters than is the axis of the pattern from time to time to the periphery of the friction wheel.

The consequence of which movement is, that the cutter chips away all the substance of the rough material, which is farther from its axis than the surface of the pattern is further from the axis of the pattern, and of course forms from the rough material an exact resemblence of the model.

Since the elemental "single-spindle" lathe Blanchard illustrated in his patent specifications could turn only short articles such as shoe lasts, certain alterations had to be made to accommodate the greater length of the musket stock. As over half his written description of the machine was devoted to a discussion of "different modes of application by placing some or all of its different parts in different positions," Blanchard had little difficulty solving the problem. In fact, he suggested at least three possible arrangements. "To turn an article which is long as a gun stock," he explained, "the frame may be lengthened, or the rough material placed above or below the model, and the cutter wheel placed above or below the friction wheel." However, in building the machines for Harpers Ferry and Springfield, the inventor chose to ignore these designs. Instead, he attached the pattern and rough stock to separate spindles on opposite sides of a swinging V frame and, in like manner, placed the cutter and tracer parallel to each other on the sliding carriage (Figure 6). Essentially the relation of the pattern to the tracer and the workpiece to the cutter on this so-called "double-spindle" machine remained the same as those on the "single-spindle" lathe for turning shoe lasts. This was extremely important since, to produce an exact copy of the master pattern, the axis of the workpiece had to be kept the same distance from the edge of the cutter wheel as the axis of the pattern was from the edge of the dumb tracer. Accordingly in both machines the tracer determined the path of the circular cutter as the pattern and workpiece revolved in unison. Once the cutter had traversed the workpiece an attendant released the holdfast nut from the screw feed, removed the finished stock from the lathe, and returned the sliding carriage to its original position by means of a rope wound by a hand-cranked roller. This completed the cycle of operation.[44]

In addition to his patented lathe, Blanchard also designed and perfected thirteen different woodworking machines to facilitate the production of gunstocks. For instance, he apparently built a machine called a "bore planer"—which grooved the stock for the barrel—at the same time he installed his rough turning lathe at Harpers Ferry. His correspondence of 1819 also indicates that he had prepared a model for cutting in the locks and mounting many months before he actually erected a full-sized machine in August

44. Specifications of Thomas Blanchard, January 20, 1820, OCO; Fitch, "Report on Interchangeable Mechanism," p. 629. Originally issued on September 6, 1819, the date of Blanchard's patent was subsequently altered by act of Congress to January 20, 1820, and twice reissued on June 30, 1834, and January 30, 1848. Contemporary descriptions of Blanchard's machinery at Harpers Ferry appear in Karl Bernhard, Duke of Saxe-Weimar Eisenach, *Travels through North America during the Years 1825 and 1826*, 2 vols. (Philadelphia: Carey, Lea & Carey, 1828), 1:187; Anne Royall, *The Black Book*, 3 vols. (Washington, D.C.: By the author, 1828), 1:285; and Joseph Martin, *A New and Comprehensive Gazetteer of Virginia, and the District of Columbia* (Charlottesville, Va.: Moseley & Tompkins, 1835), pp. 371–372. Martin's description is by far the best.

Figure 6. Blanchard's lathe adapted to gun stocks. Note the changed position of the pattern in relation to the workpiece. Reprinted from Fitch, "Report on Interchangeable Mechanism," figure 9. Drawing photographed by Frederick J. Wilson.

1820.[45] Evidently the device did not work very well because six months later the inventor notified Lee that he had made improvements: "I can cut in the whole lock with great dispatch and exactness let the variation of the plate bear [as] it may I can make a good joint to evry lock. . . . I have discovered a method by which I can vary the jig and set it to every lock . . . —it is done by a verry simple method—I am about to commense building a machine for the above mentioned purpose, and will [continue] on the same in my shop until I can do as good work as can possibly be done by hand."[46] Confident that his recessing machine could "make a closer joint in one minute than a Stocker can in one *hour*," on April 13, 1821, Blanchard offered to come to Springfield and "combat a whole battallion of stockers." While the outcome of the contest is unrecorded, the inventor by June 1822 had sufficiently developed his machinery to propose a contract to half-stock all the muskets made at Springfield for thirty-eight cents apiece. After careful consideration, Lee recommended the measure to the Ordnance Department for three reasons. First, he believed the government would save at least three cents a stock by hiring Blanchard as an inside contractor. Second, he thought it would give the inventor steady employment and, at the same time, "test the utility of the plan, & ascertain what can be saved by adopting his improvement." "But the principal object," Lee emphasized, "is to bring the Machinery

45. Blanchard to Lee, February 5, 1819, Stubblefield to Lee, August 25, 1820, Letters Received, SAR; Lee to Stubblefield, September 9, 1820, Letters Sent, SAR; Stubblefield to Wadsworth, May 31, 1819, Blanchard to Captain William Wade, November 9, 1823, Correspondence and Reports Relating to Experiments, OCO.

46. Blanchard to Lee, February 19, 1821, Letters Received, SAR. Blanchard's machinery for recessing the stock for the side, trigger, and butt plates was built on the same principle as his lock-recessing machine. According to Asa H. Waters, Blanchard's machine "for morticing into the stock . . . was mounted on a moveable frame" and "would cut on a straight line, bore a round hole, cut down and round in any direction, so that when the mortice was completed no kid glove could be fitted to the human hand so closely as was the lock to the stock" (Waters, "Thomas Blanchard," p. 257). Although the recessing machine installed at Springfield has no contemporary description, it was similar in function to one Blanchard later built for the Allegheny arsenal near Pittsburgh. See "An Estimate of the Cost of Machinery for makeing Gun Carriages for the U. States," February 2, 1830, Letters Received, OCO.

to the most perfect State." Since a private contractor could hardly be expected to underwrite such an expensive experiment, the superintendent thought the United States had to shoulder the burden and, in a sense, subsidize Blanchard's project.[47]

After well over a month of negotiations, Blanchard and Lee finally came to terms in August 1822. Furnished with shop space, water privileges, raw materials, and the general use of tools and machinery at the armory, Blanchard agreed to provide his patented machinery royalty free, hire his own workmen, and half-stock muskets at thirty-seven cents apiece. With the contract to take effect the following spring, he spent the intervening period at Millbury preparing several new machines to take to Springfield. In September he completed a new lock-recessing machine and soon afterward began work on "one to cut on *bands*" and another "to cut a square shoulder in the stock for the breech of the barrel." After arriving at Springfield in May 1823, the inventor steadily continued to elaborate and improve his designs. By November he had nine different machines in operation at the armory. During the next three years he added five more units to his mechanical repertory. These fourteen machines completely mechanized the process of stocking and eliminated the need for skilled labor in one of the three major divisions of armory production. Rarely have the contributions of one person effected so sweeping a change in such a short period of time.[48]

While Blanchard's mechanical experiments went smoothly, not everything connected with his work at Springfield proceeded

47. Blanchard to Lee, April 13, 1821, Letters Received, SAR; Lee to Bomford, June 19, 1822, Letters Sent, SAR; Lee to Bomford, June 1, 1822, Letters Received, OCO. In his letter of June 19, Lee told Bomford "what we term half stocking is to face and turn the stock, fit on the heel plate, let in the barrel, put on the bands, fit in the lock & trigger plate and bore the holes for the side & tang pin;—The other half is to let in the side plate & guards, hang the trigger, make the groove & bore the hole for & fit the ramrod, let in the band springs, smooth & oil the Stock."

48. Lee to Blanchard, July 16, 1822, to Bomford, August 7, 22, 1822, Letters Sent, SAR; Blanchard to Lee, July 25, September 2, 1822, March 13, 1823, Letters Received, SAR; Lee to Bomford, August 1, 1822, Letters Received, OCO; Blanchard to Wade, September 1, November 9, 1823, Correspondence and Reports Relating to Experiments, OCO; Blanchard, February 7, 1828, Contract Book No. 1, OCO.

trouble free. Apparently certain difficulties existed for, on July 16, 1825, Lee wrote the chief of ordnance that Blanchard "is ambitious to make all he can, hires men & boys at very low wages, that know little or nothing of the business, & is often changing hands." These circumstances not only tended to lower the quality of workmanship but also fostered ill feeling among the regular armorers, thus making Lee's job as superintendent "rather unpleasant." Convinced that Blanchard had made about "as much improvement in his machinery as he probably will" and that his own workmen could do a better job stocking muskets, Lee suggested terminating the inventor's contract for half-stocking and simply paying him a flat rate fee for the use of his machinery. With Colonel Bomford's approval, he proposed the plan to Blanchard but met with little success. In November 1825 he reported, "I have for a long time been urging Blanchard to make a definite proposition, but he seems to have a wish to evade it." With Lee offering six cents a musket and Blanchard asking as high as ten, price considerations seemed to be the major source of disagreement. Negotiations lingered on more than two years. Finally, in the fall of 1827, agreement was reached with Blanchard to grant "the free use of all his improvements . . . upon the condition that he shall be paid nine cents for each musket made at the National Armories" up through the expiration of his patent on January 12, 1834. Until then he continued as an inside contractor at Springfield on a year-by-year basis. From 1823 until his departure on December 31, 1827, Blanchard received more than $18,500 in patent fees from the national armories alone. These receipts, plus additional sales of machinery to the Boston Company, Lemuel Pomeroy, Marine T. Wickham, and the national arsenals at Pittsburgh, Washington, and Watervliet, indicate that Blanchard made a handsome profit on his patent for turning irregular forms.[49]

Even though he had originally praised Blanchard's lathe and bore planer in 1819, Stubblefield seemed undecided about introducing

49. Lee to Bomford, July 16, November 12, 1825, Letters Sent, SAR; Bomford to James Barbour, November 13, 1827, Letters Sent to the Secretary of War, OCO; Blanchard, February 7, 1828, Contract Book No. 1, OCO; *American State Papers: Military Affairs*, 5:915–916; Deyrup, p. 98.

the rest of the inventor's stocking machinery at Harpers Ferry. Undoubtedly he hesitated partly because Blanchard demanded exorbitant fees which tended to undercut any savings resulting from the use of the machinery.[50] Stubblefield was also influenced by a group of workmen, headed by his brother-in-law, master armorer Armistead Beckham, who expressed hostile feelings toward "Yankee notions," no matter how useful or ingenious. To skilled stockers, like Michael Gumpf, the introduction of Blanchard's machinery not only threatened to bastardize the art but to lower monthly wages. Indeed, the very idea of becoming a mere tender of machines seemed demeaning. Confronted with these sentiments, Stubblefield steered clear of potential strife by maintaining silence concerning the adoption of Blanchard's improvements. Consequently the decision to install a complete set of stocking machinery was not made until Lee temporarily replaced Stubblefield as superintendent at Harpers Ferry in 1827. In this case as in others, persons not permanently associated with the armory had to assume responsibility for introducing new mechanical improvements.

Drawn up on January 19, 1827, the contract stipulated that Blanchard would build nine machines and put them into complete operation at Harpers Ferry for $1600. When added to the eccentric lathe and the barrel groover erected in 1819, these units would be capable of carrying out all major operations on the stock. The government also promised to pay a royalty fee of nine cents on each musket, a commitment that was confirmed later in the year when Lee expanded the agreement with Blanchard to include both national armories. Although the machines were completed in May and shipped to Georgetown in June, Stubblefield, back in office at Harpers Ferry, postponed their installation until October. Even then the unfinished condition of millwork at the stocking shop further delayed the project. Originally Blanchard had expected to have the machinery assembled, adjusted, and in full working order

50. Since 1819, Stubblefield had repeatedly failed to pay a six-cent royalty for each musket stock turned and grooved by Blanchard's machinery. By 1823 he had fallen two years behind in these payments; by October 1825, sixteen months. Despite numerous complaints from the inventor and several reprimands by the Ordnance Department, Stubblefield continued to default on these payments until his removal as superintendent in 1829.

no later than December 1. As it turned out, the machinist he sent from Springfield did not complete the assignment until March 19, 1828, nearly four months behind schedule.[51]

Throughout this chapter considerable emphasis has been placed on various facets of cooperation between the armories. While limited communications between Springfield and Harpers Ferry were established as early as 1808, the first real impetus toward an enduring collaborative effort came after the War of 1812. Initiated by the Ordnance Department, pursued by Roswell Lee, and countenanced by James Stubblefield, both armories not only shared general administrative information but exchanged men, machinery, and raw materials as well. While everyone profited from the experience, the opening of these channels particularly favored Harpers Ferry because new technical knowledge tended to flow from Massachusetts to Virginia. In fact, only one instance of borrowing on Springfield's part was recorded between 1815 and 1829: in 1818 Lee asked to be furnished with a draft and description of a forge furnace used at Harpers Ferry. Stubblefield, on the other hand, received a considerable number of patterns and drawings from Springfield. Among them numbered plans for water wheels, triphammers, and various special purpose machines for milling, drilling, trimming, and turning iron components. Together with inspection gauges for the Model 1816 musket and the inventions of Sylvester Nash and Thomas Blanchard, these exchanges amply signify the extent to which Harpers Ferry had begun to rely on New England technology. At the same time the introduction of these improvements marked the beginning of a gradual trend away from craft-oriented methods toward the increased use of machinery at the Potomac armory.[52]

51. Blanchard to Ordnance Department, January 19, 1827, to Wade, May 25, 1827, Letters Received, OCO; Blanchard, Contract Book No. 1, OCO; Bomford to Rust, November 3, 1830, Letters Sent, OCO; Lee to Stubblefield, June 25, September 11, 1827, Letters Sent, SAR; Stubblefield to Lee, September 14, 1827, Letters Received, SAR.

52. See, for example, Lee to Stubblefield, June 6, 1816, December 23, 1818, Letters Sent, SAR; Stubblefield to Lee, January 4, 1823, Letters Received, SAR; Lee to Senior Ordnance Officer, January 10, 1818, Extracts from Colonel Archer's Report, November 25, 1823, Letters Received, OCO; "Notes of a Tour of Inspection," Harpers Ferry armory, December 1822, OIG.

As much as cooperative efforts may have induced changes at Harpers Ferry, it would be erroneous to conclude that armorers as well as managers simply shed old habits and espoused the new technology. Even though many techniques were placed at the armory's disposal, they were neither awaited eagerly nor accepted with enthusiasm. The reluctance to utilize Blanchard's machinery offers only one instance of the armory's failure to avail itself of important transitional innovations. Another striking example is the use of triphammers to weld gun barrels. There is no simple explanation why it took nearly twenty-five years for such a basic concept to take hold at Harpers Ferry. Certainly tradition-bound artisans such as George Malleory and Joseph Hoffman did not prove particularly receptive to mechanization. On more than one occasion they had voiced opinions that associated the introduction of a new machine or process with visionary schemes and charlatanism. As they were older men who frequently held supervisory positions within the armory, their disenchantment with novelties undoubtedly rubbed off on less skilled members of the labor force. Combined with the sentiments of close-knit local kinship groups that held outsiders suspect and outside ideas alien to customary practices, these inbred feelings fostered curious technological conservatism at Harpers Ferry.

Ordinarily the production worker remained a neutral element in the decision-making process at the Potomac armory. He neither introduced new techniques nor did he openly oppose them—that is, so long as such changes did not imperil his job or affect his wages. Whenever one or the other of these things occurred, factory masters could count on a confrontation either directly with the disaffected party or indirectly through his political representatives. Perhaps this factor best explains why Stubblefield seemed reluctant to replace men with machines. If so, it closely parallels a more subtle relationship. As a leading member of a small but influential family clique, Stubblefield was inextricably tied to important sources of private profit and political patronage at Harpers Ferry.

CHAPTER 6

James Stubblefield: Virginia Entrepreneur, 1815–1829

Entrepreneurial activity has many ramifications affecting both the nature of innovation and patterns of industrial development. From a purely utilitarian standpoint the ideal entrepreneur creatively molds men, machinery, and material into an efficient system of production. While the degree of success or failure of a particular venture depends on the end being sought, the decisions of the individual entrepreneur are invariably influenced by the values and anticipations of close associates as well as other groups in the community. By externalizing group goals and cultural predispositions he constantly affirms the feelings and attitudes of those from whom he derives much of his decision-making strength.[1]

One anomaly of James Stubblefield's position as superintendent is that he had two constituencies, one in Washington, the other at Harpers Ferry, whose aims and aspirations rarely coincided. As a duly appointed government official he was expected to operate the armory with cool, dispassionate efficiency. As a community leader he was expected to play a paternalistic role, providing employment for friends and neighbors, patronizing local businessmen, and otherwise seeing to it that similar interests were served. While Roswell Lee succeeded in reconciling these divergent influences at

---

1. These remarks are particularly influenced by the writings of Thomas C. Cochran and Robert K. Lamb. See, for example, Cochran, *The Inner Revolution* (New York: Harper & Row, Torchbooks, 1964), pp. 91–143; and Lamb, "The Entrepreneur and the Community," in *Men in Business*, ed. William Miller (New York: Harper & Row, Torchbooks, 1962), pp. 91–119.

Springfield, circumstances along the Potomac gave rise to a totally different configuration of events. Imbued with a strong sense of noblesse oblige and deeply committed to the maintenance of local prosperity, Stubblefield followed an entirely honest if controversial policy at the government's armory of placing community needs ahead of national exigencies. This position became increasingly evident as administrative changes at Harpers Ferry elevated the superintendent's status from a production manager of limited authority to a fully responsible entrepreneur.

Up through the War of 1812, Stubblefield's entrepreneurial role had been curtailed because he was required to share decision-making authority with the paymaster. In 1815, however, several events significantly altered the superintendent's sphere of power at the armory. After fifteen years of steady service, Samuel Annin's health had declined so much that he found it "impossible" to discharge the duties of his office with effectiveness and punctuality. The aging paymaster therefore submitted his resignation on May 15, and the secretary of war immediately appointed Major Lloyd Beall, a long-time resident of Harpers Ferry and president of the Farmers and Mechanics Bank, to fill his place.[2]

Although Annin's departure automatically gave Stubblefield seniority at the armory, an event of far greater significance soon took place. During the summer of 1815, Colonels Wadsworth and Bomford, newly arrived at their jurisdiction over armory affairs, drew up a set of regulations aimed at transforming the Virginia installation from a loosely knit group of workers into a well-disciplined organization with a clear bureaucratic chain of command. Scheduled to take effect in January 1816, these regulations completely revised, even reversed, the old administrative power structure at Harpers Ferry. Leaving no room for ambiguity, they clearly outlined the duties, accounting procedures, and spheres of authority for each officer. More important, they not only enlarged the superintendent's power as director of all operations at the armory but

2. Annin to Secretary of War, May 15, 1815, Letters Received, OSW; Alexander J. Dallas to Beall, June 7, 1815, John Smith to Stubblefield, September 24, 1810, Letters Sent Relating to Military Affairs, OSW; Beall to Colonel George Bomford, May 17, 1815, Letters Received, OCO.

specifically stated that the paymaster would be subordinate to him. Henceforth, the regulations stipulated, all correspondence related to armory matters would be sent to the superintendent. In sharp contrast with former practices, he alone had the right to make contracts, hire workmen, and supervise construction. The paymaster, in turn, simply kept a record of these transactions and acted as a disbursing agent for the armory.[3]

The administrative system at Harpers Ferry unquestionably needed to be revamped and streamlined in 1815. However, the reforms instituted by Wadsworth and Bomford consolidated control to such an extent that no means of checking the actions of the superintendent existed other than direct intervention by the chief of ordnance himself. Because communications throughout the country were slow and uncertain in 1816, even this potential source of restraint proved far from effective. Thus, while Stubblefield was nominally subordinate to military officials in Washington, he actually had a free rein at Harpers Ferry under the new system. The only specific restrictions imposed upon him stipulated that he could neither purchase land nor erect new buildings without approval from the Ordnance Department. Even so, as later records indicate, Stubblefield did construct a number of buildings—mostly dwelling houses—at the government works without the prior knowledge or sanction of the chief of ordnance.

In addition to Annin's retirement and the subsequent demotion of the paymaster to the status of bookkeeper, another important change occurred in 1815 involving the third highest office at the armory. Ever since Marine Wickham's departure for Philadelphia in 1811, Stubblefield had been without the services of a master armorer. Although he frequently pleaded for an appointment and had recommended at least three individuals for the job, the secretary of war, burdened with more pressing military responsibilities, took no action. The situation did not change until August 1815 when, after receiving a copy of the new armory regulations, Stubblefield approached Bomford about the appointment of a mas-

3. Stubblefield to Bomford, August 24, 1815, October 5, 1822, Beall to Bomford, February 6, 1816, Letters Received, OCO; Undated regulations, Letters Received, SAR. These regulations also applied to the Springfield armory.

ter armorer and submitted the name of Armistead Beckham. Bomford recommended the measure to Wadsworth who, in turn, sent the nomination to the War Department for approval. After a three-month wait, Secretary William H. Crawford announced Beckham's appointment on December 9, 1815.[4]

Ordinarily the selection of a new master armorer would call for little comment. In this particular instance, however, Armistead Beckham happened to be the brother-in-law of James Stubblefield. Even such a connection would not have been particularly noteworthy had it ended with the master armorer. But interconnecting family bonds weighed heavily at Harpers Ferry, so much so that an extended kinship group exercised almost absolute control over the destinies of the town's fourteen hundred inhabitants. Birth or marriage into one of these well-to-do "gentry" families secured powerful social and economic connections. An endorsement from one of their members was a sure guarantee of employment and comfortable security. Consequently if a young man contemplated a career as an armorer or businessman, he almost always sought the gentry's support. In like manner, it behooved prospective politicians to curry favor with these families because the way they voted influenced the political behavior of other members of the community who depended on them for jobs, credit, and other forms of patronage. Since open balloting prevailed in Virginia through the Civil War era, each man's vote became public record. Rather than risk reprisals, he tended to vote as his wealthier benefactors did. A vote for the "wrong" candidate could and often did result in dismissals, demotions, loss of business, foreclosures, and other social and economic harassments.[5]

Existing records plainly show that an inner circle of privilege did exist at Harpers Ferry and that it centered around the superintendent's wife, Mary Beckham Stubblefield (Figures 7 and 8). Besides the master armorer, she had at least three other brothers who were influential citizens and prominent in community affairs.

4. Stubblefield to Bomford, August 24, 1815, Beckham to Crawford, December 13, 1815, Letters Received, OCO.

5. John H. Hall to John C. Calhoun, March 25, 1824, Letters Received, OCO; Major William Wade to Lee, April, 14, 1827, Letters Received, SAR.

Figure 7. Oil portrait of James Stubblefield (1780–1855) by George Cook, 1822. From the collection of Mrs. Sidney Marsh Cadwell. Photograph courtesy of Mrs. Charles Green Summers.

Figure 8. Oil portrait of Mary Beckham Stubblefield (1780–1834) by George Cook, 1822. From the collection of Mrs. Sidney Marsh Cadwell. Photograph courtesy of Mrs. Charles Green Summers.

Camp Beckham, "who was not a mechanic and never present for work," held a lucrative inside contract at the armory for browning muskets. Townshend Beckham, with Stubblefield as a silent partner, operated a large tannery and oil mill on the adjacent island of Virginius (see Map). Fontaine Beckham, the most prosperous member of the family, owned a dry goods store in the center of town and a flour mill and coopers shop on Virginius. He also held a commission as justice of the peace in Jefferson County and later became the local agent for the Baltimore & Ohio Railroad.[6] In addition to these interests, he acquired even higher standing in the community by marrying the daughter of Major James Stephenson. Unquestionably one of the most respected and powerful men in the region, Stephenson boasted a distinguished political record first as a Federalist and later as an anti-Jacksonian. Having served well over a decade as a Berkeley County magistrate, six years as a member of the Virginia House of Delegates, and four terms in the House of Representatives, Stephenson had many well-placed political connections. Among his closest friends in Washington were Senator James Barbour of Virginia and Henry Clay of Kentucky. Although he maintained his legal residence in Martinsburg, Stephenson also had extensive property holdings and business interests outside Berkeley County. At Harpers Ferry he kept a "neat" inn and owned a distillery in partnership with Stubblefield. Described as "a very amiable man" who "excelled in invariably striking a quarter of a dollar with a single ball at thirty paces," the major proved to be a most valuable "silent member" of the Beckham-Stubblefield clan.[7]

6. Thomas Copeland to Wade, May 24, 1826, Hall to John H. Eaton, May 25, 1829, General John H. Wool to Bomford, May 3, 5, 1827, G. Z. to Secretary of War, July 15, 1841, Letters Received, OCO; Wade to Lee, April 14, 1827, SAR; Payrolls and accounts, Harpers Ferry armory, 1820–1826, Second Auditor's Accounts, GAO; Charles W. Snell, "A Short History of the Island of Virginius, 1816–1870" (Research report, HFNP, 1959), pp. 10–11; Millard K. Bushong, *A History of Jefferson County, West Virginia* (Charlestown, W.Va.: Jefferson Publishing Co., 1941), p. 402; Barry, *Strange Story of Harper's Ferry*, pp. 59, 84–86.

7. Nahum W. Patch to Bomford, December 14, 1825, Copeland to Wade, May 29, 1826, James Stephenson to James Barbour, November 20, 1826, G. Z. to Secretary of War, July 15, 1841, Letters Received, OCO; Bushong, p. 305; Bernhard, *Travels*, 1:185; Royall, *The Black Book*, 1:287; Edward T. Coke, *A Subaltern's Furlough* (New York: J. & J. Harper, 1833), 1:120.

Completing the circle of family relationships at Harpers Ferry was the Wager family. A niece whom Mrs. Stubblefield had raised from childhood married Edward Wager, the eldest son of John and Catherine Wager. Edward helped his mother manage the family retail business after his father's death and also served as chief clerk to the superintendent at the national armory. The Wagers, it will be recalled, originally sold the land on which the armory stood to the United States in 1796. At the same time, they reserved a six-acre plot of land and the ferry concession for themselves at the junction of the Potomac and Shenandoah rivers. As the town and armory grew, this "Ferry Tract" became an extremely valuable piece of real estate for, other than the island of Virginius which Stubblefield surreptitiously acquired in 1823, the Wager family owned all the private property in the heart of Harpers Ferry.[8] Furthermore, Catherine Wager claimed that when Tobias Lear had made the purchase for the government in 1796 he guaranteed "that the Person living on her land shall have a monopoly of the mercantile Business at the Place, & that the Persons employed about the Armory, shall be constrained to take all supplies from the Persons occupying her land and tenements." Although Anne Royall, a contemporary journalist who visited the Ferry, described Mrs. Wager as "a very fine woman," Wadsworth considered her a petty tyrant and a most "troublesome person to have any thing to do with." Apparently the elderly dowager made her monopolistic claim stick for, in 1818 and again in 1820, the chief of ordnance complained to the secretary of war:

If it be admitted that the Persons occupying the Wager property . . . shall have the exclusive privilege of supplying the people of the Armory with all they want to purchase, the persons occupying that property will be empowered to tax the Armory almost at Discretion. Mechanics constantly employed have neither the time or means to procure supplies from a distance. They must purchase on the spot at any price. I can see no

8. Henry St. George Tucker to Calhoun, July 12, 1823, Hall to Calhoun, March 25, 1824, Letters Received, OCO; Colonel George Talcott to J. M. Porter, August 17, 1843, Letters Sent to the Secretary of War, OCO; Payrolls and accounts, Harpers Ferry armory, 1817–1818, GAO; Royall, 1:287; *American State Papers: Military Affairs*, 4:143. See also Chapter 1 at note 12.

reason why the Wages should be so much higher at Harpers Ferry than at Springfield except it be that the workmen are better & cheaper supplied with the necessaries & conveniences of life at the latter place.

The price of flour for instance is always higher at Harpers Ferry than at George Town, although a great portion of the flour sold in George Town comes down the Shenandoah & Potomac passing directly by Harpers Ferry. The country on the Shenandoah is one of the most fertile & plentiful in the U. States, & provisions ought to be as cheap at Harpers Ferry as any where else, & would be, were the Trade there perfectly open & free with a full competition among the Dealers.[9]

The armory's monthly payroll averaged well over $10,000, and most of this money was spent either in the Wager store or in establishments leased from them; these facts indicate why Anne Royall attributed such great wealth to the family. By virtue of their centrally located piece of real estate, Mrs. Wager and her sons were able to control the private sector of the economy at Harpers Ferry for nearly thirty years. Prior to 1825 they alone determined who would do business at the Ferry and, through high rents and no leases, made them pay dearly for the privilege. Consequently Edward Wager's marriage to Stubblefield's adopted daughter not only expanded the inner circle to a four-family relationship but also mutually strengthened their economic ascendancy as well. Because the Beckhams, Stephensons, Stubblefields, and Wagers combined political influence and social standing with exclusive property ownership, they formed a powerful family oligarchy—or "Junto" as it was called—which literally dominated every aspect of life at Harpers Ferry—including the national armory—between 1815 and 1829. Briefly stated, their property produced wealth; their wealth guaranteed status; their status conferred privilege; and privilege insured power.

Other members of the community aligned themselves with the family clique at Harpers Ferry despite marked differences in social status. If for no other reason than the monopolistic privileges they enjoyed as businessmen, Lewis Wernwag, John G. Wilson, Samuel

9. Catherine Wager to Calhoun, September 9, 1818, Wadsworth to Calhoun, September 14, 1818, *Calhoun Papers*, 3:102, 128; Wadsworth to Calhoun, April 13, 15, 1820, Letters Sent to the Secretary of War, OCO; Royall, 1:287.

K. White, John McFarland, Philip Coontz, and the Strider brothers were just as eager as the Junto to see to it that Harpers Ferry remained a closed "company" town. Since their continued prosperity hinged on maintaining the status quo, this group closely identified its political and economic interests with those of the Beckhams, Stephensons, Stubblefields, and Wagers.[10]

In like manner, a majority of the labor force—particularly those who came from the vicinity—courted and supported the Junto. Those who did not either remained silent or risked joining the ranks of the unemployed. While workers from foremen to helpers exhibited some class consciousness, there is no indication that any pronounced class conflict existed at Harpers Ferry. This is not to say, however, that they docilely accepted subordination. Somewhat paradoxically, they presented an odd mixture of deference and defiance toward their social superiors. Although they complained about the unhealthiness of the climate and the high cost of living, these men seemed quite satisfied with their treatment at the armory. Besides being the highest paid group of armorers in the country, they were subject to few restrictions and could come and go from work whenever they pleased—a practice rarely countenanced in other establishments. They also realized that they held their positions at the superintendent's discretion and that no better situations could be found elsewhere. Therefore, as much as they may have envied or even disliked Beckham or Stubblefield, few armorers seemed willing to advocate anything that might uproot the established order at Harpers Ferry and thereby jeopardize their means of livelihood.[11]

10. Copeland to Wade, May 24, 1826, OCO; Wade to Lee, April 14, 1827, SAR; Royall, 1:287.

11. Stubblefield to James Monroe, December 9, 1814, to Bomford, April 1, 1829, Copeland to Wade, May 5, 1825, Patch to Bomford, January 25, 1826, Wool to Bomford, May 3, 1827, Benjamin Moor to Bomford, June 29, 1841, Letters Received, OCO; Asa Wood, et al. to Lewis Cass, January 1, 1833, Reports and Correspondence of Ordnance Boards, OCO; Talcott to John C. Spencer, June 25, 1842, Letters Sent to the Secretary of War, OCO. Although Springfield paid the highest wages of any of the New England armories, those at Harpers Ferry were still 12 to 20 percent higher in 1818. At the same time Lee required his armorers to work a ten-hour day while those at Harpers Ferry averaged from six to eight hours.

The highly concentrated nature of property holding and business enterprise at Harpers Ferry afforded many opportunities for exploitation. Working families, for example, had no choice but to buy from local retailers who, by carefully controlling the market and demanding exhorbitant prices for food, clothing, and other necessities, kept them in a perpetual state of debt. Government-owned housing was allocated on a discriminatory basis, sometimes even going to persons who had no connection with the armory other than friendship with the superintendent, the master armorer, or one of their kinsmen. Individuals considered "advantageous to the interests of the armory" were also given free building privileges on the public property, where they erected boarding houses, shops, and other business establishments, but rarely in competition with those already owned or leased by the Wager family.[12] Since tenants were subject to eviction, housing became a potent means of exercising social control. Often the mere threat of such a reprisal quickly caused persons who had become troublesome or indiscreet in their remarks about the Junto to hold their tongues, correct their behavior, and attempt to return to favor with the ruling families.

Yet even these practices had certain limits. Here, as always, a spirit of give-and-take came into play. This can perhaps best be seen in Stubblefield's attitude toward drinking. Most factory masters considered tippling a pernicious influence and banned it from their establishments. At Springfield Lee had absolutely forbidden "the drinking of Rum, Gin, Brandy, whiskey, or any kind of ardent spirits" under pain of immediate dismissal from the armory.[13] Yet Stubblefield winked at the practice and even tacitly encouraged it both during and after working hours. Since he owned part interest in a distillery and had at least one in-law who owned a tavern, every drop consumed increased the family fortune. At the same time, the superintendent's leniency allowed armorers to retain a long-prized freedom. As long as they were willing to pay for the privilege and not get too rowdy, he refrained from acting as a moral policeman. Nor did he ever attempt to devise work rules or any other formal regulations aimed at enforcing discipline, time-thrift, or other

12. *American State Papers: Military Affairs*, 4:143.
13. Undated regulations (ca. 1816), Letters Received, SAR.

forms of austerity. Stubblefield and Armistead Beckham were neither flogging masters nor efficient managers, but they did possess
an instinctive grasp of village culture at Harpers Ferry and were
careful not to push the armorers too far. Every penny or vote
exacted from the workingman came with reciprocal demands and
expectations. Among the most important was the maintenance of
craft customs and social conventions associated with a rural way of
life. Profit and patronage, gains and concessions, manipulation and
mollification—these subtle and often contradictory impulses became the cement that bound the community together.

Family ties, business interests, and labor relations undoubtedly
influenced Stubblefield's conduct as superintendent. In any predominantly rural society—town or plantation—where stratification
prevails, family pride, provincial attitudes, and a sense of responsibility to one's less fortunate neighbors count for much among local
leaders. Stubblefield's background and personality indicate that he
possessed all these qualities. As a "gentleman of the true Virginia
stamp," he viewed his official actions as they affected the welfare of
various groups within the community. That he did so to the satisfaction of workers as well as businessmen helps to explain his general popularity and the lack of class antagonism at Harpers Ferry
during the early nineteenth century. Acting out this paternalistic
role, however, necessitated a crucial distinction. Instead of viewing
the armory as a national establishment whose purpose was to manufacture arms as rapidly and efficiently as possible, Stubblefield
looked upon it as a source of jobs, sinecures, patronage, and profit.
In his opinion, as well as that of Armistead Beckham and many
others, the armory represented a public dole of sorts, a quasi-
welfare project in which the efficient production of firearms was a
secondary consideration. As such, the superintendent and his associates resented and resisted any attempts by outsiders to meddle in
what they considered primarily a local affair.

By far the most vexatious outsider was the chief of ordnance. At
first relations between Stubblefield and his military superior, Colonel Wadsworth, were rather smooth. The new Ordnance Regulations of 1815 certainly worked to the superintendent's advantage,

and, while not overly enthusiastic about the idea, he seemed willing to cooperate in establishing more uniform standards of manufacture at the national armories. By 1818, however, sources of conflict and ill-feeling began to emerge. One particularly annoying point of contention had to do with expenditures. Under Annin the armory had had perennial budgetary deficits dating back to 1812. To clear Annin's books and give the works a clean fiscal slate, Wadsworth released extra funds so that the superintendent could pay off these arrearages in 1816 and 1817. Thereafter he expected Stubblefield to keep within the bounds of the annual appropriations made for the armory. Yet, after the last of Annin's debts were liquidated, Wadsworth learned that Stubblefield had exceeded his budget by nearly $24,000 in 1818 and that armory accounts revealed a further deficit of $29,000 in 1819. Upset, the colonel openly expressed his displeasure in several letters to the superintendent and the secretary of war during the first part of 1820. At the same time he began to perceive that at least one reason for needless expenditures at the armory arose from the high cost of labor caused by the stranglehold a few privileged individuals exercised over the town's economy. To break this "odious monopoly" he proposed building a store on government property at Harpers Ferry that would enable armorers to purchase food, clothing, and other necessities at more reasonable prices. This measure, he argued, would then allow him to reduce wages and help correct the armory's deficit without severely distressing the workers and their families. Although Catherine Wager and her tenants lost little time in quashing the idea at the War Department, Wadsworth nonetheless insisted on carrying out a 12½ percent wage reduction during the summer of 1820. Obviously, these actions did not endear the aging chief of ordnance to Stubblefield, the Junto, or the labor force. Probably Harpers Ferry did not lament when news came of Wadsworth's death in November 1821.[14]

14. Stubblefield to Bomford, March 6, 1816, to Wadsworth, February 19, 1820, Letters Received, OCO; Wadsworth to Calhoun, April 13, 1820, Letters Sent to the Secretary of War, OCO; Wadsworth to Stubblefield, February 1, 1820, Letters Sent, OCO; Wadsworth to Calhoun, September 14, 28, 1820, *Calhoun Papers*, 3:128, 168; Stubblefield to Lee, August 1, 1820, Letters Received, SAR; *American State Papers: Military Affairs*, 2:481; Benet, *Ordnance Reports*, 1:57, 526–527.

If Wadsworth had been bothersome, his successor proved to be a thorn in the Junto's side. A graduate of West Point and an extremely capable engineer in his own right, Colonel George Bomford (Figure 9) exemplified both a new breed of professional soldier and the upper crust of Washington society. His wife, Clara, was the niece of the poet-statesman Joel Barlow and the sister of Congressman Henry Baldwin, a powerful figure in Pennsylvania politics and later an associate justice of the Supreme Court. Bomford's wide-ranging circle of friends included such military notables as Stephen Decatur, Sylvanus Thayer, and Andrew Jackson. Since 1815 he had served as Wadsworth's principal assistant, had played a leading part in formulating national armory policy, and had visited both establishments on numerous occasions. Consequently when Bomford replaced the ailing Wadsworth as chief of ordnance in June 1821, he already possessed firsthand knowledge of the state of affairs at Harpers Ferry.[15]

Strongly committed to eradicating undue local influence at the government works, Bomford kept reminding Stubblefield that his actions had to "be based upon considerations affecting the public interests *solely*; and be influenced in no degree whatever by considerations touching the interests or feelings of those individuals who may be effected by the objects proposed."[16] Immediately after assuming office, he ordered a further 15 percent reduction of wages at the armory; payroll records, however, reveal that the measure was never completely carried out, even by 1824. Like Wadsworth, Bomford continued to reprimand Stubblefield for exceeding budgets and for squandering money needlessly on dwelling houses, public roads, bridges, and other conveniences for the inhabitants of Harpers Ferry. In addition, he constantly badgered the superintendent about the nonpayment of debts, shoddy bookkeeping methods, and questionable employment practices at the armory.[17]

15. For biographical information on Bomford, see Corra Bacon-Foster, "The Story of Kalorama," *Records of the Columbia Historical Society* 13 (1910): 98–118; G. W. Cullum, *Biographical Register of the Officers and Graduates of the U. S. Military Academy*, 6 vols. (New York: Houghton Mifflin, 1891), 1:58–59.

16. Bomford to Stubblefield, April 4, 1826, Letters Sent, OCO.

17. Bomford-Stubblefield correspondence, 1821–1829, Letters Sent-Letters Received, OCO; Bomford to Barbour, April 28, 1826, to Eaton, March 23, 1830, Letters Sent to the Secretary of War, OCO.

Figure 9. Colonel George Bomford, Chief of Ordnance (1821–1842), and the foremost proponent of the uniformity system. Reprinted from the *Records of the Columbia Historical Society* 13 (1910), plate 10. Photograph courtesy of the Smithsonian Institution.

Stubblefield, in turn, tenaciously defended his right under the Ordnance Regulations of 1815 to govern the armory as he saw fit. Although disturbed by Bomford's complaints, he refrained from making any public denunciations of the chief of ordnance. Friends and relations at Harpers Ferry, however, did not exercise the same discretion. As an eccentric but able armory machinist named Thomas Copeland reported the situation to Bomford's assistant, Major William Wade,

It is not . . . my wish to complain of the conduct of Mr. Stubblefield alone—for the truth is there is least of the harm done by him—the Junto that he has about him can get him to do justice or unjustice as they please—and as to the ordnance Dept. with Col. Bomford at there head, they have no right to medle with the private arrangements the Supt. may think proper to make at this place—I have not heard Mr. Stub. say so my self—but i am purty sure he has expressed himself so—but i have heard Wager (the clerk), Rowles and several more such as them say so—and they make no privacy of saying & expressing themselves so to many others and it is and often has been said & it plainly appears they hold it as a right that the supt. is to be the proper Judge what to build up & what to pull down— & the Lord only knows he has exercised this power since I have been here with a vengence to it.

This letter, as well as other correspondence, indicates that a serious jurisdictional dispute had arisen between the local group at Harpers Ferry and the Ordnance Department and that it had become quite heated by 1826. Apparently at one point the Beckham brothers, James Stephenson, Edward Wager, and other interested persons actively sought Bomford's removal from office. This prompted Copeland to warn his friend Wade that "the words & drift of their remarks is that if Colo. Bumford dont look sharp . . . he will be displaced himself before Mr. Stubblefield."[18] Instead of inducing the chief of ordnance to ease his policy of strict accountability, the Junto's boastful threats seemed to make him even more determined that Stubblefield should follow a straight and narrow path as superintendent.

Resentment toward the outsider Bomford was increased by his insistence on placing other outsiders at the armory. Even before he

18. Copeland to Wade, May 24, 29, 1826, Letters Received, OCO.

became chief of ordnance, the Junto was indignant over the role he had played in 1819 with the appointment of John H. Hall as director of the Rifle Works, a newly organized and separate branch of the armory (see Map). Although Secretary of War John C. Calhoun made the final decision to offer Hall a contract to manufacture his patented breechloading rifles at one of the national armories, it was Bomford who first brought the gifted New Englander to Calhoun's attention, vigorously supported Hall's proposal to make arms with interchangeable parts, and in the end selected Harpers Ferry as the site for the project.[19] Despite Bomford's and Calhoun's enthusiasm about replacing common muzzle-loading rifles with Hall breechloaders, Wadsworth remained skeptical. As early as 1818 the cost-conscious Wadsworth had cautioned Calhoun that "experiments made with Arms of that description can hardly lead to any practical result, because we can never afford to introduce such Arms in to the Service." He also had questioned "whether the facility of firing will not occasion an extravagent use and waste of ammunition." Once the secretary made up his mind, Wadsworth had withdrawn his opposition; but he never became an advocate of the Hall rifle.[20]

From the outset of Hall's contract Stubblefield had complained that the New Englander's presence at Harpers Ferry threatened to "derange all the business of the armory." Both he and Armistead Beckham resented Bomford's decision to send a "visionary theorist," particularly a Yankee theorist, to the Potomac when he could just as well have ordered him to Springfield. Pointing out that the armory had been established "to make muskets & not patent rifles," the Junto claimed the Ordnance Department had no legal right to establish a private contractor at a public installation. In short, they thought Bomford had flagrantly overstepped the bounds of his

19. Hall to Bomford, January 24, 1815, April 16, 1819, to Calhoun, March 16, 1819, Correspondence Relating to Inventions, OCO; Hall to Calhoun, February 24, 1819, *Calhoun Papers*, 3:611–612; Hall to Crawford, October 28, 1816, Letters Received, OSW; E. Ward to Lee, November 4, 1818, Letters Received, SAR. The details of Hall's venture are treated in Chapters 7 and 8.

20. Wadsworth to Calhoun, November 16, 1818, to George Graham, November 11, 1816, Letters Sent to the Secretary of War, OCO. Also see Hall to Wadsworth, March 17, 1819, to Calhoun, April 26, 1819, Correspondence Relating to Inventions, OCO; Hall to Bomford, May 3, 1819, Letters Received, OCO.

authority. Why not, Stubblefield asked, avoid trouble and expense by giving Hall a regular contract under the Militia Act of 1808 and letting him make the rifles at his own establishment?[21]

Stubblefield had good reason for wanting to be rid of Hall. Having to share as much as one-fifth of his annual appropriation with the New Englander must have galled the superintendent because less money would be available for projects directly connected with the main Musket Factory. Even more serious, Hall's semi-independent status at the Rifle Works divided authority and threatened to undermine Stubblefield's local power and influence. Anne Royall clearly exposed this issue when, after visiting Harpers Ferry in 1827, she reported that "Capt. Hall . . . appears to be the principal at the Ferry, though Colonel Stubblefield has the name." A personality conflict did not help to improve relations between the two protagonists. Stubblefield thoroughly detested Hall's out-spoken, self-righteous, and egocentric mannerisms. In his opinion, the director of the Rifle Works was a money-grubbing, opportunistic adventurer who "has not only his hand, but his arm up to the shoulders (as the common phrase is) in Uncle Sam's purse." Hall, for his part, considered Stubblefield an envious, deceitful, self-aggrandizing nabob who, for purely selfish reasons, placed private interests ahead of the public welfare at Harpers Ferry.[22]

Before long these antagonistic feelings became manifest. In January 1820, Stubblefield blamed Hall for the armory's budgetary deficit—and he continued to blame cost overruns on the New Englander every year thereafter. Hall, in turn, accused the superintendent of intentionally trying to sabotage his work by withholding funds from the rifle factory and repeatedly sending him undisciplined workmen and faulty raw materials. He further complained that Stubblefield not only charged against his accounts items that were never received but also listed others at prices much

21. Stubblefield to Wadsworth, March 27, 1819, Edward Lucas, Jr., to John Cocke, January 15, 1827, "A Mechanic" (Stubblefield) to members of the U.S. Senate and House of Representatives, April 12, 1830, Letters Received, OCO.

22. Royall, 1:286; "A Mechanic" (Stubblefield) to members of Congress, April 12, 1830, Hall to Eaton, May 25, 1829, to the editor of the *Daily National Intelligencer* (Washington), September 5, 1829, Letters Received, OCO.

higher than they actually cost.[23] To make matters worse, both men feuded over matters that had little to do with armory affairs. In 1822, for example, Hall strenuously advocated and Stubblefield vigorously opposed the establishment of a publicly supported Lancastrian school for the community.[24] They again clashed in 1826 over the appointment of a postmaster at Harpers Ferry. Whatever the question, Hall could almost always count on the support of the Ordnance Department just as Stubblefield counted on the Junto. These lines of opposition only added to the enmity between the two men. Not once between 1819 and 1829 did they even come close to reconciliation. In fact, long after he was replaced as superintendent in August 1829, Stubblefield would hold a personal grudge against Hall.

Hall was not the only person to receive a cold welcome at Harpers Ferry. In 1816, for instance, the superintendent refused employment to an eminently qualified German machinist simply because he was "a forener and a Stranger."[25] Many others experienced similar treatment at the hands of Stubblefield and his kinsmen. Over the years no one had more frustrating experiences with this attitude than Springfield's Roswell Lee. On several occasions he

23. Stubblefield to Wadsworth, January 8, 1820, Letters Received, OCO; Hall to Wadsworth, October 30, December 19, 1820, to Bomford, March 1, 1823, December 9, 1826, Correspondence Relating to Inventions, OCO; Hall to Mary Hall, January 27, 1821, Hall-Marmion Letters (transcribed), HFNP.

24. Although Stubblefield sent his children to private academies in Philadelphia and Springfield, he did provide part-time school facilities at Harpers Ferry prior to 1819. Rather than concentrating on the three Rs, the curriculum consisted of instruction in "practical" subjects such as husbandry and marksmanship. Such subjects, Stubblefield believed, best suited the educational needs of armory children. As the father of seven youngsters, however, Hall held a far different opinion. "There are at Harper's ferry," he wrote Colonel Bomford, "more than four hundred children who ought to go to [grammar] school, one half of whom, it is believed, have never been or does it appear ever will go to any School unless the attempts now making to establish one on the Lancastrian plan should prove successful." Having spent much of his free time as a member of a local ad hoc school committee between 1819 and 1823, Hall was convinced that Stubblefield and Armistead Beckham "have been & . . . still are opposed to it & untill some change in their conduct decidedly favorable towards it takes place, the institution will have to struggle as it always has done, against a general impression that the principal agents of the Government at this place are against it & are gratified by opposition to it" (Hall to Bomford, November 1, 1822, Letters Received, OCO).

25. Stubblefield to Bomford, November 13, 1816, Letters Received, OCO.

sent workmen to Harpers Ferry only to learn upon their return that they had been poorly treated by both the townspeople and the armorers. When he loaned the services of a forger named David Parsons in 1817, Lee specifically enjoined Stubblefield "to protect him from any improper treatment that envy or jealousy may provoke some to show him." Evidently Parsons fared much better than some of his New England companions because he agreed subsequently to return to Virginia several times during the 1820s. Even so, by 1830 the armory had gained a fairly widespread reputation in the arms-making community for the poor reception given outsiders.[26]

One of the most serious displays of hostility toward new arrivals occurred when Nahum W. Patch, a lock filer, came to Harpers Ferry from Springfield in 1822. After a few months' employment the New Englander was temporarily designated by Stubblefield as an inspector of lock mechanisms at the Musket Factory. He must have performed his job well because the superintendent made the appointment permanent in August 1823. Soon afterward lock filers began to complain to Armistead Beckham "that the inspection was too close and their price too low to do the work" as well as Patch wanted it done. The master armorer attempted to appease these workmen by raising the piece rate on finished locks from twenty-one to twenty-five cents. At the same time he told Patch "to have the work done well, and not to receive bad work." Except for some heated verbal exchanges between Patch and several older craft-trained artisans, the uneasy truce lasted until April 20, 1825, when, amid the cheers of his co-workers, an irate filer named Henry Stipes assaulted the inspector for having rejected some of his work. Although Patch lost a tooth in the fray, his pride seemed to be hurt more than anything else. Consequently when Stubblefield summarily dismissed both men without a hearing, the discredited New Englander went to Washington to seek redress at the Ordnance Department.[27]

26. Lee to Stubblefield, March 22, 1817, Letters Sent, SAR; James Carrington to Lee, October 31, 1822, Letters Received, SAR.
27. Patch to Bomford, August 3, 1825, January 25, 1826, Letters Received, OCO.

When he learned about the circumstances of the episode, Bomford wrote the superintendent:

Mr. Patch complains, not for want of employment, but for the injury done his character; and asks only for an opportunity to establish his innocence. It seems but reasonable that this should be granted to him.

If it be, as Mr. Patch states, that he has been driven from the service in disgrace, by a combination of workmen interested in his removal, because of the strictness with which he inspected their work, the case forms a dangerous precedent. For if such results can be effected by such means, future inspectors must be more influenced by the wishes of the workmen, than by public interest. . . . It appears to me therefore, that there can be no impropriety in granting the request of Mr. Patch for an investigation of this business, which you will please to order in such a way as you may deem most conducive to ascertaining the merits of the case.[28]

"Willing to abide the result" of an "*impartial* investigation," Patch returned to Harpers Ferry only to be "treated with the most harsh and unfriendly language—sneers at my misfortunes and a denial of the right of the department to interfere." Furthermore, he reported, "Common civility is denied to me. . . . Mr. Stubblefield has exhibited hostility to me, while he exhibits friendship to those against me. His wishes in regard to the result are undisguisedly *against* me."[29]

When Bomford heard what had transpired at the Ferry, he immediately ordered, not asked, Stubblefield to hold the inquiry. The superintendent, in turn, reminded the chief of ordnance that he alone had "the right of employing Inspectors and dismissing them at pleasure," sarcastically adding that he "acted in the present case under the impression that this right still existed."

Stubblefield eventually did conduct a hearing, but without giving Patch a chance to defend himself. James Stephenson presided over the proceeding; twenty-two witnesses—all lock filers—gave hostile testimony; Patch was found guilty of "rash treatment towards them"; and, adding insult to injury, Henry Stipes was reinstated at the armory. There the matter rested.[30]

28. Bomford to Stubblefield, November 23, 1825, Letters Sent, OCO.
29. Patch to Bomford, December 14, 1825, Letters Received, OCO.
30. Stubblefield to Bomford, January 3, 1826, Patch to Bomford, January 2, 1826, James Stephenson to Stubblefield, January 3, 1826, Henry Stipes to Wil-

In the meantime, Patch had begun to provide the Ordnance Department with sworn testimony concerning undue family influence and illicit practices at Harpers Ferry. In one of his most serious charges he accused Stubblefield of keeping foreman George Malleory and several other armorers on the payroll, not because he required their services but because they were heavily indebted to the superintendent and his brother-in-law Armistead Beckham, "whose interest it is to have them thus employed." "It is well known," Patch added, that Malleory "inspects for his son and three of his brothers and they have generally done work that has not been considered of good quality." The New Englander also asserted that Stubblefield repeatedly furnished Stephenson with firewood from the armory lumberyard and even used the public team to make deliveries. Finally, he contended that the superintendent seldom visited the workshops and, as a result, knew very little about actual working conditions at the armory. Between August 1824 and April 1825, for example, Stubblefield visited the filing shop twice—"and then not to examine any thing about the work." During these lengthy absences the superintendent left the factory in charge of Beckham, a man of dubious character whose knowledge of gun making and ability as an administrator left much to be desired. This situation more than anything else, the ex-inspector declared, fostered a complete breakdown in labor morale and discipline and seriously hampered the armory's efficient operation.[31]

Nahum Patch's indictment of the superintendent and master armorer probably would have caused little notice in Washington had others been silent about administrative practices at the Musket Factory. Thus at the same time Patch lodged his complaint, a lock filer named Charles Staley made additional charges against Stubblefield and Beckham. Apparently dismissed from the armory for no other reason than his refusal to take sides in the Patch affair, Staley appeared in Washington to seek justice at the Ordnance Department. An employee at Harpers Ferry since 1812, the disgruntled armorer had many damaging things to say about his former supe-

---

liam Armstrong, January 24, 1826, Letters Received, OCO; Bomford to Stubblefield, December 17, 1825, Letters Sent, OCO.

31. Patch to Bomford, January 25, 1826 [two letters], December 14, 1825, Letters Received, OCO.

riors. Like Patch, he contended that "the superintendent retained in employ a number of drunken, lazy & disorderly men through the influence of Inspector George Malleory who is notoriously given to those vices." Staley also claimed he had "seen the public jobs of work bought and sold" mainly because Stubblefield and Beckham persisted in borrowing money from individuals upon the condition of giving them work at the armory. In the same vein, he asserted, "there are a number of men with families in the employ of government who cannot get houses—while several houses belonging to the Public are occupied by individuals having no connection with the public, who say they are not afraid of being compelled to quit them as long as the superintendent owes them money."

Staley also denounced Stubblefield's involvement in a scheme where his relatives could profit at the expense of armory employees. Worker-tenants often presented reimbursement claims for repairs they made on government-owned dwelling houses. Very seldom, however, did the armorer obtain full compensation for his expenses. According to Staley, "there has been practiced a species of speculation on a number of persons—having claims on Houses for repairs done—the superintendent stating he could not pay them for a year or two by which they have been induced to sell them to Messrs. E. Wager & F. Beckham at a great discount, who paid for them in money drawn from the paymaster a short time after the transaction." The implication was clear. While ordinary workingmen could not get Stubblefield to open his coffers, certain members of the community could and, in doing so, reaped handsome monetary returns.

Regard for special privilege seemed to pervade Stubblefield's actions as superintendent. Among other things Staley stated that in fourteen years he had "seen the public stores to wit coal, Iron, steel & etc. etc. etc. made use of for private purposes—Old guns repaired and new ones made, and the public lathes made use of in turning mill spindles etc. etc., the public teams and labourers employed for weeks at a time in filling private ice Houses—The public carts sent for individuals to Shannondale Springs for water, absent 2 days at a time, at the public expense.—This repeatedly done." Specifically, the ex-armorer cited Lewis Wernwag, James Ste-

phenson, John G. Wilson, and William Graham as frequent
beneficiaries of the superintendent's largess. For example, when
Stubblefield sold part of Virginius Island to Wernwag in 1824,
Staley testified, "the Public Sawmill was dispensed with & all articles previously sawed there, bought of Wernwag—by which he
was enabled to pay Mr. Stubblefield for his Island." Similarly,
tavern keeper Stephenson had free access to the armory woodpile;
merchant Wilson received permission to erect a private warehouse
rent free on public grounds; and businessman Graham rented the
U.S. sawmill at $100 a year even though another person had of-
fered Stubblefield $300 for the same franchise. In Staley's opinion,
these doings clearly indicated how much "malpractice and miscon-
duct" existed at Harpers Ferry.[32]

The most questionable transaction concerned the superintendent's letting of contracts for supplies and raw materials. Ordnance Department regulations specifically directed managers of
government armories and arsenals to invite proposals by public
advertisement and to award all contracts to the lowest bidder.
Stubblefield apparently paid little heed to these orders, for he made
purchases without much attention to price. According to Staley, he
continually bought pit coal at twenty-five cents a bushel from the
Strider brothers when he could have gotten the same article for
twenty cents from non-local dealers. Furthermore, the ex-armorer
argued, favoritism and private interest governed the superintendent's action on all purchases. These practices might have
been justified had the materials received at the armory been of
sufficiently high quality. Such apparently was not the case, however. Hall testified, for instance, "that in the purchase of steel, files,
and other articles for the use of the armory" Stubblefield could have
saved the government from 12 to 20 percent on the "same articles"
had he awarded contracts through competitive bidding. Instead,
the superintendent not only placed large orders solely with Jacob
Albert of Baltimore and Wickham & Company of Philadelphia, but
promised both firms "a continuance of the business for a series of
years." Although Lee knew full well the implications of

32. Staley to Bomford, n.d. [September 8, 1826], to Eaton, April 22, 1829, C.
F. Mercer to Barbour, March 17, 1827, Letters Received, OCO.

Stubblefield's actions, he refused to speculate publicly about his friend's autonomy in these matters. "How Mr. Stubblefield gets over" Bomford's "order directing 'proposals for all supplies to be advertised,' " he wrote Wickham's partner, "is not for me to determine."[33]

In addition to Staley and Hall, a group of stock sawyers from Cumberland, Maryland, bitterly complained about discriminatory practices in letting contracts at Harpers Ferry. Headed by Notley Barnard and B. S. Pigman, these lumbermen contended that Stubblefield refused to do business with them directly even though most of the walnut plank used at the armory originally came from their sawmills. In 1822, for example, Barnard and Sons had contracted with John A. Smith of Harpers Ferry for 40,000 rough-sawed gunstocks at fourteen cents apiece. Smith, in turn, sold these stocks to the national armory at nineteen cents. After Smith's death in 1823, Barnard offered to assume the contract and furnish rough stocks "on his own account" at nineteen cents each. However, Stubblefield refused the proposition and told Barnard that "no other contract would be made, until Smith's was completed; while at the same time he (Mr. S.) purchased stocks of Martin Rizer at twenty-five cents each." With the expiration of Smith's contract in 1824, Barnard evidently reached an agreement with the superintendent to furnish stocks at twenty cents apiece. Nevertheless, when he began making deliveries, Stubblefield refused to accept them under the twenty-cent contract and compelled Barnard to deliver the stocks to John Strider at nineteen cents "from whom they were received by the Govt. at twenty-five cents." In other words, Strider realized a profit of six cents per stock simply by acting the role of a middleman. This practice continued through 1829.[34]

Members of the Junto made no secret of their collusive tactics. In 1828, for instance, Wernwag informed Barnard's associate, John

33. Staley to Bomford, n.d. [September 8, 1826], OCO; Bomford to Eaton, May 12, 1830, Letters Sent to the Secretary of War, OCO; Wool to Eaton, May 26, 1829, Inspection Reports, OIG; Lee to James Baker, December 5, 1827, Letters Sent, SAR.

34. Pigman to Lee, March 27, 1827, Letters Received, SAR; Bomford to Stubblefield, January 18, 1828, to Lee, January 22, 1828, Letters Sent, OCO; *American State Papers: Military Affairs*, 3:679–680.

Mills, "that he had an understanding with Mr. Stubblefield, by which he would secure to himself the contracts for stocks; the plan it is stated, was for several persons of Mr. Wernwags family to offer proposals at different prices from 14 to 23 cents; so that they could avail themselves of the highest of their bids, which should not be under-bid by others." Under this arrangement the Cumberland sawyers as outsiders stood little chance of getting an open contract with the armory. They therefore could either sell their stock to local businessmen like Wernwag at a reduced price or do no business at all at Harpers Ferry. Although Barnard, Pigman, and Mills chose the former alternative, they did so under protest. Having encountered many rebuffs and humiliations at the hands of the Junto, they were among the most vociferous opponents of the Stubblefield administration and would play a central role in bringing about its demise in 1829.[35]

Judging from the sheer volume and pertinacity of his protest, the most undaunted critic of Stubblefield and Beckham was Thomas Copeland. A versatile mechanic who got "at the right end" of his business "by taking notice of good books & good work," Copeland began his career in Western Pennsylvania as a builder of furnaces, forges, rolling mills, and "all kinds of Machinery." While manufacturing Boulton and Watt steam engines in Pittsburgh during and after the War of 1812, Copeland, a friend later claimed, had "made the Patterns & built the Boring Mills that bored the Cannon with which the Gallant Perry won the victory on Lake Erie" in September 1813. Hall, who rarely handed out compliments, characterized Copeland as "an ingenious mechanic and an industrious man." Lee agreed that "he is certainly a man of considerable talent and possesses a good share of mechanical knowledge & ingenuity." However, the Springfield mentor added, *"He is not insensible to all this &* his loquacity & peculiarities render him at times rather tedious." As Stubblefield and Beckham soon learned, Copeland tended to mind a great deal more than just his own business.[36]

---

35. Bomford to Stubblefield, January 18, 1828, February 24, 1829, Letters Sent, OCO.

36. Copeland to Wade, July 22, 1826, Lucas to James K. Polk, July 12, 1845, Hall to Bomford, December 23, 1824, Letters Received, OCO; Lee to Wade September 21, 1825, Letters Sent, SAR.

Deeply concerned about the high loss of gun barrels at Harpers Ferry since 1819, Colonel Bomford sent Copeland to the armory to conduct an investigation and, hopefully, to propose a solution to what had become a very costly problem. After arriving in the fall of 1824, the Pittsburgh mechanic soon recognized the difficulty and reported his findings to the chief of ordnance. First of all, he had never seen "such a house full of bad Iron with such an emence scrapp heap as is at Harpers Ferry." "It would have been well spent money" had Stubblefield taken the time to visit various furnaces and select his iron more carefully. "At any rate," Copeland concluded, "very little of what is understood to be good Juneata Iron has went to Harpers Ferry for these several years." The armory's principal supplier, Peter Shoenberger of Huntingdon, Pennsylvania, "mostly endeavoured to have the Iron for H. Ferry neatly drew but very seldom paid much attention to the most particular part, that of having good Pig Iron & having it carefully Refined." The armory's accumulation of over $100,000 worth of faulty materials since 1819 convinced Copeland that Stubblefield had not been very vigilant in the execution of his duties. While he refrained from accusing the superintendent of outright dishonesty in the dealings with Shoenberger, he nevertheless observed "that Mr. Stubblefield is a poor soft headed simple & unfit man for the post he fills & that Shoenberger is a cunning Deep Designing Crafty man—that betwixt the Knave & the Fool the Government of the U.S. pays dear for the two as poor Richard says."[37]

Convinced that the Musket Factory suffered greatly from careless management, Copeland proposed a twofold solution to the gun barrel problem. First and foremost, the armory absolutely should procure iron from forges other than those owned by Shoenberger and Henry P. Dorsey, also of Huntingdon. Having just completed a lengthy tour of the Juniata district, he assured the Ordnance Department that "there is no dependence in neither of them for none of them has any command of any Blast Furnace that will make good pig Iron—they mix them with bad & at best they make but medling Iron." Much to Stubblefield's annoyance, Copeland recommended that the armory discontinue patronage of Shoenberger and

37. Copeland to Bomford, December 29, 1824, to Wade, January 22, 24, October 6, 1825, July 31, 1826, Letters Received, OCO.

place orders for barrel iron with Gloniger & Anshutz of Tyrone and Valentine & Thomas of Bellfonte. Second, the armory needed "a man of experience" to make, improve, and keep "all and every part of the machinery in good order at the least possible expense." Especially the equipment in the barrel-forging shop urgently needed renovation as both tools and techniques had become sadly antiquated. Accordingly the Pittsburgher proposed updating the barrel-welding process by installing "4 to 6 tilte hammers" in place of traditional hand labor. At the same time he asked permission to experiment at Harpers Ferry with a new method of preparing barrel skelps and welding them by means of rollers.[38]

Impressed by the thoroughness of Copeland's report and the feasibility of his suggestions, Bomford sanctioned the proposals and on January 5, 1825, issued orders to Stubblefield to make the necessary facilities available at the armory. Stubblefield, astonished "at an attempt being made to rowel gun bariels in two pieces," wanted to know "why he was troubled with it." The two men soon clashed over who would supply the iron for the experiment, the superintendent still insisting on purchasing it from Shoenberger. Although Stubblefield threatened to withold funds from the project, Copeland prepared a set of drawings and went to Pittsburgh to have the rolls, guides, and housing cast by Kingsland, Lightner & Company and finished by George Lewis. Nearly three months elapsed before the machinery was completed, shipped to Harpers Ferry, and installed at the factory. In the meantime, Copeland urged Stubblefield "to be a little patient" and to give his experiment a fair trial. The superintendent, however, had no more use for Copeland than he did for Hall or, for that matter, any other uninvited interloper who attempted to set up shop at the armory. Indeed, he particularly detested the Pittsburgher and considered him a tattling lackey of the Ordnance Department.[39]

---

38. Copeland to Wade, January 22, May 5, 19, 1825, to Stubblefield, April 25, 1825, Letters Received, OCO.

39. Copeland to Wade, January 24, 1825, to Stubblefield, May 2, 1825, Letters Received, OCO; Payrolls and accounts, Harpers Ferry armory, 1825, GAO. Despite a thorough search of archival materials, Copeland's drawings have not been found; he probably based his ideas about rolling gun barrels on British sources, particularly encyclopedias. See Copeland to Wade, July 22, 1826, Letters Received, OCO.

Relations between Copeland and Stubblefield reached a crisis by May 1825. Complaining that the superintendent "had objections to anything" he proposed, Copeland informed the Ordnance Department that Stubblefield not only continued to purchase large quantities of barrel iron from Shoenberger but adamantly refused "to have aney thing more dun towards having the barrels welded" by triphammers. The craft-oriented forgers, supporting their superintendent in the controversy, resented Copeland's presence and scoffed at his improvements. In fact, Copeland believed the superintendent and master armorer had intentionally tried to sabotage his machines by allowing several of these workmen to tamper with them during his absence from the armory. "From the first commencement of this experiment," he declared, "Mr. Stubblefield and all his connections set their factory against it, and of course against me."[40]

Realizing that the situation had become tense if not hopeless, Bomford arranged for Copeland to continue his experiment at Springfield during the summer of 1825. In contrast to the stormy relations with Stubblefield, Copeland's association with Roswell Lee was splendid. After spending July and August supervising the manufacture and proof of eighty-seven barrels, Copeland returned to Harpers Ferry completely "satisfied with the manner in which the experiment on his Iron was made." Lee, however, did not give the process an unqualified endorsement. He had reservations about Copeland's method of welding barrels with two seams instead of one and considered the concept of welding with rollers somewhat premature. Nevertheless, he did think "the experiment made by Mr. Copeland furnishes evidence that good Iron can be made by that process" and that "his plan for making Iron by Rolling may do well." Writing to Bomford, Lee concluded:

The plan of Rolling Scalps for barrels is not a new process, it has long been in use in England & has been a subject of conversation here for many years . . . Should it be found that Iron prepared by rolling from the blooms after being twice heated will answer a good purpose, it will be advisable to have the rolling done at the Armories, where particular attention will be paid to the Gauges; for it is very difficult to obtain Iron rolled

40. Copeland to Wade, May 5, 19, 1825, OCO.

to the exact size at the Iron Works & we have frequently to reduce it under the hammer or suffer a loss by having it reduced too thin, this is one among a great many reasons why a Rolling Mill will be useful at the National Armories.[41]

While Copeland's demonstration convinced Lee that a rolling mill was needed at Springfield, it totally failed to achieve the same effect at Harpers Ferry. Statements made by the chief of ordnance had led Copeland to believe that work on a full-scale mill would soon begin because, after completing his experiments, he stayed on at the armory as a machinist. As each month passed the likelihood of such an undertaking ebbed away, and by July 1826, Copeland had become convinced that, whatever Bomford's wishes were, Stubblefield had no intention of building a mill at the Ferry. To make matters worse for Copeland, not only was his project abandoned, but his character was assailed and his family mistreated by members of the local Junto. Embittered and frustrated, he wrote Major Wade, "I am kept here at high wages doing nothing. Lawyer Macabe a respectable man at this place heard three of the Beckhams & Edward Wager tell . . . their acquaintances in Mr. Stevenson's tavern that the Dept. was retaining me here and that i was of no benefit to the establishment. . . . It would be veary desirable to me," Copeland concluded, "if Col. Bomford would give such orders as either to have me usefully employed—or discharged."[42]

Long before leaving the works Copeland made clear his intention "to take my time and choose a proper place to get Redress." Between September 1825 and October 1826 the indignant mechanic bombarded the Ordnance office with long letters elaborating on the link between family privilege and illicit practices at the armory. In many instances he repeated charges already made by John Hall, Nahum Patch, and Charles Staley. He asserted, for example, that he had frequently rendered services at the privately owned mills of the Beckhams and Wernwag on Virginius Island "for which he was paid on the public rolls." He also noted that a number of armorers

41. Copeland to Wade, July 9, 1825, Letters Received, OCO; Lee to Bomford, August 2, 1825, Correspondence and Reports Relating to Experiments, OCO; Lee to Wade, September 21, 1825, Letters Sent, SAR.
42. Copeland to Wade, July 24, 1826, Letters Received, OCO.

such as John Rochenbaugh and John Resor "could throw a good deal of light" on the fraudulent activities of the Stubblefield administration but refused to do so for fear of losing their jobs.[43]

Copeland's most extensive criticism concerned practices at the barrel-forging shop. While providing a running commentary and statistical tables on the high loss of gun barrels at the works since 1824, he asserted that the barrel welders—many of whom were "largely in debt to Mr. Wilson, or Wager & Beckham"—repeatedly tried to cover up their mistakes by consigning ruined skelps to the forge before they could be discovered by inspectors. He also accused the superintendent of giving Wernwag many privileges at the armory, among them free scrap iron and castings from the government's storage bins.[44] More serious were charges regarding purchases of iron at Harpers Ferry. Copeland seemed baffled "that there should be such a wish on the part of Mr. Stubb$^{d.}$ to get Iron from no other Ironmaster but Mr. Shoenberger." While he refrained from impugning the honesty of the superintendent, Copeland expressed his suspicions about the motives of Armistead Beckham and inspector Joseph Hoffman. Both men, he asserted, had accepted expensive presents from Shoenberger and, in return, had recommended his iron to the gullible superintendent. Furthermore, Copeland contended that whenever the Pennsylvania ironmaster visited the armory he regularly distributed "considerable sums of money among the barrel welders" to gain favor with them. Shoenberger, in response, vehemently denied ever having attempted to bribe the armorers. Rather, he insisted, "I have been in the habit of giving the workmen a treat occasionally when I would visit the Armory for 12-15 years back in Mr. Annin's time." Whether this practice was construed as a bribe or a treat, Copeland denounced it as highly irregular and prejudicial to the "Publick trust." "I will just state," he wrote Wade, "that there is & allways will be too much loss sustained in all the component parts of the Muskets" so long as "these people manage this Armory."[45]

43. Copeland to Wade, May 5, 19, September 4, 1825, Letters Received, OCO; Wade to Lee, April 14, 1827, Letters Received, SAR.
44. Copeland to Wade, June 16, December 30, 1825, May 24, July 22, September 4, 1826, Letters Received, OCO.
45. Shoenberger to Lee, July 22, 1829, Letters Received, SAR; Copeland to Wade, July 22, 31, 1826, to Bomford, August 19, 1826, Letters Received, OCO.

Having received many reports about absenteeism, mechanical backwardness, the nonpayment of debts, and the misallocation of funds at Harpers Ferry, Bomford had long considered Stubblefield an ineffective administrator. The accusations of Copeland, Hall, Patch, and Staley only served to bolster his conviction that something had to be done to restore order, discipline, and efficiency at the government works. Well aware of the Junto's political influence with Secretary of War Barbour and other ranking members of the Adams' administration, Bomford knew he stood little chance of getting Stubblefield removed from office. He therefore chose a more subtle means of instituting reforms at the Ferry.

Since the Springfield armory not only had experienced remarkable gains since 1815 but also had emerged as one of the most progressive manufacturing establishments in the United States, Bomford turned to Roswell Lee for assistance. His plan was shrewd but simple: Stubblefield and Lee would exchange positions temporarily until the latter had transferred certain "improvements at Springfield" to the Virginia installation and set it on proper footing. In a letter dated November 23, 1825, Bomford informed Lee that "the Harpers Ferry armory needs your services." "I am aware," he added, "that this proposition may be unpleasant to you, on account of the inconvenience to which it may subject yourself and family; and I would not willingly make it, but from a full conviction of its great importance to the public interests." Nevertheless, he tactfully concluded:

The Armory at Springfield is now so well organized that your continued service there, is less essential to its prosperous management than formerly. And it is but justice, on this occasion to state that the present highly improved condition of that armory, is, by the Dept. attributed solely to the energetic zeal and fidelity with which its affairs have been administered since they have been confided to you. And it is with a well grounded hope that a like improvement of the Harpers Ferry Armory, will be effected by the measure now proposed.[46]

Lee had recently returned from a lengthy tour of inspection connected with the establishment of a national armory on the

---

46. Bomford to Lee, November 23, 1825, Letters Sent, OCO; Samuel Lathrop to Lee, December 8, 1825, Letters Received, SAR.

"Western Waters" of the Ohio and, physically exhausted, tried his best to avoid going to Harpers Ferry. Even under the best of circumstances he considered it a loathsome assignment. Now, with the armory in a state of turmoil, the situation was far from normal, and Lee emphatically did not want to become involved in the heated jurisdictional dispute between his friend Stubblefield and the Ordnance Department. Pleading ill health, family responsibilities, and a backlog of work at Springfield, he petitioned the secretary of war to countermand Bomford's order. When Secretary Barbour denied the request Lee cynically informed Major George Talcott, a friend at the Ordnance office, that "I am ordered to Harpers Ferry to take charge of that Establishment as a *reward* (I presume) for the arduous services I have rendered at Springfield, and Mr. Stubblefield is to be *punished* by taking my place at that Armory." Essentially Talcott sympathized with his friend's plight.

I cannot see why Col. Bomford should wish to remove you to Harpers ferry. Is it not intended merely to drive Stubblefield out of his situation? He surely cannot leave his plantation and negroes to the care of overseers, and I would advise you to go to Harpers ferry with a stipulation that you should return to Springfield at the end of a year or two at most. . . .

Although I do not think your own interests would suffer materially by your being absent a year from Springfield. . . . I hate this sort of proceeding, of whipping Stubblefield over your back—If he is not a proper man for the situation why not drop him at once instead of squeezing him and you in the same Vice—but perhaps they very much need your services at H. Ferry—If so let them compensate you handsomely for the sacrifices you must make and quit this policy which they often pursue of "making a man work for nothing & fine himself."[47]

Miffed about being "driven from 'pillow to post' " by the Ordnance Department, Lee threatened to resign and accept a supervisory position at a cotton mill in Springfield rather than go to Virginia. However, since he actually preferred not to leave government employment and since Bomford could ill afford to lose his valuable services, a compromise was reached after Lee suggested deferring the transfer between Stubblefield and himself until the fall

47. Lee to Talcott, March 20, 1826, Letters Sent, SAR; Talcott to Lee, April 23, 1826, Letters Received, SAR.

of 1826. Agreeing to the postponement, Bomford assured Lee that he could return to Springfield no later than August 1, 1827. Although Lee continued to grumble about the hardship and "pecuniary disadvantage" of the arrangement, he nevertheless packed his bags and headed south on the appointed date. After stopping in Washington for a brief conference with the chief of ordnance, he proceeded up river and assumed command of the armory around November 15.[48]

By the time of Lee's arrival, Stubblefield had already left Harpers Ferry but not for New England. Stricken with an attack of the bilious fever that had killed his son-in-law, Edward Wager, and debilitated 90 percent of the labor force, the superintendent first had gone to Berkeley Springs to take the medicinal waters there and, since mid-October, had been resting at his Berry Hill plantation in Jefferson County. By January he had sufficiently recovered to resume work, but still did not move to Springfield. Although the planned transfer was never officially abandoned, Stubblefield managed to avoid the assignment and remain in Virginia. During the next five months he divided his time between town and country, an interested if not uneasy onlooker.

Lee's sojourn at Harpers Ferry proved to be a personally difficult and trying experience. Attempting to conciliate the various factions at the armory left him little time for instituting thorough reforms. He did manage to introduce a series of labor regulations based on the Springfield model and, as previously noted, made arrangements for the installation of Thomas Blanchard's new machinery. In addition to his regular duties as superintendent, Lee hosted a three-man commission assembled by the Ordnance Department to examine Hall's machinery in December 1826. Most of his attention was necessarily devoted to matters having little to do directly with the manufacture of arms.

Bomford's solution of temporarily replacing Stubblefield at Harpers Ferry failed to placate the superintendent's enemies. As observed, the Cumberland lumbermen numbered among the most

48. Lee to Bomford, April 29, October 5, 1826, to Wool, October 30, 1826, Letters Sent, SAR; Wade to Lee, October 13, 1826, Letters Received, SAR; Bomford to Lee, October 13, November 15, 1826, Letters Sent, OCO.

vociferous critics of his administration. By March 1827, Barnard and Pigman had carried their complaints to the halls of Congress. Consequently when several members of the Maryland delegation demanded an investigation of Stubblefield's conduct as super-intendent, Secretary of War Barbour had little choice but to order a court of inquiry held at Harpers Ferry. With Inspector General John E. Wool presiding, the court convened on April 26, 1827. Eight days later the hearing ended, and Wool issued a long report to the secretary of war which noted a "few instances of neglect of minor importance" but nothing seriously impugning the super-intendent's reputation. "The charges against Mr. Stubblefield," he concluded, "are generally unsupported by evidence and appear to have been entertained on information proceeding from malicious sources." With this vindication, Barbour notified Lee that Stubblefield would be reinstated on June 1 and that he could then return to Springfield. While warning the Virginian not to repeat the "few irregularities" noticed by Wool, the Secretary gave him full authority to dismiss all "persons employed in the public service who have disseminated false and malicious reports or charges to the prejudice of that order and subordination which is necessary at so important an establishment."[49] Needless to say, Copeland's name headed the list of those fired during the weeks that followed.

The outcome of the Wool inquiry produced mixed reactions among those who had followed the proceedings. Members and friends of the Junto naturally celebrated the court's pronouncement as a great moral victory over the centralist tendencies of the Ord-nance Department. Jubilant over the prospect of returning home much sooner than expected, Lee wrote to his master armorer, Jo-seph Weatherhead, that, in view of the gravity of the charges, "I was much gratified as well as surprised that so little was proved against him." Caught between his loyalties to the Ordnance Depart-ment and those to his friend Stubblefield, he could say little more.[50]

49. Extracts from the Proceedings of a Court of Enquiry convened at Harpers Ferry, April 26, 1827, Letters Received, SAR; Wool to Bomford, May 3, 1827, Letters Received, OCO; Bomford to Lee, May 8, 1827, to Stubblefield, May 8, 1827, Letters Sent, OCO.

50. Lee to Weatherhead, May 16, 1827, Letters Received, SAR.

On the other hand, George Bomford had been conspicuous for his silence during the whole affair. Other than relaying information to and from the secretary of war, he took no part in the proceedings; nor did he venture an opinion once they had ended. His assistant, however, was not so reticent. Several weeks before the inquiry began, Wade informed Lee, "I had supposed that the family influence which has hitherto controlled everything at Harpers Ferry, and monopolized all the patronage, power, profits, and emoluments of that place, had been somewhat lessened; but I now perceive that it has been too deeply planted to be easily eradicated." The inspector general's report only served to strengthen this conviction. According to Wade, the secretary of war had ordered the inquiry merely to pay lip service to Congress. In his opinion, Barbour never intended it to be a full-scale or, for that matter, candid investigation of Stubblefield's conduct as superintendent. Although he did not "mean to infer that there was any improper bias on the minds of any of the members of the court," several things "appeared strange" to Wade:

One, was on the charge for neglect, in not causing barrels to be proved according to regulations. It was proved that the Supt. sent the regulations to the Master Ar. as soon as received; and also that they did not reach the Inspector for two or three years afterward; yet no neglect was found. Another instance was, that, it was alleged that waggoners bills were permitted to be hawked about and sold at a discount while public funds were at the armory: the Inspector who sold and the workmen who purchased were both named in the specifications: yet neither of them were examined upon the point, and of course the fact was not found.[51]

Wool's report clearly left many questions unanswered. Why, Wade asked, did the board fail to make adequate use of armory records during the course of the inquiry? Why, after reporting "Mr. Stubblefield has discharged his duties as Superintendent with fidelity," did Wool tell General Jacob Brown six months later that the Harpers Ferry armory had, "undoubtedly, been badly managed"? Why, if Stubblefield had been innocent of any serious neglect did Stephenson, Wernwag, and several others have to make restitution for several thousand dollars worth of iron and lumber

51. Wade to Lee, April 14, August 6, 1827, Letters Received, SAR.

they had received on demand from the superintendent? The answers seemed abundantly clear to another critic, Charles Staley. "Mr. Stubblefield was cleared" simply because the board "never examined half the witnesses." Those that were called to testify, he intimated, were known friends and associates of the superintendent. In Staley's opinion, the investigation had been a sham and mockery of justice from the beginning.[52]

In many respects the Wool inquiry tended to confuse rather than elucidate the situation at Harpers Ferry. While the inspector general noticed some minor instances of neglect, he did not specify them in detail nor did he adequately explain why more serious charges were "generally unsupported by evidence." It also appears that many witnesses were, in fact, not examined under oath by the committee. Even Wool hinted that perhaps he had not been quite so vigilant as he might have been. That the board intentionally selected favorable witnesses is highly doubtful. Nevertheless, Copeland, Staley, and other malcontents insisted that Armistead Beckham and a group of local thugs had threatened and physically prevented those with negative testimony from coming forward. As a result the final report gave no clear indication of what had or had not been proved against the superintendent.

The events of 1827 apparently failed to disturb Stubblefield. Throughout the hearings he maintained a calm, self-assured composure and, once they ended, he made no overt effort to change his standard policies or assume a more active role in everyday armory affairs. Much to Bomford's chagrin, the superintendent continued to delegate authority to Beckham, exceed annual appropriations, let contracts without inviting proposals, and turn defective muskets into store. Soon these and other practices again became potent ammunition in the hands of Stubblefield's enemies.

As early as January 1828, Congressman Michael C. Sprigg of Cumberland, Maryland, renewed Barnard's charges against the superintendent of favoritism in letting contracts for coal and gun stocks. He succeeded in getting a resolution passed by the House of

---

52. Wool to Brown, November 16, 1827, Inspection Reports, OIG; Wool to Bomford, May 3, 1827, Staley to Eaton, April 22, 1829, Letters Received, OCO; Lee to Bomford, June 21, 1827, Letters Sent, SAR.

Representatives calling upon the secretary of war to furnish a complete statement of accounts relative to the purchase of these materials at Harpers Ferry between January 1, 1820, and December 1, 1827. Meanwhile Copeland had not forgotten his vow to get even with the superintendent. Now a machinist at the Washington Arsenal, he persuaded Sprigg to present a memorial to Congress complaining of his discharge from the armory and accusing Stubblefield of mismanagement. Barnard, Copeland, and other discontents had learned an important lesson since 1827. Instead of acting independently, they began to pool their information and political resources to mount a concerted attack on the Stubblefield administration. Their strategy was effective, because by January 1829, Sprigg had joined forces with Chauncey Forward of Pennsylvania and at least ten other congressmen to demand another, more thorough investigation of the Harpers Ferry armory.[53]

If political partisanship had favored Stubblefield in 1827, it weighed just as heavily against him in 1829. Having staunchly supported the unsuccessful candidacy of John Quincy Adams in the election of 1828, the superintendent realized that he and his in-laws would soon be bankrupt of influence with the new administration. With the exit of James Barbour, William Wirt, and other members of Virginia's "Old Guard," rough times lay ahead. Writing to Stubblefield on January 27, 1829, Roswell Lee warned *"Our time is nearly out—rely upon it."*[54]

Actually Lee's gloomy prediction applied to his friend more than to himself. No sooner had Andrew Jackson entered office than Sprigg, Forward, and Edward Lucas, Jr., of Shepherdstown appeared at the War Department "to stir the matter" of Harpers Ferry with the newly appointed secretary, John H. Eaton. In general they contended "that the means furnished by the Govt. for the fabrication of arms, at that armory, are misapplied; that materials

53. Lee to Stubblefield, February 8, 1828, January 13, 1829, to Isaac C. Bates, January 21, 1829, Letters Sent, SAR; Stubblefield to Lee, February 17, 1829, Wickham to Lee, February 18, 1829, Letters Received, SAR; Bomford to P. B. Porter, January 24, 1829, Letters Sent to the Secretary of War, OCO; Forward to Bomford, February 21, 1829, Letters Received, OCO; Bomford to Stubblefield, April 18, 1829, Letters Sent, OCO.
54. Lee to Stubblefield, January 27, 1829, Letters Sent, SAR.

of bad quality are procured; that unskillful workmen are employed; that the arms made and turned into store, are of bad quality, and have to undergo expensive repairs before they can be issued for service; that the discipline of the establishment is loose; and that its general management is inefficient and injudicious; whereby the arms made are unusually expensive." Since Sprigg, Forward, and Lucas were loyal party men, Eaton could hardly deny their request for an official inquiry into Stubblefield's conduct. Accordingly he ordered John E. Wool not only to inspect the Harpers Ferry armory but also to conduct a "rigorous scrutiny into its management." At the same time he invited Congressmen Forward and Sprigg to assist Wool in making the investigation. Clearly Eaton expected them to keep a watchful eye on the proceedings and to make sure that the inspector general submitted something more than a perfunctory report.[55]

With witnesses assembled and a reluctant Lee again appointed acting superintendent of the armory, the second Stubblefield investigation got underway on May 7, 1829. Although he kept in constant touch with the participants, Bomford again remained in the background. Finally, after a week of taking testimony "day and night," Wool broke silence to inform the secretary of war that "thus far, nothing has appeared of a criminal nature against Major Stubblefield, though much to satisfy me that he has not been as vigilant, and as energetic in the discharge of his duties as his highly responsible situation required." In any event, Wool concluded, "I am not without apprehension that the public interest will require his removal."[56]

When the inspector general submitted his final report on May 26 nothing had occurred to substantially alter his opinion. Stubblefield had been extremely negligent but not willfully dishonest. Specific questions had arisen about discretionary employment practices, absenteeism, favoritism in letting contracts, needless expenditures, and the misuse of government property at Harpers Ferry. However, very little could be proven for or against the

---

55. Lucas to Copeland, March 13, 1829, Letters Received, OCO; Eaton to Wool, March 27, 1829, Letters Sent to the Secretary of War, OCO; Bomford to Sprigg, March 27, April 30, 1829, Letters Sent, OCO.
56. Wool to Eaton, May 14, 1829, Letters Received, OCO.

superintendent because his records were too muddled and incomplete to be used as evidence. This factor alone stood as a serious indictment of the man's administrative competence and ability.[57]

Wool's secretary, Lieutenant John Symington, probably came closest to the truth when he characterized Stubblefield as "an honest good hearted Man who has been too easy perhaps in dealing with scoundrels" at the armory. Certainly scoundrels abounded at Harpers Ferry. The board clearly ascertained, for example, that inspector Malleory had frequently given pay vouchers to armorers for work they did not do; that Fontaine Beckham, in collusion with Edward Wager, had speculated in private claims against the armory; and that many trees on government property "were cut down for no other purpose than to procure bark" for Townshend Beckham's tannery on Virginius Island. The most shocking revelation, however, had to do with the purchase of iron at Harpers Ferry. Much to Copeland's gratification, the board reported that Shoenberger had in fact distributed large sums of money among the barrel welders in 1825 and 1826 and that he did so with the full knowledge and approbation of the master armorer. Furthermore, as the trial progressed, it became clear that the real culprit at Harpers Ferry was not Stubblefield but his brother-in-law, Armistead Beckham. The master armorer's name seemed to be synonymous with corruption, turmoil, and intrigue at the Ferry. He had taken bribes, falsified records, intimidated workers, played favorites, condoned the use of violence against those who threatened to expose his activities, and even attempted to have Copeland jailed on fictitious charges to prevent him from testifying before the Wool committee. Nevertheless, because the superintendent had assumed ultimate responsibility for the affairs of the armory, he had to suffer the consequences.[58]

57. Wool to Eaton, May 26, 1829, Inspection Reports, OIG. Wool later indicated that he had tried to word his report so as to give the least possible umbrage to Stubblefield. "Anxious to know public sentiment as well as that of the [Stubblefield] family," he assured Lee that "no person could have exerted himself more than I did to do justice to all concerned, and no person knew my sentiments as well as you, and certainly no person was half as capable of judging my actions and conduct as yourself" (Wool to Lee, August 7, 1829, SAR).

58. Symington to Major Rufus L. Baker, May 23, 1829, Letters Received, AAR; Wool to Eaton, May 26, 1829, OIG.

Stubblefield's reaction to these disclosures was one of dismay and disbelief. In a futile attempt to defend his administration he addressed a fourteen-page letter to the Board of Investigation asserting that he had always acted in good conscience and that he had never intentionally sought to defraud the government or shirk his duties as superintendent. Yes, he admitted, federal funds had been used to construct private roads and bridges at Harpers Ferry, but these were necessary for the "correct" operation of the armory. Yes, defective muskets had been turned into store since 1815, but this was due to "looseness both in the execution of work and the inspecting" caused in part by the necessity of hiring inexperienced workmen and the recurrence of bilious fever epidemics "almost annually." Yes, large amounts of faulty barrel iron had been purchased since 1819, but the superintendent positively denied "knowing or suspecting" any foul play. Yes, supply contracts had been given without advertising for proposals, but "such had been the practice since he came into office."[59]

When presenting his case, Stubblefield was obviously under great emotional stress. Influential political friends who had formerly offered encouragement and support were now either out of office or curiously silent. The calm self-assurance he had displayed in 1827 had given way to grim pathos. His letters were strained and desperate. The arguments adduced were weak and, in some cases, even contradictory. The more he tried to justify his position, the more authorities became convinced that he had indeed been neglectful and had outlived his usefulness at Harpers Ferry.

By the end of May it seemed only a matter of time before the secretary of war would remove Stubblefield from office. If for no other reason, Lieutenant Symington thought political expediency dictated the measure. As much as Lee regretted the idea, even he acknowledged "that Mr. Stubblefield cannot remain for any considerable time with advantage to the public or comfort or Honor to himself." Convinced that an abrupt dismissal would "prostrate every well regulated Establishment in the country," however, he asked Bomford to restore Stubblefield "for a short time and let him

59. Stubblefield to Board of Investigation (Wool), May 25, 1829, Letters Received, OCO.

resign in, say two or three months." Bomford accepted this advice and evidently convinced Eaton of its wisdom. Consequently the Virginian went to Washington and submitted his resignation on June 1, 1829. With the understanding that he would be relieved of office at the end of two months' time, Eaton reinstated him as superintendent and ordered him to adjust all claims and close his accounts through June 30. On August 1, Stubblefield relinquished his duties to Lieutenant Symington of the Ordnance Department and retired to his Berry Hill plantation. On the same day, Eaton announced the appointment of Thomas B. Dunn, manager of the nearby Antietam Iron Works, as his successor. Ironically Armistead Beckham—the one person who deserved to be removed from office—refused to resign as master armorer and remained at Harpers Ferry until May 1830. Even then, instead of issuing a forthright dismissal, the War Department ordered Beckham to exchange positions with Benjamin Moor of the Allegheny arsenal in Pittsburgh.[60]

Stubblefield's departure from Harpers Ferry ended an erratic twenty-two-year association with the federal government. Between 1807 and 1815 he did his job well and might even be classified as an innovator insofar as he introduced the division of labor, piece rates, and the increased use of machinery. However, all this changed with the reorganization of the national armories under the Ordnance Department, Samuel Annin's retirement, and the appointment of Armistead Beckham as master armorer. After 1815, when significant mechanical developments were taking place in New England, he became more wary of outsiders and more conservative in his approach to technological change. Having consolidated control at Harpers Ferry, acquired a large country plantation, and developed outside business interests, he began to dissociate himself from everyday armory affairs and delegate managerial authority to lesser

60. Symington to Baker, May 23, 1829, AAR; Lee to Bomford, May 29, 1829, Stubblefield to Eaton, June 1, 1829, Sprigg to Bomford, June 29, 1829, Dunn to Bomford, August 12, 21, 1829, Gerard B. Wager to Sprigg, March 10, 1830, Letters Received, OCO; Bomford to Stubblefield, June 8, August 3, 1829, Letters Sent, OCO.

officers, most notably the master armorer. This action probably would not have been detrimental to the efficient operation of the works had he chosen his subordinates more carefully. But Stubblefield was not a good judge of men, and the Wool investigations clearly revealed the damage that Beckham, Malleory, and others had done. Although cooperative efforts between the two national armories had succeeded somewhat in updating production, the superintendent could point to few if any positive accomplishments undertaken at his own initiative during the 1820s. Fraught with internal strife and mired in administrative confusion, Harpers Ferry seemed totally incapable of closing the technological gap that existed between it and the New England branch of the arms industry. A representative of the Ordnance Department indicated just how wide this gap had become when, upon visiting the premises in 1832, he reported that "this armory is far behind the state of manufacture elsewhere and the good quality of their work is effected, at great disadvantage, by manual labor." Contrasting Springfield with Harpers Ferry, he noted that "there is so little machinery at the latter place that no fair comparison of prices can be made."[61]

As much as he may have believed in the rightness of his actions, James Stubblefield had sacrificed national goals for community convenience at Harpers Ferry. Repeated misjudgments and carelessness naturally brought criticism from George Bomford and other utilitarian-minded government authorities who considered him too closely wedded to local precepts to act objectively as an administrator. When the forces of "Jackson and Reform" finally succeeded in displacing Stubblefield from office, the local Junto suffered a blow from which it never fully recovered. Although the Beckhams, Stephensons, and Wagers continued to have substantial power and influence in the community, they never regained the same control over both town and armory as they possessed between 1815 and 1829. After 1830 the stakes of power at Harpers Ferry rested more

61. Talcott to Bomford, December 15, 1832, Reports of Inspections of Arsenals and Depots, OCO; Talcott and Major Henry K. Craig to Secretary of War, April 9, 1833, Reports and Correspondence of Ordnance Boards, OCO.

with professional politicians than with established family oligar-
chies. Even then, however, provincial attitudes and extended
bonds of kinship continued to affect policies and procedures up to
the coming of the Civil War.

CHAPTER 7

John H. Hall: Yankee in
the Garden, 1819-1841

Although the early history of the Harpers Ferry armory illustrates how much personal interests and cultural affinities can
retard innovation, government operations do not by their very nature discourage mechanical creativity. Talented individuals, working under government auspices, can achieve important technical
results. A case in point is the career of John H. Hall (1781-1841), a
spirited New Englander who labored more than twenty years
under special contract at Harpers Ferry. The story of Hall's efforts
to introduce a patented breechloading rifle into military service and
to standardize production through the use of precision techniques
testifies to more than the virtues of individual genius; it also clearly
depicts the trials and tribulations of rendering an effective mechanical synthesis in an age of developing industrialism.

Little is known about Hall's early years, except that he de-
scended from old Yankee stock and grew to manhood in and around
the thriving seaport of Portland, Maine. His father, Stephen, was a
Harvard divinity graduate and native of Westover, Massachusetts,
who had moved to Portland and married Mary Cotton Holt, a
twenty-four-year-old widow of a former Harvard classmate, in
1778. From that time until his death in 1794, Stephen Hall successfully managed his father-in-law's tannery in Portland and played an
active role in local politics, ardently supporting the separation of
Maine from Massachusetts during the 1780s. There John, the second of six children, was born in January 1781.[1]

1. Brief, somewhat tentative treatments of Hall's family background are provided by Charlotte J. Fairbairn and C. Meade Patterson, "Captain Hall, Inven-

Judging from his family background and his skill in expressing himself, Hall received a better-than-average education. He probably had contemplated a college career as an adolescent, but, with his father's unexpected death, had abandoned the idea in order to participate in the family business and perhaps even serve an apprenticeship to some auxiliary trade. Whatever the case, around 1808 he opened a shop near Richardson's Wharf in Portland where he carried on a general woodworking business as a cooper, cabinetmaker, and boatbuilder. In the same year his mother died, and Hall became one of the executors of the moderately well-to-do family estate. Five years later, in 1813, he married Statira Preble of York, sister of the Honorable William Pitt Preble of Portland and niece of Captain Edward Preble, the American naval hero of the Tripolitan War (1801–1805).

Statira Preble Hall was born in 1788. Like her spouse, her ancestry went back to the very beginnings of the Massachusetts Bay Colony. Described as a "tall, elegant woman" whose facial features were "stamped with benevolence," Statira added a marvelous sense of balance to her husband's brilliant but somewhat eccentric temperament.[2] Their union produced seven children: John Jr., William Augustus, Lydia Ingraham, Willard Preble, Annie, Mary, and George. From the outset Statira was Hall's closest confidante, the person whose judgment he trusted most. During twenty-eight years of marriage he made no important decision without first consulting her.

By the time of their marriage Hall had already turned his attention from woodworking to making firearms. Early in 1811 he had designed a rather bizarre-looking breechloading rifle which, according to one writer, "was a long step from the beaten path." Several years later Hall told then-Major George Bomford that "I invented the improvement in 1811, being at that time but little acquainted with rifles, and being *perfectly ignorant of any method whatever* of loading guns at the breech." He also maintained to have developed his breechloader without any knowledge of earlier designs by an

tor," *The Gun Report* 5 (Oct. 1959):6–10; and R. T. Huntington, *Hall's Breechloaders* (York, Pa.: George Shumway, 1972), pp. 1–2, 345–348.

2. Royall, *The Black Book*, 1:286.

Italian, Giuseppe Crespi, and a British army major named Patrick Ferguson. Unlike the mechanisms of Crespi and Ferguson, the "receiver" of Hall's rifle consisted of a hinged block that rose above the breech end of the barrel when released by a spring-operated catch (Figure 10). Its operation was simple. By pressing the spring release located under the breech, the shooter raised the forward part of the breech block or receiver, charged the chamber with gunpowder, and pressed the bullet into position with his thumb. After closing the breech and priming the flashpan, the rifle was ready for firing.[3]

When Hall sought to patent his invention in the spring of 1811, the Superintendent of Patents, Dr. William Thornton, immediately challenged the claim and informed the New Englander "that a similar improvement had already been made." After making further inquiries, Hall received a letter from Thornton "declaring *himself* the inventor," though adding "that he was nevertheless desirous of sharing the invention . . .& hoped there would be room enough for both of us." Perplexed, Hall went to Washington to meet the superintendent and examine his model. When Hall arrived at the Patent Office, Thornton promptly exhibited a Ferguson rifle but had prepared neither drawings nor specifications of the gun he supposedly had invented as early as 1792. Nevertheless, Thornton maintained, the mere "thought of a plan" resembling Hall's rifle authorized him to claim the invention. "In short," Hall declared, "the gun as he represented it was so very different from mine that had it ever been made, we might each have obtained patents for our respective improvements without any risque of ever interfering with each other."

Since Thornton refused to issue a patent "unless in connexion

3. William Gould, *Portland in the Past* (Portland, Me.: B. Thurston, 1886), pp. 448–449; Hall to Major George Bomford, January 24, 1815, Correspondence Relating to Inventions, OCO; Patent specifications of John H. Hall and William Thornton, May 21, 1811, Restored Patents, RPO (hereafter cited as Hall-Thornton patent). In addition to greater accuracy and firepower, Hall claimed that his rifle held several distinct advantages over conventional muzzle-loading weapons: it could be loaded and fired more rapidly; it could be loaded virtually in any position; the barrel did not foul as easily; and soldiers could not over-load their weapons, thereby lessening the problem of bursting which frequently occurred during the heat of battle. For an informed discussion of the ballistics, operation, and different designs of Hall's weapon, see Huntington, pp. 181–265.

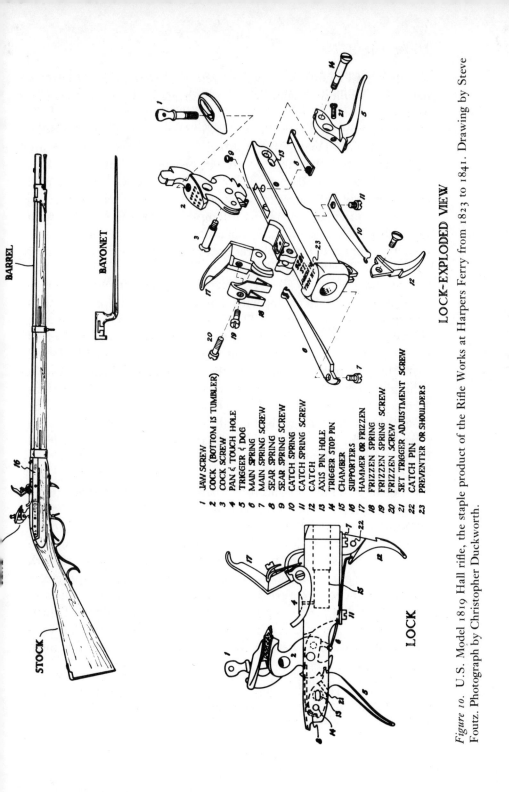

BARREL

BAYONET

STOCK

LOCK

LOCK-EXPLODED VIEW

1 JAW SCREW
2 COCK (BOTTOM IS TUMBLER)
3 COCK SCREW
4 PAN & TOUCH HOLE
5 TRIGGER & DOG
6 MAIN SPRING
7 MAIN SPRING SCREW
8 SEAR SPRING
9 SEAR SPRING SCREW
10 CATCH SPRING
11 CATCH SPRING SCREW
12 CATCH
13 AXIS PIN HOLE
14 TRIGGER STOP PIN
15 CHAMBER
16 SUPPORTERS
17 HAMMER OR FRIZZEN
18 FRIZZEN SPRING
19 FRIZZEN SPRING SCREW
20 FRIZZEN SCREW
21 SET TRIGGER ADJUSTMENT SCREW
22 CATCH PIN
23 PREVENTER OR SHOULDERS

Figure 10. U.S. Model 1819 Hall rifle, the staple product of the Rifle Works at Harpers Ferry from 1823 to 1841. Drawing by Steve Foutz. Photograph by Christopher Duckworth.

with himself," Hall called upon Secretary of State James Monroe to have the conflicting depositions adjudicated by law. However, Monroe seemed adverse to conducting an inquiry and, in Hall's words, had advised that "it would be more to my interest to be connected with Doct. Thornton even at the expense of half my right than to have it wholly to myself, because his influence in that case would be exerted in my favor but otherwise would be exerted against me, in the one case he would be friendly & disposed to render assistance in case of any attempts of others to interfere with my rights by attempting the obtainment of patents for the same invention connected with alterations, an event frequently occuring with patents likely to prove important."[4]

Cautioned by the secretary of state, Hall had little choice but to take out joint letters patent with Thornton for an improvement in firearms on May 21, 1811. At the same time both men signed a subsidiary agreement that Thornton would receive one-half of all proceeds on the sale of the patent rights to private individuals and that Hall, in turn, would have the exclusive privilege of establishing factories for the patent rifles. Although the stipulations seemed beneficial on the surface, Hall soon deeply regretted having "foolishly associated" with the wily commissioner of patents. No sooner had the patent taken effect than the two partners disagreed on how the rifle should be marketed. For obvious reasons Thornton wanted to sell the rights, collect his royalty fees, and be done with the business. Hall, on the other hand, wanted to restrict usage solely to the patentees. Consequently when the New Englander attempted to enlarge his operations by obtaining a government contract in 1811, Thornton used every means in his power to thwart the measure which, "if once effected, would do away with all hopes of selling patent rifles" on a concession basis and thereby ruin his prospects of realizing a lucrative profit on the venture. Thus, instead of promoting Hall's interests in Washington, Thornton ironically became one of his staunchest opponents between 1811 and 1820.[5] His part in the whole patent rifle affair is most

4. Thornton to Hall, April 20, 1811, Hall-Marmion Letters (transcription), HFNP; Hall to Bomford, January 24, 1815, OCO.

5. Ibid.; Hall-Thornton patent, May 21, 1811, RPO; Major J. B. Crane to Bomford, September 13, 1817, Hall to Colonel Decius Wadsworth, November 20, 1816, Correspondence Relating to Inventions, OCO.

questionable. Since he is often cited as a prolific inventor in his own right, one naturally wonders just how many other forced associations he made during his lengthy tenure as superintendent of patents.

To manufacture and market his rifles, Hall had to make large outlays for labor, raw materials, tools, and shop facilities. Initially he met these expenses by investing his personal savings as well as much of his mother's fortune in the business. Once these funds became depleted, he resorted to borrowing money from several Portland businessmen, using the family homestead as collateral. By May 1817 he had spent nearly $20,000 and had accumulated personal liabilities in excess of $6,000. Operating on a narrow margin and barely able to make interest payments, let alone a profit, Hall stood precariously near bankruptcy. He urgently needed further credit, credit that would be forthcoming if he could negotiate a government contract.[6]

Since 1798 many of the largest arms makers in the United States had depended almost exclusively on the federal government for contracts which provided much-needed working capital in the form of monetary advances. Between 1798 and 1809, for example, Eli Whitney had received generous subsidies enabling him to build a factory, complete a long-delayed contract for 10,000 muskets, liquidate outstanding debts, and, at the same time, pursue lawsuits in the South over his cotton gin.[7] Although the young republic depended on armorers such as Whitney, Asa Waters, and Simeon North for technical creativity, they, in turn, depended on the public for patronage. The fortunes of these inventor-entrepreneurs, therefore, varied not only with fluctuations in the economic market but also with shifts in domestic politics. Such conditions critically affected Hall's career, for without government support his financially troubled enterprise would have languished in Portland.

As early as 1811, Hall had approached President James Madison

6. Portland Scrapbook (Photostats from the archives of the First Parish Church, Portland), Maine Historical Society, Portland, Maine; Crane to Bomford, September 13, 1817, Hall to John C. Calhoun, November 12, 1818, Correspondence Relating to Inventions, OCO.

7. Deyrup, *Arms Makers*, pp. 41–42, 46–47; Robert S. Woodbury, "The Legend of Eli Whitney and Interchangeable Parts," *Technology and Culture* 1 (1960):235–253.

about introducing his rifle into military service; he was referred to the secretary of war's office, where little encouragement was received from William Eustis. Since the latter had already made arrangements with nineteen contractors for the delivery of 85,200 muskets under the Militia Act of 1808, he could hardly be expected to adopt an obscure breechloading rifle whose utility had yet to be proven. This consideration, plus Thornton's opposition and a long-standing preference among the military for conventional muzzle-loading weapons, prevented the New Englander from being awarded a contract. Refusing to allow his invention to be consigned to oblivion, Hall badgered the war department until Eustis' successor, General John Armstrong, consented to give the rifle a trial and ordered Bomford to investigate its merits. Accordingly, between December 1813 and November 1814, Hall delivered eight breechloaders—five rifles at $40 each and three muskets at $30—to Bomford at Albany, New York. After conducting a lengthy series of field tests, the future chief of ordnance "reported in favor of their adoption" and, with the secretary of war's authorization, placed another order with the Portland inventor for an additional two hundred rifles on December 23, 1814. Unfortunately, Bomford's timetable called for their delivery by April 1, 1815. Since each rifle had to be fashioned by hand, Hall had neither the work force nor the facilities to complete such a large order in so short a time. He therefore thanked the war department but declined the offer.[8]

Clearly disturbed about missing an opportunity to introduce his rifles into the service, Hall spent most of 1815 at Portland improving their design and training "not very expert workmen" to make them "in the best possible manner." After completing these preparations, he announced his readiness to manufacture "any number" of rifles for the government. At the same time he assured the secretary of war, "I have spared neither pains nor expence . . . in getting my

8. Eustis to Hall, March 4, 1811, Letters Sent Relating to Military Affairs, OSW; Hall to Secretary of War, November 21, 1814, to Bomford, November 21, 24, 1814, January 24, February 24, 1815, April 16, 1819, to William H. Crawford, January 23, 1816, Crane to Bomford, September 13, 1817, Correspondence Relating to Inventons, OCO; Hall to Bomford, January 16, 1827, Letters Received. OCO.

tools & machinery into the best possible order for executing the work with accuracy." Continuing, he confidently asserted:

Only one point now remains to bring the rifles to the utmost perfection, which I shall attempt if the Government contracts with me for the guns to any considerable amount viz., to make every similar part of every gun so much alike that . . . if a thousand guns were taken apart & the limbs thrown promiscuously together in one heap they may be taken promiscuously from the heap & will all come right. This important point I conceive practicable, & although in the first instance it will probably prove expensive, yet ultimately it will prove most economical & be attended with great advantages.[9]

Whether Hall intended it or not, this statement hit a responsive chord at the Ordnance Department. Having already committed themselves to the introduction of more uniform manufacturing standards at the national armories, both Bomford and Wadsworth exhibited keen interest in any proposal that might help bring the system into effect.

Hall evidently expected to receive a contract for one thousand rifles at the same price Bomford had offered in December 1814, $40 a stand. Admittedly these were generous terms, but Hall believed "the advantages to be derived from labor-saving machinery in manufacturing the rifles" with interchangeable parts would more than compensate for their high cost. In his opinion, "a *favorite* & important part of the American small arms would then be at the height of perfection, & would vastly excell those of any other nation." Since the United States would ultimately benefit most from the achievement, he thought it only right and just that the public should foot the bill. In Hall's mind the price of progress came high. The Ordnance Department, however, thought otherwise.

When Hall entered negotiations with Captain John Morton in June 1816, he soon discovered that the Ordnance Department had no intention of paying $40,000 for his rifles. Even Bomford, who fervently hoped the arm would be adopted, considered the price far too high. While he lacked authority to make a firm commitment,

9. Hall to Bomford, January 18, 1816, to Crawford, January 23, April 15, 1816, to Captain John Morton, June 22, 1816, Correspondence Relating to Inventions, OCO; Hall to Crawford, October 28, 1816, Letters Received, OSW.

Bomford did promise the inventor that he would "give every facility in his power" to accept the latter's proposal if made "at sufficiently reduced prices." Faced with the alternative of either lowering his price or forgoing a contract, Hall countered by offering to manufacture five hundred rifles at $25 apiece. "In all probability," he noted, this figure would "amount to less than the expences which would be incurred by me in making the guns." Even so, Wadsworth still seemed dissatisfied with the proposal. Urging "great Deliberation and Caution" before deciding on their permanent adoption, he convinced the secretary of war that no more than one hundred rifles would be required "to institute a pretty extensive Experiment."[10]

Such a meager order came as a severe disappointment to Hall. It not only dashed his hopes for building labor-saving machinery and manufacturing the arms with standardized parts, but increased his financial anxieties as well. Clearly the government struck a hard bargain. Desperate for work and "very desirous" to have his rifles "introduced into public use," Hall finally came to terms with Wadsworth in December 1816. Specifically the agreement called for one hundred patent rifles at $25 each. Although he was given a year to make delivery, Hall completed the arms at Portland and shipped them to Captain George Talcott at the U.S. arsenal in Charlestown, Massachusetts, by November 1817. Since Talcott had previously "entertained and expressed very unfavorable opinions of the invention," Hall anticipated little encouragement from him. However, much to the inventor's surprise the ordnance officer "inspected and received them for the United States, and after testing the arms, reported upon them in very favorable terms." Indeed, Talcott wrote Morton, "I cannot sufficiently praise them. Every thing (except the Bayonets) is as perfect as can be wished. I am decidedly of the opinion that they ought & most eventually [will] supercede the common Rifle,—but their principles are such that

10. Hall to Crawford, October 28, 1816, OSW; Hall to Morton, June 22, 1816, to Bomford, September 28, 1816, to James Monroe, March 21, 1817, Wadsworth to Secretary of War, November 11, 1816, Correspondence Relating to Inventions, OCO.

they must always be constructed with great care and attention as respects the individual parts."[11]

Given this endorsement, Hall pressed hard for a larger contract. Upon learning of Wadsworth's intention to send his rifles to an outpost in Missouri for extensive field trials before deciding on their adoption, he complained bitterly to President Monroe that "if eighteen months more are suffered to pass, as formerly, my business must stop, my workmen must disperse, & my ruin ensue." With his creditors demanding payment and his patent running out, Hall wanted to know why the introduction of his guns should "be delayed any longer than is necessary to obtain full proofs respecting them?" "With all due deference" to the chief of ordnance, he considered it extremely shortsighted and unfair to object to his rifle merely because its price exceeded that of a common rifle. After all, he argued, "it is impossible in the commencement of such a business to execute the work to such advantage & with such facility & dispatch as after it has been long established." For this reason, he concluded, "It would be absurd to reject guns & use bows merely because the latter cost less. Would it not be as improper to reject a species of firearms which will do twice as much execution as common ones merely because they cost more? . . . Permit me to suggest that there ought to be a difference observed between contracting with an inventor & a mere manufacturer . . . of allowing an inventor (in cases of contracts) . . . any price he may ask."[12]

As so often in government affairs, Hall's ability counted less than his influence. Astutely recognizing this fact, he worked assiduously during 1817 and 1818 to form a lobby at the nation's capital favorable to his cause. The group included not only his influential brother-in-law, William Preble, but also the congressmen from Maine, George Bradbury, Albion K. Parris, Enoch Lincoln, Prentiss Mellon, and Ezekiel Whitman. To further his case, the inventor also enlisted the support of Isaac Ilsley, Robert Ilsley, Wood-

11. Hall to Wadsworth, November 20, 1816, to Monroe, March 21, 1817, Correspondence Relating to Inventions, OCO; Talcott to Morton, December 9, 1817, Hall to Bomford, January 16, 1827, Letters Received, OCO.

12. Hall to Monroe, March 21, 1817, OCO.

bury Storer, John L. Storer, Josiah Paine, and General William King—all of whom had political connections in Washington and were friends of the Hall-Preble families in Portland. As much as anything else, the activities of this pressure group explain why Secretary of War John C. Calhoun decided to conciliate Hall temporarily by inviting him "to repair to the National Armory at Harpers Ferry" to supervise the construction of a few patent rifles for the purpose of improving the model and testing them.[13]

Although Hall received Calhoun's invitation in March 1818, he did not arrive at Harpers Ferry until August. There, with the assistance of several armorers, he made four sample weapons: two that corresponded in caliber and barrel length with the regulation musket and two that embodied standard rifle dimensions. After the samples were completed in October, James Stubblefield and Armistead Beckham—both of whom felt inconvenienced by the inventor's presence and resentful of his brusque manner—subjected the arms to a rigorous trial and examination. In addition to taking one-third the time to load, the breechloaders were found to be at least as accurate and powerful as the common rifle. As a result of these findings, Hall noted, Stubblefield and Beckham begrudgingly "yielded their prejudices" against his rifles and "made favorable reports concerning them." However, when Stubblefield submitted his final returns to the Ordnance office, he also observed that the arms cost nearly $200 apiece "in Materials, Labor & loss of time." This prompted the economy-minded Wadsworth to doubt their utility and especially to question "whether the facility of firing will not occasion an extravagent use and waste of ammunition." In any event, he told the secretary of war, "experiments made with Arms of that description can hardly lead to any practical result, because we can never afford to introduce such Arms into the Service."

Wadsworth's remarks immediately led to another confrontation with Hall, who, by November 1, 1818, had arrived in Washington

13. Ibid.; Hall to Calhoun, November 12, 1818, Correspondence Relating to Inventions, OCO; Hall to Calhoun, December 30, 1822, to Bomford, January 27, 1827, Letters Received, OCO; Hall to Calhoun, February 24, 1819, *Calhoun Papers*, 3:611–612; Hall to Calhoun, March 16, 1819, Misc. Letters Received, Third Auditor's Accounts, GAO.

with his newly finished specimens. Irked with the chief of ordnance for treating him with "coldness and neglect," he addressed a
long letter to Calhoun which, in effect, repeated earlier complaints
about the government's indifference to his invention. At the same
time he detailed the merits of the rifle and urged the secretary to
take affirmative action. Like his famous contemporary Whitney,
Hall never underestimated his own abilities, asserting that "at one
& the same time I have had to invent a new trade & bring it to
perfection—teach it to others & invent the tools for executing the
work." With firm resolve, he reminded Calhoun that if the government failed to adopt his invention, the loss would not only be his
own but also that of the nation.[14]

Though Wadsworth considered Hall's claims bold and extrava-
gant, Calhoun evidently thought otherwise because he ordered the
sample rifles sent to Greenleaf's Point in Washington for a new
series of experiments by a board of officers headed by Colonel
Nathaniel Towson. After a three-month trial the Towson committee issued a report testifying to "the advantage in favor of the
new one over the common rifle as 2 to 1." Furthermore, the members recommended the introduction of Hall's rifles into regular
service. As to endurance, the board reported, "the firing of these
new arms was continued from time to time until the musket had
been discharged 7,061 times, and the rifle 7,186 times, this appear-
ing to the board a fatigue of at least equal to what these pieces
would be exposed to in 14 or 15 campaigns, and probably more
than they would be required to undergo in a lifetime." In their
summation the officers stated that "the advantages of these [Hall's]
guns over the common ones now in use are, *first*, the celerity and
ease with which they may be loaded in all situations. It is of great
consequence in the rifle; for the difficulty of loading is the greatest
objection to its more general introduction into service; *second*,
greater accuracy and less recoil; *third*, less weight."[15]

14. Hall to Calhoun, November 12, 1818, OCO; Wadsworth to Calhoun, November 16, 1818, Letters Sent to the Secretary of War, OCO.

15. Hall to Calhoun, February 24, 1819, *Calhoun Papers*; Hall to Bomford,
January 16, 1827, OCO; U.S., Congress, House, Committee on Military Affairs,
Report on the Petition of John H. Hall, 24th Cong., 1st sess., 1835–1836, H. Rept.
375, p. 6.

Although Hall considered the document "very guarded," the findings of the Towson committee prompted Calhoun to offer the New Englander a contract for one thousand rifles. Hall responded that the Ordnance Department had led him to believe that, once the government finished the tests, "a purchase of a right to make at least 10,000 would be immediately effected." Nevertheless, he quickly added, "I submit in silence to the painfull disappointment." Three days later, on March 19, 1819, he signed the contract. Made with Wadsworth, the agreement called for the manufacture of one thousand breechloading rifles for which the inventor would receive a royalty of $1 for each weapon produced. The document also stipulated that instead of being paid a piece rate or hourly wage, Hall would receive a set salary of $60 a month.

In making his initial proposition, Calhoun gave Hall the option of either preparing the arms at his own shop in Portland or using facilities at one of the national armories. Since financial problems had forced the dismissal of most of his workmen during the summer of 1818, Hall chose the latter alternative. He also asked to be sent to Springfield, not only because it was nearer his home in Maine but also because he was mindful of the antagonism he had encountered in Virginia. The chief of ordnance, however, adamantly refused the request primarily because he wanted to keep a sharp eye on Hall's operations. Consequently the contract of March 1819 specifically stated that Hall would serve at Harpers Ferry and "perform the Duty of an Assistant Armourer in instructing and directing the Workmen, to be employed in fabricating the Firearms above Specified."[16]

For eight years Hall had struggled to perfect a rifle design that could be produced by existent techniques and to convince government authorities to depart from customary preferences for muzzle-loading weapons. That he risked a small fortune, encountered firm opposition, and experienced numerous setbacks is without doubt. Yet, his surmounting of these difficulties indicates he possessed not only mechanical ability but also the necessary tenacity and influence to sell his product in a reluctant postwar market. In sign-

16. Hall to Calhoun, March 16, 1819, GAO; Contract Book No. 1, OCO.

ing his contract Hall went one step beyond contemporaries such as North and Whitney because the federal government subsidized him not with monetary advances but as a private manufacturer at a public armory. In this sense the contract of 1819 was unique.

Hall assumed his official duties at Harpers Ferry early in April, though he did not move his family to Virginia until the following summer. Other than making engagements for gun stocks and welding some barrels at the armory forge, he apparently accomplished very little in 1819, primarily because the Ordnance Department lacked appropriations for the rifle project and had "no money to spare." In fact, Hall did not occupy a separate workshop until February 1820. Even then the building he received consisted of a small woodframe structure which had been used as a sawmill since the 1790s and badly needed repairs. What became known as the Rifle Works—a separate division of the Harpers Ferry armory—was not officially recognized until April 1820 when the paymaster submitted Hall's first payroll to the Second Auditor's Office. (See map.) Prior to that time the inventor conducted his business at the larger Musket Factory and, except for the assistance of Robert Blanchard, Timothy Herrington, and Peletiah Thompson, he worked alone.[17]

Hall at first had expected to complete his contract no later than September 1821. However, work on the rifles proceeded haltingly. When in August 1821, Roswell Lee inquired about the inventor's progress, Stubblefield replied, "Capt. Hall has not finished any of his Guns, but is moving on slowly with the component parts." About the same time Whitney visited Harpers Ferry to scrutinize the "New System" being adopted there. While frequently interpreted as a reference to Hall's work, the so-called "New System" Whitney mentioned did not concern interchangeable parts but rather a method of inventorying stock distributed to armory workmen. Whitney, no doubt, did inspect the Rifle Works but, consider-

17. Hall to Calhoun, April 26, 1819, to Wadsworth, October 30, 1820, Correspondence Relating to Inventions, OCO; Hall to Bomford, May 3, 1819, Stubblefield to Bomford, May 15, 1822, Letters Received, OCO; Philip R. Smith, "History of Lower Hall Island, 1796–1848" (Research report, HFNP, 1959), pp. 1 ff.; Payrolls and accounts, Harpers Ferry armory, 1820, Second Auditor's Accounts, GAO.

ing the date, it is likely he did not see much of interest. If he did, he made no mention of it in his correspondence with Lee, his friend and former employee. In fact, Hall, by his own admission, did not finish his tools and begin work on machinery until the summer of 1822. Though he sent twenty rifles to the Ordnance office in the summer of 1823, he did not complete the entire parcel until December 1824.[18]

Reasons for the delay are not difficult to find. A major factor, mentioned earlier, was the personal animosity between Hall and Stubblefield and the resulting confrontations over allocation of appropriations, shop facilities, workers, and raw materials. Furthermore, bilious-fever epidemics during the summers of 1820 and 1821 seriously impeded operations at the Rifle Works. Of prime importance, however, was Hall's preoccupation with the idea of making his rifles with interchangeable parts and, as a result, his postponement of manufacture until tools and machinery were constructed that could accomplish the task. This factor not only deferred completion of the rifles by three years but substantially increased their cost as well. Aware of the problem, the inspector general remarked upon visiting Harpers Ferry in 1822 that "Mr. Hall . . . is too fond of projects, too much of an innovator ever to have been entrusted with publick means to complete machinery of his own invention. The 1000 Rifles which he is about making will cost the Government more than 50 dollars each before they are finished."[19]

Despite the criticisms of Stubblefield and the inspector general, Hall refused to be swayed from his purpose. Having devoted more than half of his time to the preparation of tools and machinery, he readily admitted to the secretary of war in 1822 that the apportionment of expenditures "cannot be considered as too favorable for the Rifles." Nevertheless, he argued, "the principles upon

18. Stubblefield to Lee, August 23, 1821, Whitney to Lee, August 23, 1821, Letters Received, SAR; Lee to Bomford, June 9, 1821, to Whitney, June 13, 1821, Letters Sent, SAR; Bomford to Stubblefield, 1821, Hall to Calhoun, May 15, 1822, December 21, 1824, to Bomford, December 30, 1822, March 3, 1823, January 16, 1827, Letters Received, OCO.

19. "Notes of a Tour of Inspection," Harpers Ferry armory, December 1822, OIG.

which my tools & machines have been constructed are applicable to every species of small arms & have for their object the production of perfect uniformity with the least possible expence." While costly in the short run, he emphasized that "when completed, . . . the machinery will answer as well for one hundred thousand of the guns as for one thousand." These reasons, he believed, sufficiently warranted the indulgence of his superiors in Washington.[20]

Fortunately Hall received encouragement and support from Bomford and Calhoun. As chief of ordnance and secretary of war respectively, two more important sympathizers could not have been found.[21] Both men recognized the potential benefits of Hall's mechanical experiments and, having visited Harpers Ferry on several occasions between 1821 and 1825, both possessed ample proof that the New Englander's work was no mere hoax or visionary pipe dream. Well aware of their interest in the project, Hall kept Bomford and Calhoun informed of his progress and often sought their advice. Consequently when the first glimmerings of success appeared during the winter of 1822–1823, he lost no time writing the secretary of war:

I have succeeded in an object which has hitherto completely baffled (notwithstanding the impressions to the contrary which have long prevailed) all the endeavors of those who have heretofore attempted it—*I have succeeded in establishing methods for fabricating arms exactly alike, & with economy, by the hands of common workmen, & in such a manner as to ensure a perfect observance of any established model, & to furnish in the arms themselves a complete test of their conformity to it.*

One of the many advantages of this perfection will be to render the fabrication of each part of a gun totally independent of all the other parts & thus to prevent a great loss of time which is constantly occurring & at the same time to facilitate such a complete division of labor as will, ultimately, reduce the expence of manufacturing arms to its lowest possible amount.[22]

20. Hall to Calhoun, May 15, 1822, OCO.
21. Bomford, as stated, succeeded Wadsworth as chief of ordnance in June 1821. For an appraisal of Calhoun's many accomplishments as secretary of war (1817–1825) see White, *The Jeffersonians*, pp. 246–250.
22. Hall to Calhoun, December 30, 1822, OCO.

While the letter betrays Hall's unabashed egotism, he at least never shunned inquiry nor acted clandestinely in these matters. Quite to the contrary, he welcomed public scrutiny and opened his shop to all comers in hope of gaining national recognition. When the first thousand rifles were almost completed in 1824, he not only invited the secretary of war and chief of ordnance to inspect the finished products but also asked them to conduct a full-scale investigation of his manufacturing operations as a matter of public record. On the first score, Calhoun and Bomford went to Harpers Ferry and witnessed "the manner in which the several parts, promiscuously taken, came together, fitted and adapted to each other." Thoroughly impressed with what they saw, both men approved the extension of Hall's contract to a second thousand rifles in July 1824. Bomford further recommended an extensive trial of Hall's rifles at Fortress Monroe, Virginia, in 1825. Though he did not question the outcome, he wanted "their superiority over other arms fully and satisfactorily proved" before deciding on their permanent adoption by the army. He doubtlessly thought that the publicity generated by these tests would enhance the reputation of the Ordnance Department as well as the inventor's. By the time Bomford made the proposal, however, Calhoun had retired as secretary of war and his successor, James Barbour, took no action. There the matter rested until May 20, 1826, when the House of Representatives passed a resolution introduced by John H. Cocke, chairman of the committee on military affairs, calling for detailed information "concerning the fabrication, cost & utility of the patent Rifles."[23]

The impetus behind Cocke's resolution came not from Hall but, interestingly enough, from his adversaries. Stubblefield, as observed, bitterly resented having to share his annual appropriations with the New Englander, and strong feelings existed in the local Junto that the armory should make muskets, not Hall rifles. In a

23. Hall to Calhoun, December 21, 1824, February 19, 1825, Thomas Copeland to Major William Wade, February 19, 1826, Edward Lucas, Jr., to Cocke, January 15, 1827, Letters Received, OCO; Contract Book No. 1, OCO; Bomford to Barbour, September 28, 1825, July 3, 1826, Letters Sent to the Secretary of War, OCO; Hall to Bomford, December 9, 1826, Correspondence Relating to Inventions, OCO.

determined effort "to put a stop to the Manufactory of the patent Rifles altogether," Stubblefield, Edward Wager, James Stephenson, and Wager's cousin, Edward Lucas, Jr., pooled their political influence and worked primarily through Congressman William Armstrong of nearby Romney, Virginia, to persuade Cocke to introduce the resolution. Once "the waste & extravagance of the Publick money on the Patent Rifle" came to light, they expected Congress would withhold further appropriations from the project and thereby end Hall's association with the Harpers Ferry armory.[24] However, the scheme backfired. Ironically, the inquiries that followed thoroughly vindicated Hall and drew further attention to Stubblefield's unsatisfactory conduct as superintendent.

Upon official notification by the committee on military affairs, Secretary Barbour authorized Bomford to investigate every aspect of Hall's operations at Harpers Ferry. The chief of ordnance acted promptly, issuing orders on July 7, 1826, to the commandant of Fortress Monroe to supply two companies with Hall rifles; they were to begin "a course of practice in opposition to an equal number of men armed with new pattern Springfield Muskets" and another company armed with Model 1814 Harpers Ferry rifles. These trials lasted nearly five months, terminating in December 1826 when a military board consisting of the staff of the Artillery School of Practice submitted a long report to Bomford who, in turn, relayed it to the Cocke committee.

In evaluating the general utility of Hall's breechloader, the board expressed "its perfect conviction of the superiority of this arm over every other kind of small arm now in use." "This opinion," the members emphasized, "has been formed after having seen two companies armed with them for five months, performing all the duties to which troops are liable in garrison, and contrasting them in a variety of ways with the Common Rifle and Musket—in all which trials their great and general superiority has been manifest." The examining officers repeatedly praised the "uniformity of all its component parts," noting that, in contrast to the other arms, Hall's rifle was unique in this respect. This consideration weighed heavily

24. Lucas to Cocke, April 27, 1826, Copeland to Wade, May 29, 1826, undated letter (1826), Letters Received, OCO.

with the committee because the rifle's parts could be mutually exchanged with one another, thus greatly simplifying the task of making field repairs. Finally, the members concluded, "the Hall Rifle, after having sustained 8,710 discharges, appears in a fit condition for service."[25]

In addition to the actual performance tests, Cocke's resolution had also called for information about the manufacture of Hall's rifles. Since one of the most serious allegations against Hall concerned the wasteful extravagance of his tools and machinery, Congress particularly singled out these items for critical examination. Bomford was personally qualified to make such a report, but he evidently wanted to maintain some semblance of neutrality as chief of ordnance and, so far as possible, remain aloof from the proceedings. For this reason he delegated the question to his administrative assistant, Major William Wade.

Wade began making inquiries about the selection of members for an examining committee in October 1826. Recognizing that the complexities of Hall's machinery could hardly be understood let alone assessed by the technically uninitiated, he immediately established two guidelines for conducting the investigation. "In the first place," he noted, "it should be made by persons who are . . . practical judges of machinery; and have some acquaintance with the manufacture of arms, or at least a knowledge of works in iron." Second, "they should be persons who from their established reputation as machinists or Engineers, are known to the Public, in order that their report and opinions may command proper respect."[26]

With these criteria in mind, Wade asked Roswell Lee to serve as one of the examiners and to assist him in finding two other candidates for the assignment. On October 9 he wrote the Springfield superintendent, "I do not see how we can well get along without you." Lee, however, refused the appointment because he had already "formed an opinion in relation to the subject" and feared that

25. Bomford to Barbour, July 3, 1826, OCO; Major J. B. Crane (report), December 11, 1826, Reports and Correspondence of Ordnance Boards, OCO; Reports of a Board of Officers at Old Point Comfort on Hall's Patent Rifle, December 11, 1826, Reports of Tests and Experiments, OCO.

26. Wade to Lee, October 9, 1826, Letters Received, SAR.

it would lead to "trouble & difficulty which might otherwise' be avoided" with Stubblefield. In any event, he responded, "Were I to make the selection I would appoint the following persons (viz.) Maj. Wm. Wade, James Carrington, & Joseph Weatherhead Esquires." Although an excellent suggestion, Wade obviously could not serve, and Weatherhead could not be spared from his duties as master armorer at Springfield during Lee's temporary transfer to Harpers Ferry as acting superintendent. Thus, only Carrington remained a possibility.[27]

Carrington possessed impressive credentials. In Lee's opinion, few persons in the United States could equal his qualifications and experience. Widely known as an armorer as well as a machinist, he had served as foreman of Eli Whitney's celebrated "Gun Manufactory" near New Haven since 1799. Although Whitney rarely handed out compliments, even he considered Carrington "a man of strict integrity" and "very competent." In addition to his regular duties as shop foreman, Carrington also had the rare privilege of inspecting Whitney's arms for the government. On several occasions he had acted as an intermediary in disputes between the Ordnance Department and private arms contractors. After Whitney's death in 1825, Carrington resigned his position as foreman and established his own foundry, coffee mill, and razor strap factory at Wallingford, Connecticut. Even then, he continued to inspect arms made by Philos and Eli W. Blake at Whitneyville. If anyone commanded the respect of the arms-making community it was James Carrington. Consequently, with Wade's wholehearted approval, Lee asked his friend to head the committee on Hall's machinery. On October 19, 1826, Carrington accepted.[28]

27. Lee to Wade, October 17, 20, 1826, Letters Sent, SAR. Weatherhead had served as master armorer at Springfield since December 1825 and would continue in that position until December 1833. Prior to his arrival at the national armory in 1817, he had worked as an armorer-machinist at Simeon North's Staddle Hill factory in Middletown, Connecticut.

28. Whitney to Bomford, July 3, 1821, Lee to Bomford, July 27, 1824, Letters Received, OCO; Bomford to Lee, December 14, 1825, Letters Sent, OCO; Lee to Phineas Blair, March 2, 1822, to Carrington, October 17, 1826, to Wade, October 20, 1826, Letters Sent, SAR; Carrington to Lee, January 14, 1823, January 7, October 19, 1826, Lemuel Pomeroy to Lee, July 30, 1824, Letters Received, SAR; Henry Dearborn to Whitney, December 10, 1808, Misc. Letters Sent, OSW.

While waiting for Carrington's reply, Wade explored the possibility of adding Luther Sage to the committee. When asked for his opinion of Sage, Lee characterized him as "a practical mechanic, faithful & industrious, prompt in his duty & firm in his decisions," adding that he had "full confidence in his integrity & ability." In short, Lee concluded, "I think Mr. Sage will answer very well for one of the examiners."[29]

Between 1815 and 1823, Sage had served as an assistant armorer under Lee at Springfield, spending most of his time as an inspector of contract arms. Well liked and respected by contractors, he had traveled to virtually every important armory in New England and had had ample opportunity to observe the introduction of new techniques. His familiarity with the technically sophisticated works of Simeon North, Nathan Starr, and Robert Johnson at Middletown, Connecticut, would serve him well as a member of the investigating committee. Transferred to the Frankford arsenal near Philadelphia in April 1823, Sage had also inspected arms made by Marine T. Wickham, Henry Deringer, John Rogers, and Brooke Evans. With such broad-ranging experience he possessed ample qualifications to judge the novelty of Hall's machinery. Consequently Wade appointed Sage as the second-ranking member of the examining committee in October 1826. As a government employee he had no choice but to accept.

Since Carrington and Sage were New Englanders, Wade wisely decided to balance the board by choosing a third commissioner from one of the southern states. It took him several weeks to find a suitable person for the assignment, but he eventually recommended Gideon Davis, a machinist and millwright from Georgetown, D.C. With Davis' acceptance, Bomford announced that the committee would convene at Harpers Ferry on December 11, 1826. At the same time he notified the commissioners, "your attention should be directed to the operation of the respective machines so much in detail as to enable you to arrive at just conclusions respecting the general merits of each when contrasted with the several machines hitherto in general use for the manufacture of small arms."

29. Lee to Wade, October 17, 1826, to Bomford, January 15, 1823, Letters Sent, SAR.

However, two days before the scheduled date of the inquiry, Davis informed the Ordnance Department that he had been unexpectedly detained in Richmond on business and would be unable "to assist in examining Mr. Hall's machinery." Since Carrington and Sage had already set out for Harpers Ferry, it was too late to postpone the investigation. Upset at the news, Bomford tried to improvise by instructing acting superintendent Lee "to supply the vacancy." "If there be no person at the Armory who would, in your opinion, be altogether suitable for this duty," he added, "you will please to request Messrs. Carrington & Sage to proceed in the business, by themselves."[30]

Acting with typical celerity, Lee had a replacement for Davis by December 12. At first he had considered adding his trustworthy clerk, Elizur Warner, to the board. But, "as Capt. Hall seemed inclined to have some person entirely disconnected with the National Armories," Lee decided on Colonel James Bell of West Brook, Frederick County, Virginia. While "highly spoken of as a military man and of very respectable character," Bell admittedly knew little about machinery. Nevertheless, Lee thought the unusual circumstances warranted his appointment. If nothing else, Bell could bear witness to the competence of his fellow committeemen as well as the objectivity of their report.[31]

Having arrived at Harpers Ferry on the evening of December 11, Carrington and Sage began their duties the following morning. Bell joined them three days later. Together they spent over three weeks examining Hall's works and, according to Carrington, had a very agreeable time. In conjunction with "contrasting the new with the old machinery," the commissioners diligently studied the costs of construction and repair, the quantity of work performed by each machine in a given time, the relative portion of manual labor required in each case, and the quality of work performed. These tasks completed, Carrington assembled his notes and prepared a seventeen-page report entitled "On Hall's Machinery." On January 6,

30. Bomford to Messrs. Carrington, Sage, and Davis, December 2, 1826, to Lee, December 9, 1826, Letters Sent, OCO.

31. Lee to Bomford, December 12, 1826, Lucas to Cocke, January 15, 1827, Bell to Alfred H. Powell, February 25, 1827, to Bomford, February 25, 1827, Letters Received, OCO.

1827, the committee officially concluded its proceedings, Bell returned to his home at West Brook, and Carrington and Sage accompanied Lee to Washington to present their findings to the chief of ordnance.

It is doubtful whether Hall himself could have composed a more laudatory document. The members of the Artillery School of Practice had praised his rifle, but the Carrington committee found his machinery unparalleled in contemporary practice. "Capt. Hall," they declared, "has formed & adopted a system, in the manufacture of small arms, entirely *novel* & which no doubt, may be attended with the most beneficial results to the Country, especially, if carried into effect on a large scale." Continuing, they observed that "his machines, for this purpose, are of several distinct classes, and are used for cutting iron & steel & for executing wood work; all of which are essentially different from each other & differ materially from any other machines we have ever seen, in any other establishment."

Although Carrington and Sage briefly described these machines, they thought their general merits, since they contrasted so sharply with techniques used elsewhere, could "be better understood by pointing out the difference of the results produced by them." An infallible test of the accuracy and effectiveness of Hall's machinery could therefore be found in the arms themselves. "It is well known," the commissioners stated, "that arms have never been made so *exactly similar* to each other by any other process, as to require no *marking* of the several parts & so that those parts on being changed would suit equally well when applied to every other arm." Yet, they noted, "the machines we have examined, effect this with a certainty & precision, we should not have believed, till we witnessed their operations."

If uniformity, therefore, in the component parts of small arms is an important desideratum, . . . it is, in our opinion, completely accomplished by the plan which Hall has carried into effect. By no other process known to us (& we have seen most, if not all that are in use in the United States) could arms be made so *exactly alike*, as to interchange & require no *marks* on the different parts. And we very much doubt, whether the best workmen that may be selected from any Armory, with the aid of the best machines in use *elsewhere*, could in a whole life make a hundred rifles or muskets that would, after being promiscuously mixed together, fit each other, with that exact nicety that is to be found in those manufactured by Hall. . . . We would however further observe, that in point of *accuracy*, the quality of the work is greatly *superior* to any thing, we have ever seen or expected to see, in the manufacture of small arms & cannot with any degree of propriety, be compared with work executed by the usual methods, and it fully demonstrates the *practicability* of what has been considered almost or totally *impossible* by those engaged in making arms, viz., *of their perfect uniformity.*

Finally, the self-effacing members concluded, "we were not fully sensible, when we commenced this examination, of its importance & feel our incompetency to do that justice to the subject it requires & wish it had been confided to those, who were more able to report the merits of the *machinery* and the *inventor*, who we trust will receive that Patronage from Government his talents, science & mechanical ingenuity deserve." These were humble words indeed coming from individuals such as James Carrington and Luther Sage.[32]

In view of Carrington's intimate association with Whitney and Sage's familiarity with the Middletown contractors, their report paid remarkable tribute to the magnitude of Hall's achievement. Both men emphatically agreed that, by producing a large parcel of firearms with interchangeable parts, Hall had done something that his contemporaries had not. Interchangeability had completely eluded Whitney during his lifetime and had been only partially accomplished by Simeon North by 1827. Under Lee's leadership the Springfield armory—undoubtedly the most progressive of government installations—had introduced the concept of serial machine production of highly similar parts as early as 1818; but, even there, components had to be individually marked, filed, and fitted in a soft state before final assembly. Marking components so that they can be distinguished from others in the same batch is unnecessary in the manufacture of truly standardized parts. The Carrington

32. James Carrington, Luther Sage, and James Bell to Bomford, January 6, 1827, Reports and Correspondence of Ordnance Boards, OCO (hereafter cited as the Carrington Committee report).

report as well as existent specimens of Hall rifles themselves indicate that Hall had obviated this requirement as early as 1824. Herein lay the novelty of his work: rifle components after careful gauging and machining could be assembled in a case-hardened state. Such a result convinced Cocke and other members of Congress that Hall's project was indeed worthwhile, despite its expense.

Nevertheless, criticism by the Junto continued. No sooner had Carrington and Sage left Harpers Ferry than Edward Lucas, Jr., warned Cocke not to take their findings too seriously, since they came "from a paternal roof and friendly fireside." Fancying himself as a "discomfiter of visionary theorists," Lucas seized upon the hasty appointment of Bell to level charges of favoritism against the chief of ordnance. Specifically, he intimated that Bomford had permitted Hall to hand pick all three commissioners with the purpose "of getting up . . . a favorable report." No one took his accusation seriously.[33]

Two years later Stubblefield renewed the attack on Hall. Having previously scoffed at the novelty of the New Englander's machinery, in 1829 he distributed a printed circular in Washington that admitted their utility but denied Hall as the inventor. Provoked by the superintendent's belittling remarks, Hall immediately penned a devastating rebuttal in a letter to the editor of the *National Intelligencer*. At the same time Bomford countered Stubblefield's claim that "these tools and machines were previously well known and used in other armories" by curtly referring to the Carrington committee's declaration to the contrary in 1827. While Stubblefield admitted that he had never read the Carrington report, he nonetheless continued his attacks through August 1830. But like Lucas, he utterly failed to generate any congressional support.[34]

Even after Stubblefield's resignation in 1829, the problem would continue under his successors. Jealous of their prerogatives and

33. Lucas to Cocke, January 15, 1827, OCO.
34. Stubblefield to Bomford, July 11, October 8, 1829, Hall to the editor of the *Daily National Intelligencer* (Washington), September 5, 1829, "A Mechanic" (Stubblefield) to members of Congress, April 12, 1830, Letters Received, OCO; Bomford to Stubblefield, September 16, 1829, Letters Sent, OCO.

resentful of having to share appropriations with the semi-autonomous Rifle Works, time and again superintendents George Rust, Jr., and Edward Lucas, Jr., were to clash with Hall over the allocation of shop facilities, workmen, and raw materials. Hall, for his part, did little to alleviate the situation. Haughty, self-righteous, and outspokenly critical, he proved to be an extremely difficult person to get along with.[35]

Despite the antagonism and hard feeling at Harpers Ferry, the investigations of 1826 did result in Hall being given a third contract on March 8, 1827. Reflecting the government's positive sentiment, the new agreement more than doubled his salary to $1450, increased the yearly allotment of rifles to three thousand, and continued the royalty of one dollar for each gun produced. Significantly Hall received this fee not for the rifles themselves but rather for the use of certain machines "for cutting metallic substances" he had patented the day before signing the contract. In addition to superintending the production of three thousand arms annually, he explicitly agreed to improve "the methods of conducting the business" and to perfect "the machinery therefor." This alteration of Hall's services at Harpers Ferry emphasized the experimental nature of his work. Contrary to what arms collectors and military historians often assume, his primary function after 1826 was not to make rifles per se but to refine and further develop mechanized techniques for their manufacture.[36]

After the release of the Fortress Monroe and Carrington reports the demand for patent rifles mounted rapidly. Between 1827 and 1829 no less than ninety-six congressmen asked for specimens to exhibit in their respective districts. While similar requests came from two foreign governments, private individuals, and the Marine Corps, the largest orders by far flowed in from different state governments. In 1808 Congress had authorized an annual appropria-

35. See, for example, Hall to Bomford, March 6, 1835, March 2, 1836, July 20, 1837, Rust to Bomford, November 14, 17, 1836, Lucas to Bomford, August 1838, September 6, 1839, Letters Received, OCO.

36. Contract Book No. 1, OCO; Leggett, *Index to Patents*, 2:922; Hall to Bomford, March 6, 1835, to Lewis Case, October 15, 1834, Benjamin Moor to Bomford, October 12, 1835, Rust to Bomford, November 25, 1834, Letters Received, OCO.

tion of $200,000 to arm and equip the militia. Under the provisions of this act each state received a yearly quota and could stipulate its needs for military supplies. Consequently, when the governors of six states issued calls for Hall rifles in lieu of muskets, the Ordnance Department faced an unexpected supply problem. At first Bomford refused to honor these requests primarily because he deemed it unwise to place such expensive weapons in the hands of inexperienced troops. As demands from the states mounted, however, he was soon overruled by the secretary of war and therefore had to find some means of filling these orders.

Although Hall believed he could easily supply both federal and state needs if provided adequate shop facilities, a legal technicality required all arms made under national armory appropriations to be "reserved solely for the use of U. States troops." Since the War Department defrayed Hall's expenses from these funds, it meant that his rifles could not be issued to state militias. Congress had traditionally reserved all appropriations for arming and equipping the militia to private contractors. To furnish the states with Hall rifles, it therefore became "necessary that some of the contractors should commence the fabrication of them." Faced with this need, Bomford recommended Simeon North of Middletown, Connecticut.[37]

North's widespread reputation as an innovative arms maker undoubtedly prompted Bomford to offer him a contract for the patent rifles. Yet, when Hall was asked to communicate his views on the subject, he seemed very apprehensive about allowing "a contractor to attempt making them." "The amount of capital *must* be large & the risk great," he cautioned, "for if the contractor should fail of full and complete success, his arms must all be rejected and he will be ruined, as the introduction of the Rifles into the service in so defective a state as not to admit exchanging all their parts with each

37. Bomford to Hall, November 17, 1828, Letters Sent, OCO; Bomford to Peter B. Porter, December 11, 1828, Letters Sent to the Secretary of War, OCO. For literature on Simeon North (1765–1852), see Merritt Roe Smith, "John H. Hall, Simeon North, and the Milling Machine," *Technology and Culture* 14 (1973): 574– 577; S. N. D. North and Ralph H. North, *Simeon North, First Official Pistol Maker of the United States* (Concord, N.H.: The Rumford Press, 1913); and *Dictionary of American Biography*, s.v. "North, Simeon."

other, and with those made here would totally defeat the great object for which so much expense has been incurred." Hall estimated that the production cost per rifle would be around $16.68, adding "that a contractor ought in justice to himself, to get as much more in addition to this amount as will compensate for the deterioration of his property and the interest of his capital while getting his manufactory under way." While he did not oppose giving North a contract, he was skeptical that any private contractor could achieve and maintain the necessary standards for manufacturing his rifle with interchangeable parts.[38]

Apparently North possessed the capital and was willing to assume the risk because, on December 15, 1828, he contracted with the Ordnance Department for five thousand rifles at $17.50 apiece. The pact stipulated that deliveries would begin on July 1, 1829, at the rate of one thousand annually.[39] Almost immediately, however, North encountered production bottlenecks, not the least of which was an oversight at the Ordnance office that forced him to build tools and equipment from several regular-issue rifles rather than from specially prepared pattern pieces and a set of Hall's gauges. This threw the New England armorer off schedule by more than a year.

When finally piecemeal deliveries were made during the summer of 1830, numerous inspection problems arose. Having received instructions from Bomford to rigorously scrutinize the quality and character of North's work, Hall took the chief of ordnance at his word and conducted extremely critical inspections. Determined to maintain high standards of uniformity, he found numerous faults in the rifles made at Middletown. When called upon for an explanation, North stated that "the defects specified may have arisen from the imperfections of the patterns which had been furnished him." Overly sensitive, Hall considered this intimation "ungenerous & unjust" and refused to have the blame "thrown upon" himself. More than likely, he told Bomford, the discrepancies occurred be-

38. Hall to Bomford, December 5, 1828, Letters Received, OCO.

39. Contract Book No. 1, OCO. The $17.50 contract price included the fee for the use of Hall's patented machinery. Whether Hall ever received payment from North cannot be ascertained from existent records.

cause North had used the pattern rifles "in making tools and machinery for executing the contract" rather than preserving one of them "from all injury & unimpaired, for a model, as it ought to have been."[40]

No doubt both men had cause for complaint. To an extent, the Ordnance Department deserved censure for neglecting to order adequate patterns and gauges prepared for North in the first place. After Bomford made these arrangements, difficulties over inspection procedures eventually disappeared. By 1834, Hall with pride could observe that components of the rifles made by North not only exchanged well with each other "but equally well with those made under" his "immediate supervision at Harpers Ferry."[41] For the first time fully interchangeable weapons were being made at two widely separated arms factories. Although another decade would pass before the practice became common throughout the firearms industry, the feat marked an important stage in the evolving pattern of precision production. No two individuals played a more important role in this development or as machine-tool innovators in general during the early nineteenth century than did North and Hall.

Hall worked at a less hurried pace after 1829 than during his early years at Harpers Ferry. He slowed down, however, by necessity not by choice. Although the Ordnance Department extended his 1827 contract on April 22, 1828, to nine thousand rifles, Bomford's plan for increasing annual output at the Rifle Works to three thousand stands had completely dissolved by the following spring. In fact, Hall had just finished forging components for the new rifles when he received word that certain appropriations would have to be curtailed to make up former deficits accumulated by the Stubblefield administration. This news came as a severe disappointment to Hall, for it shattered his hopes for expanding pro-

40. Bomford to Hall, August 1, December 17, 1830, to Otis Dudley, December 17, 1830, Letters Sent, OCO; Hall to Bomford, June 3, July 26, September 28, October 5, December 27, 1830, April 25, 1831, Dudley to Bomford, December 7, 1830, Letters Received, OCO.

41. Hall to Cass, October 15, 1834, OCO. Also see Hall-North correspondence, 1830–1832, Letters Sent-Letters Received, OCO.

duction and forced him to lay off a number of valuable hands. At the same time, congressional refusal to appropriate funds for the construction and repair of several buildings at the Rifle Works proved equally frustrating. Not only was the erection of a new smith's shop delayed but also Hall was prevented from taking over a grinding mill and tilt-hammer shop on the Shenandoah which had been promised him in 1827.

With operations seriously restricted by a lack of adequate shop space, Hall continued to experience the same "inconvenience and embarrassment" noted by the Carrington committee several years earlier. Constantly having to move one machine out of the way to make room for another precluded the possibility of manufacturing all parts of the rifle simultaneously. More than once the inspector of arsenals and armories had pointed out that Hall had to finish "the entire quantity of some parts," then completely rearrange his tools and machinery before proceeding to another set of components. Hall opined that it took longer to set up the machinery for each operation than it actually did to make the components. In short, inadequate facilities forced him to relocate and adjust his machinery each time he changed from making receivers to stocks and stocks to barrels. In view of this expensive and time-consuming handicap, it is easy to see why he needed seven years to fulfill his obligations under the contract of 1828 and why the cost of his rifles never fell below $14.50 apiece during the 1830s.[42]

While far from pleased with circumstances at Harpers Ferry, Hall evidently thought it best to remain silent lest he incur displeasure at the War Department and chance having his appropriations further reduced by the retrenchment-minded Jackson administration. In the fall of 1834, however, two things occurred that he could neither overlook nor accept.

Having committed himself to maintaining "unity of command" both within the army and at all military installations, Secretary of War Lewis Cass on September 14 directed that "the Rifle Factory at Harpers ferry shall be hereafter considered as a branch of the

42. Talcott to Bomford, December 15, 1832, Reports of Inspections of Arsenals and Depots, OCO; Hall to Bomford, March 6, 1835, Letters Received, OCO. See Table 2.

National Armory at that place and the director of that Factory will be considered as a Master Armorer and as such be subject to the Ordnance Regulations adopted on the 1st May 1834." Hall was infuriated. As he interpreted the provision, the Rifle Works would not only lose its independent status but also be placed under the immediate supervision of superintendent Rust, a vigorous opponent of the project. Under these conditions Hall felt certain that Rust would downgrade the business and eventually cause its ruin. Equally repugnant was the idea of being "put on the footing of a mere Master Armorer." This demotion, he told Bomford, added insult to injury by allowing enemies to display "envy and malice toward me" and "render my situation so irksome & mortifying as to compel me to quit the public service." Already, he noted, "numerous attempts . . . have been made to impress upon the minds of the community, generally, a conviction that my degradation is certain." Hall warned the chief of ordnance that "nothing but the interposition of the Government can now protect me."[43]

The question of Hall's status under the new Ordnance regulations occurred at a most inauspicious time since his contract also expired in October 1834. Although Bomford had no qualms about renewing it, his proposition failed to satisfy Hall. He offered the New Englander a four-year contract on the same terms as the one signed in 1828, but reduced the annual quota of rifles to one thousand. The agreement further stipulated that Hall would permanently surrender the use of his machinery to the United States and any of its assignees. Needless to say, the inventor flatly refused these terms and even accused the chief of ordnance of trying to undermine his patent rights. He also reminded Bomford that it would be impossible to realize economies of scale and thereby reduce the cost of his rifles if the government persisted in limiting their production to a paltry thousand stands yearly. Indeed, he argued, "Many of the expenses would remain quite as large . . . as they would be, if manufacturing thrice as many. . . . Not only all the advantages, in point of economy, arising from the minute subdivision of Labor in the Rifle business . . . will be sacrificed, but, in

43. Rust to Bomford, October 27, 1834, Hall to Bomford, October 27, 1834, March 10, 1835, to Cass, October 15, 1834, Letters Received, OCO.

consequence of the subdivisions not being adopted to business on a small scale—positive loss would be produced—and to a great extent. Such is the inevitable effect of attempting to execute on a small scale, business previously prepared by means of its machinery, for being executed on a large scale."[44]

Evidently Bomford agreed with Hall, but he could do nothing to alter the proposed contract because it had originated with Cass. For this reason Hall again turned to the "Members from Maine," particularly Ether Shepley, Rufus McIntire, and Francis O. J. Smith, for assistance in pleading his case with the secretary of war. On February 26, 1835, the entire delegation, two senators and eight representatives, sent a sharply worded letter to Cass in which they expressed "surprise and regret" at the "attempt to impose on Mr. Hall a contract to relinquish to the government his right to his inventions *forever* & for all other inventions anticipated to be made by him under a contract for four years only and absolutely without consideration beyond the trifling compensation on 4000 rifles." They also decried placing the Rifle Works "specifically under the control of a man conscious of having attempted to deprive Mr. Hall of his rights." Certainly, the members wisely conceded, "we know this attempt could not knowingly be sanctioned by you & we hope not by the Head of any Bureau of the Department." Nevertheless, they concluded, "The attempt has been made with the sanction (inadvertent we trust) of the Ordnance Department by a subordinate [George Rust] and that one under whom immediate superintendence it is proposed to place Mr. Hall as an ordinary Master Armorer. Will the Department subject Mr. Hall to the trial and wholly disregard that sensitiveness peculiar to a man of his inventive powers? Can it possibly conduce to the *interest* of the United States to do so, to say nothing of the justice or liberality which might with propriety be urged?"[45]

With conflicting pressures brought to bear by Rust and his sup-

44. Hall to Bomford, March 6, 1835, OCO. Also see Rust to Bomford, November 25, 1834, Letters Received, OCO; Bomford to Cass, March 14, 1835, Letters Sent to the Secretary of War, OCO; Hall to Statira P. Hall, February 25, 1835, Document File (transcription), HFNP.
45. Maine Delegation to Cass, February 26, 1835, Letters Received, OCO.

porters at Harpers Ferry on the one hand and by Hall and the Maine delegation on the other, Cass sought to avoid resentment toward himself by referring the question to a special body of officers knowns as the "Ordnance Board" for review and settlement. After several days of deliberation the president of the board, General Alexander Macomb, submitted a report on March 17, 1835. In effect, the officers endorsed an earlier recommendation of Bomford which advised that Hall should retain his semi-independent status as director of the Rifle Works but should be employed at Harpers Ferry on a noncontractual basis. Accordingly the board suggested that "his compensation should be at the following rate per annum, viz., one thousand dollars for his personal services, and sixteen hundred dollars for the use of his inventions and improvements in machinery." By this arrangement the government reserved the right to determine the number of rifles produced each year and agreed to pay Hall a flat royalty fee of $1600 regardless of whether he manufactured four thousand arms annually or none at all. Most important, the agreement "impliedly guaranteed" the use of his machinery "not only at Harpers Ferry but elsewhere." Although Hall considered these terms a poor substitute for a regular contract, friends cautioned that the secretary of war had made up his mind and nothing better could be expected. Thus, acknowledging the compromise for what it was, the inventor notified the War Department of his acceptance around March 20, 1835.[46]

Between 1836 and 1840 Hall renewed this agreement annually. During the same period he retooled the Rifle Works for the manufacture of carbines, experimented with methods of drilling cast-steel gun barrels, and spent a great deal of time in a fruitless attempt to petition Congress for compensation on his inventions. Since he had completed his tools and machinery by 1834, these deeds seemed anticlimactic in retrospect. By 1836, Hall realized that his breechloading system was no longer a novelty, for in that year John W. Cochran, Samuel Colt, and Baron Hackett succeeded

46. Extract from the proceedings of a Board of Officers, March 17, 1835, Correspondence Relating to Inventions, OCO; Bomford to Cass, March 14, 1835, Talcott to Joel R. Poinsett, March 17, 1840, Letters Sent to the Secretary of War, OCO; William A. Hall to Major Henry K. Craig, April 26, 1841, Letters Received, OCO.

in having their designs tested in competition with his rifle and the standard smooth-bore musket. Although the results of trials conducted at Washington and West Point proved Hall's weapon superior to those of his three competitors, the examining officers were by no means convinced that breechloaders made the best arms for military service. In fact, the board prefaced its report by skeptically remarking that "an arm which is complicated in its mechanism and arrangement deranges and perplexes the soldier."[47]

The popularity of the patent rifle began to wane in 1837 and 1838 when a number of reports reached Washington criticizing its fragility in service. Almost invariably these reports came from militia commanders who readily admitted that "the weapon would probably answer better when used by disciplined soldiers."[48] Nevertheless, certain faults did exist. As a Dragoon officer serving in Florida put it, "One company . . . was armed by me; and at the first discharge, five carbines were rendered perfectly useless; and, on the second, eight more were badly broken, the splinters flying in every direction—one of which passed near my face."[49] At first the Ordnance Department suspected that such guns were defective in their barrels and receivers. But after inspecting and re-proving them at Harpers Ferry, Hall found these parts perfectly sound. Instead, he reported, the real problem lay not in poor workmanship but in the use of steamed walnut gunstocks that became brittle with age and tended to splinter with regular service. Although Hall and Bomford's assistant, Colonel George Talcott, considered this problem easily remedied by redesigning the stock and avoiding the use of steamed walnut, Secretary of War Joel Poinsett thought other-

47. U.S., Congress, Senate, Committee on Military Affairs, *Report on Improvements in Fire-Arms*, 25th Cong., 1st sess., 1837–1839, S. Document 15, p. 3; *American State Papers: Military Affairs*, 7:468. For Hall's experiments with cast or crucible steel gun barrels, an innovation commonly attributed to the Remington and Whitney companies, see Lucas to Bomford, February 6, 26, March 1, 2, 12, 1839, to Talcott, March 21, 1840, Letters Received, OCO. On Hall's petitions for compensation, see Hall to Francis O. J. Smith, January 1836, F. O. J. Smith Papers, Maine Historical Society; Petition of John H. Hall (HR26A–HR27A), HR; Petition of John H. Hall (S24A–S25A), USS.
48. See, for example, Captain H. R. Peyton to Adjutant General R. Jones, December 20, 1838, Reports of Ordnance Boards, OCO.
49. Quoted by R. T. Huntington, "Hall Rifles at Harpers Ferry," *Guns Magazine* 7 (April 1961):45.

wise. Convinced that the rifles had outlived their usefulness, he informed the chairman of the House committee on military affairs early in February 1840, "I would not have adopted them and shall make little use of them hereafter in the regular service." For all intents and purposes this sounded the death knell of the Hall breechloader.[50]

Time had taken its toll in more ways than one. Beginning in the summer of 1837 a chronic illness (probably tuberculosis) forced Hall to restrict his activities at the Rifle Works and turn over direction of its affairs to his second eldest son, William. By 1840 his condition had so deteriorated that he took an extended leave of absence and, with his wife, went to Missouri to visit their third son, Willard Preble Hall. He failed to regain his health and died in Huntsville, Randolph County, on February 26, 1841, at the age of sixty. During the next few weeks Statira Hall received several letters from former co-workers and associates extolling her husband's genius and acknowledging his cumulative achievements as a watershed in the mechanical arts. She doubtlessly appreciated these eulogies, but as a wife and mother she harbored bitter memories of her spouse's work. "No one can know as I do," she wrote Talcott, "the great sacrifices both of comfort and interest he has made. No one but myself can imagine his days of toil and nights of anxiety while inventing and perfecting his machinery. Never did he for one minute hesitate to sacrifice his own interest when he thought it would interfere with the interests of the government. Had he in 1820 listened to the proposals of foreign governments, he might now be enjoying health and prosperity, yet he refused, all because he thought by doing so he should benefit his own government."[51] In Statira Hall's opinion, the public had been a harsh and unappreciative master. It robbed her husband not only of rightful compensation but of life itself.

---

50. Poinsett to John W. Allen, February 4, 1840, Petition of John H. Hall (HR26A–D15.1), HR; Poinsett to Martin Van Buren, December 5, 1840, Letters Sent to the President, OSW. See also Hall to Bomford, August 4, 16, 1838, to Poinsett, January 3, 1839, Letters Received, OCO; Captain J. F. Lee to Talcott, January 6, 1844, Correspondence and Reports Relating to Experiments, OCO.
    51. Statira P. Hall to Talcott, October 7, 1840, Letters Received, OCO.

CHAPTER 8

Hall and the American System, 1824–1840

The early nineteenth century witnessed a remarkable transformation from craft to factory production in the firearms industry. At the heart of this phenomenon stood the division of labor, an organizational format made increasingly possible by new techniques that completely reordered previous modes of production, replaced hand skills with the precision of the machine, and noticeably altered the nature of work itself. The origins of this change undoubtedly were in Europe, particularly in France and Great Britain of the late eighteenth century. Nevertheless, America made some boldly original innovations of its own and, in doing so, sowed the seeds for a system of interchangeable manufacturing that eventually bore its name. While countless individuals contributed to the new technology, John H. Hall stood foremost among those who combined inventiveness with entrepreneurial skill in blending men, machinery, and precision measurement methods into a workable system of production. The achievement formed the taproot of modern industrialism.

Even if sufficient data existed, the significance of Hall's contribution cannot be meaningfully evaluated by economic indicators alone. The cost of the first thousand rifles delivered in 1824 totaled $20,592.50, or $20.59 apiece. This figure included expenditures for appendages, labor, raw materials, and patent fees, but not interest or depreciation on capital. With the completion of the second parcel in 1827, the cost per stand decreased to $15.93 and included a charge for interest. Between 1832 and 1839 the expense for rifles and carbines made under Hall's supervision averaged $16.32 apiece. Costs, of course, varied inversely with the number of arms annually produced. In 1832, for instance, 4,360 rifles averaged $14.50 each, the lowest unit cost achieved by Hall during his twenty-one years at Harpers Ferry. With the delivery of only 970 stands two years later, the cost per rifle jumped to $21.13. Compared with Simeon North's contract price of $17.50, the Harpers Ferry product was far more expensive. When queried on this point in 1835 even Colonel Bomford admitted that "there appears, thus far, to have been no saving."[1]

Hall's failure to achieve significant reductions in the cost of his rifles did not ostensibly bother the chief of ordnance. From the outset he had considered the Rifle Works an experimental venture, the primary purpose of which was to develop and test new mechanical ideas. While he attempted to rationalize the project before an economy-minded Congress by stressing the *potential* savings in the adoption of mechanized techniques, he nonetheless felt that process refinement was the important product to be derived from Hall's labors. Economy, according to Bomford's strategy, became a subtle machination for wooing congressional support and throughout the 1820s and 1830s served as a guise for working out the engineering ideal of uniformity in the manufacture of firearms.

As much as he understood the anomalies of the situation, Hall was disturbed by Bomford's attitude because he felt it compromised honesty and impaired his reputation. Time and again he pointed out that erratic cost fluctuations had to be expected if the government persisted in restricting output to one or two thousand rifles a year. In his opinion, it seemed self-defeating to equip a plant with expensive machinery and then force it to operate at half capacity in inadequate shop facilities. These limitations proved doubly frustrating in Hall's case because they prevented him from demonstrating the large-scale economic benefits of his system and from amassing a personal fortune through the collection of patent fees.[2] In short, the experimental emphasis given Hall's work must be recognized. His contracts as well as special inspection reports and

1. Benet, *Ordnance Reports*, 1:305. See Table 2.
2. See, for example, Hall to Bomford, March 6, 1835, Letters Received, OCO.

general Ordnance office correspondence corroborate this point. Even armory payrolls reveal how much process refinement dominated Hall's efforts.

In 1835 the master armorer at Harpers Ferry prepared a table detailing the distribution of labor assignments at the Rifle Works. This document discloses that, with the exception of four years between 1819 and 1835, Hall employed at least as many men on the construction of tools and machinery as he did for actually making firearms. Such a labor ratio denotes an uncommon preoccupation with machine making. Whereas Hall employed a machinist for each armorer at his works, the armorer ratio at the nearby Musket Factory stood at about twenty to one. The same document also reveals that total expenditures at the Rifle Works amounted to $432,899.30 between 1819 and 1835. Of this sum well over one-third ($149,489.89) was invested in the preparation of tools and machinery. This figure is nearly three times greater than similar allocations of capital made at the main armory at Harpers Ferry and at Springfield during the same period. Both the apportionment of labor and the large expenditures on machinery at the Rifle Works highlight the exploratory character of Hall's venture. Ultimately his reputation stands or falls on the contributions he made to developing the mechanized production of interchangeable parts.[3] For this reason something more specific must be said of his tools and machinery.

The Carrington committee's highly favorable opinion of Hall's machinery, together with the Fortress Monroe trials of 1826, played an important part in the inventor's career for two reasons. They provided official testimony to his success as a first-rate mechanical innovator, and they represented an incontrovertible piece of evidence to which he could refer whenever skeptics criticized his work or questioned the originality of his inventions. He often used the reports to advantage in skirmishes with the Junto as well as the War Department. For the most part, however, Hall's admirers outnumbered his adversaries. Many persons, in addition to Carring-

3. Benjamin Moor to Bomford, October 12, 1835, Letters Received, OCO; Payrolls and accounts, Harpers Ferry armory, 1821–1840, Second Auditor's Accounts, GAO; *American State Papers: Military Affairs*, 5:915–922.

ton, Sage, and Bell, expressed amazement at the magnitude of his achievements. For instance, Inspector General John E. Wool after visiting the armory in 1829 singled out Hall's "highly important" and "very perfect" machinery. Without mincing words on the subject, Wool considered "the principles as applied to the Lock . . . of so much consequence that it ought if possible be extended to the Manufacture of Locks for the common Musket."[4]

Three years later Colonel George Talcott paid a similar visit to Harpers Ferry. Finding the Musket Factory "deficient in many useful machines" and "far behind the state of manufacture elsewhere," the inspector of arsenals and armories discovered that Hall's works presented a marked contrast. "This manufactory," he reported, "has been carried to a greater degree of perfection, as regards the quality of work and uniformity of parts than is to be found elsewhere—almost everything is performed by machinery, leaving very little dependent on manual labor." Talcott's opinion did not change with the passage of time. If anything, his enthusiasm and admiration grew. Returning to the armory in 1836, he expressed "deep astonishment at the immense quantity of machinery in operation" at the Rifle Works. Indeed, he observed, "The first conception, or idea, of such surprising machinery, its construction and use, could have been affected by no ordinary mind. I am bound to add that the plans of the inventor have been crowned with signal success and he deserves a pecuniary recompense, far greater than any sum that has been suggested, as a suitable reward for his labors."[5]

Military inspectors were not the only persons who recognized the significance of Hall's work. In 1874, William Wade, the former Ordnance officer who had since become a leading Pittsburgh iron founder, recalled that the New Englander "introduced the system of making all the parts [of his rifle] interchangeable," adding that "we all regarded it at the time as a memorable feat, and as marking

4. Wool to General Alexander Macomb, October 31, 1829, Inspection Reports, OIG.

5. Talcott to Bomford, December 15, 1832, Talcott, "Inspection of Hall's Rifle Factory at the Harpers ferry Armory," October 31, 1836, Reports of Inspections of Arsenals and Depots, OCO.

an important epoch in progressive improvements." Several years later Asa Holman Waters, an arms contractor from Worcester, Massachusetts, paid tribute to the inventor as "the first to perfect and carry [the uniformity system] into practical operation, and probably to conceive of it as now practiced." Even Eli Whitney Blake, while defending the labors of his uncle, Eli Whitney, and arguing that any self-respecting mechanic could produce arms with interchangeable parts if allowed to alter his model as difficulties arose, acknowledged "the truly original and inventive genius of Mr. Hall." Writing to the editors of the Washington *National Intelligencer* on July 1, 1835, he mentioned that "Mr. Hall has done much (perhaps I should say EVERYTHING) in the department to which his efforts have been directed, and he deserves well of his country. Indeed, the cup both of his merit and his fame is full—so full that its contents cannot be increased by pouring from the recepticles which belong to others [namely, Eli Whitney]." The most generous compliment paid Hall came from the chairman of the Senate committee on military affairs, John Tipton of Indiana. Speaking before that body in December 1837, he referred to Hall's machinery as "the greatest improvement in the mechanical arts ever made by one man." These and other remarks helped fill Hall's "cup of fame," but they did little to compensate him for his effort.[6]

While historians have overlooked the experimental character of Hall's work, almost all of them place the transplanted New Englander in the vanguard of those who pioneered the American System during the early nineteenth century. After interviewing a number of persons who had actually worked in the antebellum armories and completing a discerning essay on "The Rise of a Mechanical Ideal," the engineer-historian Charles H. Fitch wrote one of his collaborators in 1883, "The Ferry may well claim priority in Capt. Hall's work." More recently, Robert S. Woodbury has stated that "John Hall's methods can be fairly clearly established, at least suffi-

6. Wade to Colonel S. V. Benet, March 10, 1874, in *Ordnance Notes, No. 25* (Washington: Ordnance Office, 1874), p. 138; Waters, "Thomas Blanchard," pp. 259–260; Blake to the editor, *Daily National Intelligencer* (Washington), July 8, 1835; Isaac N. Coffin to Edward Stanley, June 6, 1842, Petition of John H. Hall (HR27A–G124), HR.

ciently for us to be sure that modern interchangeable manufacture derives far more from his inventive genius at Harpers Ferry than from Eli Whitney's manufactory at Mill Rock." Other scholars have voiced similar opinions.[7] Yet, despite these accolades, no one has attempted to delineate the extent of Hall's mechanical contributions or place them in proper historical perspective since Fitch submitted his informative "Report on the Manufactures of Interchangeable Mechanism" to the Bureau of the Census in 1881. Indeed, twentieth-century historians have relied almost exclusively on this document, the *American State Papers*, and the published correspondence of the Ordnance Department in sketching the outlines of Hall's inventive activities between 1819 and 1841. The authenticity of these sources cannot be questioned nor can their importance for assessing Hall's techniques be ignored. But, taken alone, they leave many gaps and can easily be misinterpreted without a more intimate analysis of day-to-day operations at the Rifle Works.

To attain a clearer grasp of the nature and extent of Hall's mechanical synthesis at least five basic questions must be answered:

(1) What types of tools and machines did the New Englander use at Harpers Ferry?

(2) Did he learn from the experience of his contemporaries? If so, what ideas did he retain in his own mechanical designs?

(3) What innovations did he introduce? How did they differ from techniques used elsewhere?

(4) Did he build his own machinery?

(5) What influence did Hall exert on his contemporaries? What lasting contributions did he make to the mechanical arts in America?

Such queries are not easily answered since the Patent Office fire of 1836 and the demolition of the Rifle Works in 1861 destroyed many records bearing directly on Hall's activities. Among the most valuable items lost were detailed drawings and descriptive specifications of his patented machinery for cutting metallic substances. Two sets

7. Fitch to Burton, January 13, 1883, James H. Burton Papers, Yale University Archives; Woodbury, "The Legend of Eli Whitney," p. 251. The most recent treatment of this topic is Rosenberg's *American System of Manufactures*, pp. 1–86.

of drawings that Hall provided for reference purposes to the office of the chief of ordnance have also disappeared from departmental files. A fruitless search for these materials in a number of public repositories suggests that their recovery is very unlikely unless they are in private hands. It is therefore difficult to specify exactly what machinery existed at the rifle shops, for the materials that do exist are sadly deficient for making such a determination. Nevertheless, a reasonably reliable account can be reconstructed from an examination of Hall's work returns, parts of his correspondence, official inspection reports, and other contemporary sources related to arms manufacture during the middle decades of the nineteenth century.

While Hall had proposed making his rifles with interchangeable parts as early as 1816, his correspondence indicates that he made components by hand and had not devised any other means prior to moving to Harpers Ferry in 1819. In fact, he did not complete new tools for the purpose until the winter of 1822–1823. Besides hand implements commonly used by armorers, these items included three sets of case-hardened gauges. Hall distributed one set to his workmen, another set to inspectors, and the third, or master set, he kept in his office for the purpose of detecting any deviations in the working sets caused by wear, warping, and other hazards of constant usage. These instruments were intended, he asserted, "not only to enable those who use them to do their work right but so as to preclude the possibility of their doing it essentially wrong." By using them as guides, errors could be limited to those that were remediable. Such methods, he believed, held the potential of approaching the "utmost limits of perfection . . . within any assignable bounds."[8]

The rigor of Hall's inspections played an important part in the manufacture of arms with interchangeable parts. In addition to conducting on-the-spot inspections of work in progress, he double-checked each parcel for defects. Accordingly, after a number of rifles had been put together and fully completed, they were once more disassembled and their parts exchanged before turning them

8. Hall to John C. Calhoun, December 21, 1824, Letters Received, OCO; Hall to Joel R. Poinsett, February 21, 1840, Petition of John H. Hall (HR26A), HR.

over to the armory storekeeper. This procedure, Hall declared, represented "an Inspection of Inspections—and an infallible test of the intelligence and attention with which previous Inspections have been conducted—as well as of the complete accomplishment in all respects of the accuracy aimed at."[9]

The extent to which Hall gauged work in 1822 is not known, but he evidently continued to improve the method through 1826. By the time of the Carrington report, his inspection devices not only outnumbered but also greatly surpassed those used for the manufacture of common muskets at the national armories. In 1882 John H. King, one of the ablest mechanics at the Rifle Works, testified "that Capt. Hall used over 63 gauges on his breech-loader." He further stated that the inspection instruments prepared at Harpers Ferry for the new Model 1840 musket were patterned after Hall's, though they were not nearly so accurate. Similarly North's superintendent, Selah Goodrich, recalled that when Hall's inspectors arrived from Virginia in 1830, they brought "more numerous and exact gauges than had ever before been used" in the Connecticut Valley.[10] As previously noted, Hall and North eventually adopted the same gauges and manufactured rifles so that the output of one plant would exchange with the output of the other—an important technical achievement for both men. This substantiates the presence of two elements of modern interchangeable manufacture—precision instrumentation and uniform standards of measurement—at Harpers Ferry and Middletown by 1834.

Hall's concern for maintaining accurate standards also extended to machining operations. By 1828 he had completely replaced filing jigs and other less durable hand tools with machines which he had fitted with work-locating and work-holding fixtures; these, in effect, supplied the same sort of guidance hand filers experienced with jigs. In addition to various viselike clamps and setscrews, provisions were made for adjusting the machines whenever the need arose. Such rectifications assured greater accuracy and helped guard against what the inventor characterized as "minute errors . . .

9. Hall to Bomford, April 25, 1831, October 13, 27, 1834, Letters Received, OCO.

10. Fitch to Burton, November 11, 1882, James H. Burton Papers, Yale University Archives; Fitch, "Rise of a Mechanical Ideal," pp. 520, 522–523.

in the work from such alterations as are unavoidably produced by changes of temperature, by abrasion, and by every other similar cause that might operate, injuriously, in the intervals between the rectifications."[11]

The correct positioning of a workpiece in relation to the cutting tool assumed critical importance during machining operations, as Hall explained to the secretary of war in 1840:

In making a part of an arm like a prescribed model, the difficulty is exactly the same, as that which occurs in making a piece of Iron exactly square. In such a case, a man would Square the 2d. side by the 1st, the 3d. by the 2d., and the 4th by the 3d., but on comparing the 4th side with the 1st, it will be found that they are not square; the cause is that in squaring each side by the preceding side, there is a slight but imperceptible variation and the comparison of the 4th with the 1st gives the sum of the variations of each side from a true square. And so in manufacturing a limb of a gun so as to conform to a model, by shifting the points, as convenience requires, from which the work is gauged & executed, the slight variations are added to each other in the progress of the work, so as to prevent uniformity. The course which I have adopted to avoid this difficulty, was to perform & gauge every operation on a limb, from one point called a *bearing* so that the variation in any operation could only be the single one from that point .

In manufacturing the receiver or breech block of Hall's rifle, for example, a bearing located on the right side of the workpiece determined its relative position for all subsequent machining operations. From this point the workmen secured it in special fixtures to be successively drilled, morticed, milled, and bored. In like manner, each component part of the rifle had a similar bearing point. The use of this technique is particularly noteworthy because, next to his patented machinery, Hall regarded it as his most important mechanical contribution. "This principle," he declared, "is applicable in all cases where uniformity is required."[12]

Hall generally classified his machinery under two headings: "Primary machines," those with which other machines were made, and "secondary" machines, those used for the actual production of com-

11. Hall to Calhoun, December 21, 1824, OCO.
12. Hall to Poinsett, February 21, 1840, HR; Hall, "Nomenclature of Hall's Rifle, Shewing the component parts, materials, dimensions, etc.," [1835], Correspondence and Reports Relating to Experiments, OCO.

ponent parts. Very little is known about the former except that Hall considered them very expensive for the limited amount of work they performed. Yet, these primary machines were necessary because he could not adequately equip his shop without them. Such units included engine lathes, drill presses, and other mechanical contrivances commonly associated with early nineteenth-century tool building. There is no evidence that Hall ever used a planer at his factory, however. He therefore encountered many of the same technical problems contemporaries did in producing true machine surfaces by means of hammers, cold chisels, and files. Such a constraint necessarily impeded the degree of precision that could be achieved in manufacturing interchangeable parts.[13]

That Hall built most of his own machinery is evidenced by repeated statements that he not only made the machines used at the Rifle Works but also the tools with which they were constructed. When the House of Representatives called for information on the subject in 1826, the inventor responded that he had not yet purchased a machine from outside sources. Since the armory lacked facilities for casting beds, headstocks, carriages, and other metal machinery parts, Hall necessarily had to arrange for this work at private foundries. Prior to 1827 he procured his castings from McPherson & Brien's forge near Harpers Ferry and from John Mason's Columbian Foundry in Georgetown. After that date he placed most of his orders with William Miller & Company of Baltimore, William Barker & Son of the same city, and James D. Paxton of Adams County, Pennsylvania. In all cases Hall followed the same procedure. He supplied patterns and drawings to the founders, who cast the respective parts and then shipped them to Harpers Ferry where they were finished and assembled into complete machines. Most noteworthy of this work is the large number of cast-iron frames listed on the purchase vouchers. Most machine tools used at private and public armories prior to 1830 rested on wooden framework; apparently Hall was one of the earliest manufacturers in the United States to introduce machinery of total metal construction.[14]

13. Hall to Bomford, December 5, 1828, Letters Received, OCO.

14. James Stubblefield to Colonel Decius Wadsworth, January 8, 1820, Hall to Calhoun, May 15, 1822, December 21, 1824, to Bomford, December 30, 1822,

Not every machine housed at the Rifle Works was designed and constructed by Hall. Though he claimed to have made "great alterations in it," he did use a rack-and-pinion rifling machine patented by James Ruple and William Parkinson of Washington, Pennsylvania. The machine had originally been installed at Harpers Ferry in May 1819 for rifling barrels of the muzzle-loading Model 1814 rifle. With the discontinuance of the model in 1820, James Stubblefield transferred the apparatus to Hall's works. Like most machines of the period, it rested on a massive wooden frame measuring nearly eighteen feet long and nineteen inches wide. The unit operated by a water-powered pulley which actuated a sliding carriage by means of cranks and arms fastened to the end of a revolving axle which passed through the pulley and rested near the top of the frame. A cutting rod attached to an indexed leader produced a spiral groove of one turn every nine feet, or one-third of a turn per barrel. While capable of rifling forty barrels a day, the machine usually operated at two-thirds capacity. For its use, Ruple and Parkinson received twelve cents for each finished barrel.[15]

In addition to rifling equipment, Hall purchased primary machines on at least two different occasions. In October 1828 he bought a "gage lathe" from the Savage Manufacturing Company, a textile firm located near Baltimore. Eight years later he acquired a "screw engine lathe" from the Ames Manufacturing Company of Cabotville, Massachusetts. Whether he made other outright purchases cannot be determined from existent records. If he did, they were very few in comparison with those built at Harpers Ferry.[16]

When James Carrington, Luther Sage, and James Bell inspected the Rifle Works in 1826, they paid particular attention to three distinct classes of machines which not only differed essentially

---

January 16, 1827, Letters Received, OCO; Payrolls and accounts, Harpers Ferry armory, 1821–1836, GAO.

15. Ruple & Parkinson to Stubblefield, December 29, 1826, to the paymaster at Harpers Ferry, February 16, 1827, M. McKennon to Bomford, April 25, 1832, Hall to George Rust, May 30, 1834, Letters Received, OCO; Moor to Rust, January 2, 1833, Correspondence Relating to Inventions, OCO; Bomford to Lewis Cass, April 30, 1834, Letters Sent to the Secretary of War, OCO.

16. Hall of Bomford, October 18, 1828, Letters Received, OCO; Bomford to Hall, October 18, 1828, Letters Sent, OCO; Nathan P. Ames to Hall, [1836], Misc. Letters Sent, James T. Ames Papers, Bingham Collection.

from one another but materially from any they had ever seen. Specifically the committee mentioned equipment for stocking, forging, and cutting components which, in their opinion, represented Hall's most significant mechanical innovations.

Contrary to statements made by Asa H. Waters, Hall did not use Thomas Blanchard's patented eccentric lathe or, for that matter, any of the Millbury mechanic's ingenious gunstocking machinery. Blanchard undoubtedly could have adapted his techniques to fashion the stocks of Hall's rifles, but the extra expense of preparing new patterns coupled with the costliness of the inventor's patent fee evidently dictated against the measure. Instead, Hall devised several special-purpose machines for the task. Although he had the essential movements worked out by 1826, he continued making improvements through 1835. When completed, they comprised a stable of five different machines.

On the basis of information provided by a former machinist at Harpers Ferry, Fitch provided a fairly detailed description of Hall's stocking machinery in his 1883 report on interchangeable manufactures. Hall's principal machine consisted of a "gang of circular saws set together on a spindle" and "arranged so that the teeth broke the joint irregularly" thus preventing the wood from splitting with the grain.

This saw gang was used in mill-planing or profiling, the gang being placed between the two sides of the slide bearings of a jig frame, in which the rough stock was clamped, the spindle of the gang being underneath and in direction perpendicular to the slide bearings, and the jig-frame having the outline of the profile of a gun stock, and being moved over the slide bearings, which were on a level with the cutting edges of the gang. The saw-gangs used were from a quarter to half an inch thick and 3 to 7 inches in diameter. . . . By its use the stock was brought to a square-edged profile, top, bottom and sides.

Among Hall's stocking machinery was a circular saw which slabbed the face of the stock and produced the surface in which the barrel was subsequently bedded. Once faced and straightened, this surface became the bearing point for subsequent operations of jigging or profiling. Hall employed another circular saw to crosscut the ends of the rough stock off to their proper dimensions. He also devised a fourth machine for roughing out the groove for the barrel. This apparatus carried a revolving cutter positioned on a horizontal axis and performed a profiling operation. Finally, Hall designed a machine to spot-groove the bed for the barrel. According to Fitch,

This machine carried a spindle (horizontal) say about the length of the barrel, the surface of which was turned off so as to have standing at intervals in its length of say 3 to 4 inches, narrow belts about three-sixteenths of an inch wide, which were cut with diagonal teeth of shallow depth, the spindle being in the first place turned and finished to the exact longitudinal profile and diameter of the barrel. This spindle was of necessity provided with intermediate supports to prevent it from springing when in use. The gun stock, previously rough-grooved-out for the barrel, being secured in proper position in a sliding frame, was lowered down to the slowly-revolving cutter-spindle until arrested by a properly adjusted 'stop.' The result was that the roughly-grooved bed for the barrel was 'scored' at intervals of 3 or 4 inches of the exact diameter of the barrel at corresponding points in its length. These 'scores' became the guide for the hand-workmen in planing out the bed for the barrel, and were the means of saving much time in fitting by hand.[17]

Although Hall's profiling machines represented a notable engineering achievement, the rest of his stocking operations lacked mechanization. Once the workpiece had been spotted and scored, much remained to be done with hand tools. In addition to rounding and shaping the stock with drawknives, spokeshaves, and files, finishers had to recess areas for the receiver, bands, guard, and butt plate by hand. While Hall's procedures included the use of gauges for determining cross sections at various points and even succeeded in making the stocks so as to require no fitting during assembly, they nonetheless seemed primitive and incomplete alongside those of Blanchard. For this reason the government gradually abandoned

17. Fitch to Burton, July 23, 1881, James H. Burton Papers, Yale University Archives; Fitch, "Report on Interchangeable Mechanism," pp. 630–631. Cf. Waters, "Thomas Blanchard," pp. 259–260; Hall to Bomford, March 24, 1831, March 9, 1835, Letters Received, OCO; Edward Lucas, Jr., to John Robb, April 5, 1839, Letters Received, SAR.

Hall's techniques and adopted the Blanchard method at the Rifle Works between 1843 and 1845.[18]

To contemporary observers, Hall's "great" and "small" forging machines appeared particularly impressive even though die forging with drop hammers was not new. The principle, a possible derivative of pile driving, had been attempted by the French armorer Blanc during the 1780s. By the 1830s, if not earlier, the method had fairly widespread usage among American metalworkers, most notably braziers, tinsmiths, and clockmakers (Figure 11). The novelty of Hall's machines lay in their massive construction and the expeditious manner in which they compressed and shaped iron, a much less malleable substance than brass, lead, or silver.

Applied to every iron component except the barrel, Hall's "great" and "small" drop hammers completely eliminated hand forging at the rifle factory. Both machines operated by means of an endless chain which passed over toothed pulleys situated at opposite ends of a vertical column. Actuated by a crank on the lower pulley, a spur or hook in the drop engaged the chain which lifted the weight between iron guideways until it reached the desired height. At this point the operator placed the article to be forged on a steel die located in an anvil block embedded in lead at the base of the column. By pulling a cord attached to a lever at the top of the frame, he released the drop and the massive weight fell downward, compressing the iron workpiece in the die with tremendous impact and thereby producing the desired form.[19]

Hall experienced greater difficulty perfecting his dies than building his drops. He thought the defects—though few in number—in his first thousand rifles "occurred almost wholly in the forged work and, principally, in consequence of inadequate allowance in some of the swedges for the shrinking of the metal, a point, which, in

18. Carrington committee report, January 6, 1827, Hall to Bomford, January 16, 1827, OCO; Hall to the editor, *Daily National Intelligencer* (Washington), June 26, 1835; Payrolls and accounts, Harpers Ferry armory, 1843–1844, GAO.

19. James Baker to Major Rufus L. Baker, May 13, June 17, 1833, Letters Received, AAR; Talcott, "Inspection of Hall's Rifle Factory," October 31, 1836, OCO; Hall to Bomford, August 20, 1830, Letters Received, OCO; Fitch to Burton, July 23, 1881, James H. Burton Papers. Yale University Archives; Fitch, "Report on Interchangeable Mechanism," p. 636.

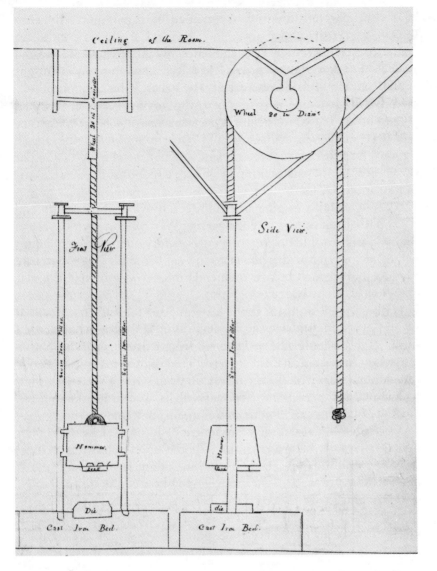

Figure 11. A primitive drop press used at a Philadelphia lamp factory in 1833. Such presses were commonly used throughout the United States during the early nineteenth century and probably influenced the design of Hall's heavier drop forges. Photograph from the National Archives, Letters Received, AAR.

new work, can only be fully ascertained by experience." By 1826 Hall had corrected this problem, for the Carrington committee made special mention of a device which quickly ejected the forging from the swedge, "however deep the latter may be or however tight it may stick." "By means of this apparatus," the committee observed, "the machine is made to apply, advantageously, to every species of forging which requires great *accuracy* of *form* & *proportion*, & where it becomes necessary to extricate suddenly the article forging, from the swedge, to prevent the hot metal from taking out its temper, by its long continuance in it, which would soon make it soft and unfit for use."

Although Hall's forging machines combined "great *stability* and *strength*" with accuracy of construction, Carrington and associates thought utility could be greatly increased and much drudgery eliminated by applying water power to the endless pulleys. Apparently the inventor heeded this recommendation because he had the machines working by water in 1830.[20]

Hall received inquiries about installing his forging equipment at several private foundries and machine shops. Whether he honored these requests cannot be determined from existent records. Nor is it absolutely certain that the government adopted his techniques at the national armories. Nevertheless, in drop forging with dies, Hall had few if any peers before the late 1830s. In Fitch's opinion, he "was greatly in advance of his contemporaries of the North."[21]

After they left the forge, components had to be dressed to their final dimensions before assembly. These operations were critical since close tolerances had to be maintained to insure interchangeability as well as the proper operation of the finished rifle. The machining of the receiver or breech mechanism required particular care since it had to fit snugly with the barrel to perform safely. Consisting of twenty-three parts, it was a far more complex mechanism to manufacture than the common musket lock,

20. Hall to Calhoun, December 21, 1824, Carrington committee report, January 6, 1827, OCO; Payrolls and accounts, Harpers Ferry armory, 1830, GAO.

21. Fitch to Burton, July 23, 1881, Burton Papers; William Barker & Son to Bomford, February 18, 1828, Letters Received, OCO; Wade to S. V. Benet, March 10, 1874, *Ordnance Notes, No. 25*, p. 138.

requiring a number of deep and irregular cuts that could not be done expeditiously in filing jigs (see above Figure 10).

To meet these needs Hall employed a variety of machines designed to cut, shape, and smooth the exterior dimensions of metal components. Installed at the Rifle Works during the early 1820s, these contrivances performed with cutters and saws "work usually done elsewhere with grindstones, chisels & files." The principles on which they operated, however, did not originate with Hall. For instance, the Carrington committee mentioned a trimming machine used for cutting off the surplus metal of forgings. While constructed "to operate with extreme *Steadiness* & almost *unlimited* power," Hall's model definitely resembled a machine used at Springfield for similar purposes since 1818 except that his workmen operated it by hand.[22]

Hall also built a number of drilling-and-milling machines for hollow milling screw pins and for drilling, reaming, and countersinking holes in components. Again this design was not new since, as previously noted, vertical-spindle drilling-and-milling machines had been used at both national armories before the War of 1812. However, Hall did introduce several improvements which substantially increased their versatility. One was the use of a screw-adjusted worktable which could be moved both horizontally and vertically in relation to the cutting tool and firmly secured once the correct position had been determined. Another was the addition of "*conical* and *perfectly similar* & *concentric* sockets" to the table where a variety of fixtures could be attached for positioning the workpiece at right angles to the revolving cutter. "A great variety of implements . . . are adapted to the mandrils of this machine," the Carrington committee observed, "and each one suits equally well, in both the *upper* and *under* mandril."[23]

22. Carrington committee report, January 6, 1827, Hall to Calhoun, May 15, 1823, December 21, 1824, to Bomford, December 17, 1824, Roswell Lee to Wadsworth, January 10, 1818, OCO; Fitch to Burton, July 23, 1881, James H. Burton Papers, Yale University Archives.

23. Carrington committee report, January 6, 1827, OCO; Fitch to Burton, July 23, 1881, James H. Burton Papers, Yale University Archives. A clear-cut distinction between true milling, hollow milling, and other machining processes is provided by Battison, "Eli Whitney and the Milling Machine," pp. 16–23.

In addition to standard units for hollow milling and trimming components, Hall patented several other machines in 1827 for cutting metallic substances. Because drawings of these devices have yet to be found, their specifications remain hazy. Nevertheless, the Carrington committee designated them as *Straight cutting, Curved cutting & Lever cutting Machines,*" indicating that they formed the largest portion of Hall's production equipment.[24]

Of the three, Hall's straight-cutting machine is most interesting and historically important because it represents an early version of plain milling as practiced today. Designed to produce a combination of straight, flat, fluted, or ribbed surfaces, it could easily be applied to the production of a great variety of regular and irregular shapes. The second or curve-cutting machine served "to produce surfaces of either *double* or *single* curvature of any of the regular curves," though it had been fitted to produce only the latter at the time of the Carrington report. Years later James H. Burton, a former machinist and master armorer at Harpers Ferry, identified the device as a bridge-milling machine or a lathe with a former.[25] Finally, Hall's "lever" machine executed work that could not be conveniently done by either of the other machines. Applied principally to the breech mechanism of the rifle, in December 1826 its functions included boring the pan and cutting recesses in the receiver for the battery, cock, and main spring. In 1881, Fitch described it as "similar in principle to some forms of hand-milling machines now in use." Other than these few facts, nothing is known about this particular device.[26]

24. Leggett, *Index of Patents*, 2:922. Unless indicated otherwise, the following account of Hall's milling machinery is based on the Carrington committee report, January 6, 1827, OCO.

25. Fitch to Burton, July 23, 1881, James H. Burton Papers, Yale University Archives. Bridge milling machines employed a hinged table which held the workpiece in position on its upper side and a copy of the required pattern underneath. The pattern rested on a fixed fulcrum or "bridge" situated directly below the table and in line with the revolving mill cutter. When set in motion the table carrying the workpiece traversed the cutter and bridge simultaneously. The bridge, in turn, traced the pattern and imparted a rising and falling motion to the table, thereby transferring the desired movement to the object being cut.

26. Fitch to Burton, July 23, 1881, James H. Burton Papers, Yale University Archives; Payrolls and accounts, Harpers Ferry armory, 1827, GAO; "Nomenclature of Hall's Rifle," (1835), OCO.

All three machines shared a number of features. Powered by water, they operated by leather belts which, the Carrington committee observed, "by increasing their *tension & width*," overcame "any resistance, however great, arising from their number or the velocity of their motions." Furthermore, the numerous pulleys that transmitted power from the belts revolved "without any of that *shaking & trembling* which frequently accompanies their motions & which often proves injurious, especially when the motions are very rapid." Hall remedied any defects in the pulleys by balancing them after they had been turned in a lathe. "This," the commissioners reported, "is done by *loading* the *light side* with lead or any heavy substance till an *equilibrium* takes place, by which means there is no tendency in their revolutions, however rapid, to wear more on one side of their Journals or gudgeons, than on the other." Balanced pulleys enabled Hall to run his machines much faster than usual without sacrificing steady accurate performance. In 1829, General Wool remarked that they not only operated well at 3,000 revolutions per minute but were constructed so the speed of their moving parts could be either increased or decreased at will.[27]

Another factor enhancing the effectiveness of Hall's cutting engines was their size and weight. Nearly every person who inspected the Rifle Works during the 1820s and 1830s saw the relationship between their great stability and accurate performance. On more than one occasion ordnance officers spoke highly of their "massive style of construction." Even Fitch in 1884 recognized the significance of Hall's "excessively solid and heavy" machines, although he considered them very clumsy by contemporary standards. Based on information provided by former machinists at Harpers Ferry, he estimated that one of Hall's milling machines contained enough metal to make three or four modern units.[28]

One clue to the great weight of Hall's machines lies in the types of operations they performed. The sheer size and depth of cuts that had to be taken on certain components—particularly the receiver—

---

27. Wool to Macomb, October 31, 1829, OIG.

28. Talcott to Bomford, December 15, 1832, Talcott, "Inspection of Hall's Rifle Factory," October 31, 1836, OCO; Fitch, "Rise of a Mechanical Ideal," p. 519; Fitch, "Report on Interchangeable Mechanism," p. 641.

necessitated toothed cutters capable of doing much heavier, more demanding work than was required of earlier milling machines equipped with rotary file-type cutters. This produced stresses and tensions that could be compensated for only by heavy, rugged machine construction. Several references as well as finished components themselves indicate that "saws" capable of removing a substantial chip of iron were employed on Hall's machines. Since the inventor also used formed cutters to execute irregularly contoured work, this too would indicate the need for machining stability. Lacking formal engineering training, Hall, like most empirically trained mechanics of the period, attempted to cope with this problem by overbuilding his machines. His mechanical designs were intimately related to the types of mill cutters used. Both innovations were transitional in that they served as a halfway house between the rotary filer prototypes of the pre-1820 period and the fully evolved plain milling machines that appeared in the arms industry during the 1840s.[29]

Hall's cutting engines doubtlessly appeared clumsy and crudely proportioned to persons living in the 1880s. Considering the vast number of improvements in machine design since 1840, this is understandable. To Hall's contemporaries, however, they seemed far from primitive. Their durability and utility are well attested to by armory records. When the last allotment of patent rifles was delivered in 1844, the Rifle Works housed forty-nine cutting machines excluding those used for trimming, drilling, and holow-milling operations. Acting at that time under orders from the superintendent, master armorer Benjamin Moor and several other workmen readjusted their counterpulleys and fitted them with new holders and cutters adapted to the production of components for the new Model 1841 percussion rifle. Except for five curve-cutting units broken up for scrap in 1852, these machines continued in operation at Harpers Ferry up through 1855 and possibly as late as the Civil War. This fact not only documents their remarkable longevity of service but also corroborates Hall's claim that they pos-

29. More detailed treatments of this subject are provided by Battison, pp. 26–33; and Smith, "Hall, North, and the Milling Machine," pp. 573–591.

sessed universal application and could be used for the manufacture of common arms as well as patent rifles.[30]

The most distinctive feature of Hall's cutting engines is that he employed "common hands," "even boys," to tend them without the slightest fear of jeopardizing standards of accuracy or the quality of work performed. In 1834, for example, he informed Colonel Bomford that "the best person, decidedly so, that has ever worked with the cutting machines is a boy of but eighteen years of age, who never did a stroke of work in his life previous to commencing with them a few months since." Two years later the inspector of arsenals and armories reported that "most of the machines are attended by boys with a few men to supervise and keep the tools in good order," and he noted, "The number of skillful mechanics required is therefore quite limited for the manufacture of Rifles."[31]

The organization of labor at the Rifle Works illuminates the character of Hall's manufacturing operations. Essentially the system was a variant of the inside contracting method used throughout the arms industry during the antebellum period. In a letter to Daniel Bedinger, the paymaster at Harpers Ferry, Hall described its features:

Many of the operations in the Rifle business are executed to more advantage & more economically, by the joint efforts of several persons—to one of whom only—the *Principal*, are the articles to be operated upon, issued & charged—*he* is responsible for them—that they are returned in good order & uninjured—and to *him* payment is made. . . . He, the Principal, works by the side of these assistants, and watches their operations, and applies corrections to each as they become necessary—this is more particularly the case with the operations executed by the aid of the Machinery invented for facilitating the fabrication of those Rifles in the perfect manner mentioned—a large portion of *these* operations—with the machinery— is effected with the aid of Boys, each of whom can perform as much & as

30. Major John Symington to Talcott, September 30, 1845, December 9, 1847, Major Henry K. Craig to Talcott, September 10, 1844, Colonel Benjamin Huger to Craig, September 20, 1852, Letters Received, OCO; Hall to Poinsett, February 21, 1840, HR.
31. Hall to Bomford, December 30, 1822, December 17, 1824, October 11, 1834, to Calhoun, December 21, 1824, Letters Received, OCO; Talcott, "Inspection of Hall's Rifle Factory," October 31, 1836, Reports of Inspections, OCO.

good work, with those machines, under the immediate direction and eye of the Principal, as a man—these Principals hire those boys, and pay them—themselves—for their services, at such rates as are mutually agreed upon between those boys & themselves.

Such a system would have been impossible without self-acting machinery. "To keep a machine in operation," the Carrington committee noted, "*activity* is more necessary than *judgement*, for the machines, after the work is put into them, go thro' with the operation without any further *aid* from the boy, and when the operation is completed, give notice to the boy, who has been employed during the operation, in putting in and taking out work from other machines." In other words, Hall's cutting machines not only functioned without any manual guidance but also ceased operation once the workpiece had been finished. This enabled a laborer to tend as many as three or four machines simultaneously. Since their management required no mechanical skill, any reasonably alert individual could learn the job within a short period of time. At the height of his career Hall boasted "that one boy by the aid of these machines can perform more work than ten men with files, in the same time, and with greater accuracy."[32] While attempting to discredit their utility, even Edward Lucas, Jr., one of Hall's sternest critics, acknowledged that one machine "completed as much work in a day, as three men could file by hand."[33] The Carrington committee confirmed this estimate, adding that "the cost of the cutters & saws does not amount to *half that* of files used to do the same quantity of work."

According to master machinist Burton, the Hall Rifle Works housed "not an occasional machine, but a plant of milling machinery by which the system and economy of the manufacture was

32. Hall to Bedinger, October 29, 1834, payrolls and accounts, Harpers Ferry armory, 1834, GAO; Hall to Poinsett, February 21, 1840, HR.

33. "But," Lucas hastily added, "this operation is not a fair test, because the receiver, upon which the trial was made is necessarily forged much larger, for the saws of the machine to act upon, than the same limb would be if forged for the purpose of being filed by hand." Here Lucas unwittingly contradicted himself, for rather than exposing the supposed inefficiency of Hall's engines, he indicates that they were capable of removing more iron in less time than could be accomplished by hand filing. See Lucas to John Cocke, January 15, 1827, OCO.

materially altered."[34] Featuring solid metal construction, advanced
cutting tools, variable speed control, and automatic stop mechanisms, Hall's cutting machines represented a significant step forward in milling iron. Combined with sophisticated gauging techniques, they enabled him to dispense entirely with the use of filing
jigs and develop a system of production capable of manufacturing
arms with interchangeable parts. In doing so, he far surpassed the
legendary work of Whitney and further elaborated upon that of
North and other New England contemporaries. Yet all this would
have little significance if, as some scholars contend, Hall's improvements languished at Harpers Ferry.[35]

Determining the extent of Hall's influence is not a simple task.
Essentially it hinges on whether or not other arms makers adopted
his techniques. On this question Hall's sentiments are poignantly
clear. Beginning in 1834 he repeatedly complained that his patented machinery had been introduced at the Springfield armory
without his permission. "Moreover," he asserted, "by order of the
Ordnance Department, they have been and still are used by Col.
[Simeon] North, in manufacturing arms for the United States by
contract." While Springfield officials and North neither affirmed
nor denied these claims, Hall's lawyer, Isaac N. Coffin, and his
immediate family made similar allegations. Significantly, the chief
of ordnance and the chairmen of the House and Senate committees
on military affairs confirmed their validity.[36]

Surely ample opportunity existed for assessing and copying
Hall's techniques. Between 1821 and 1835 many gun makers vis-

34. Fitch to Burton, July 23, 1881, James H. Burton Papers, Yale University
Archives.

35. See, for example, George H. Daniels, "The Big Questions in the History of
American Technology," *Technology and Culture* 11 (1970):13n; Paul J. Uselding,
"Henry Burden and the Question of Anglo-American Technological Transfer in
the Nineteenth Century," *Journal of Economic History* 30 (1970):317.

36. See, for example, Hall to Cass, October 15, 1834, Statira P. Hall to Talcott, [1840], Letters Received, OCO; Hall to Cass, March 18, 1835, Correspondence Relating to Inventions, OCO; Bomford to Cass, September 16, 1834, Letters Sent to the Secretary of War, OCO; Bomford to James I. McKay, February 8,
1836, Letters Sent, OCO; William A. Hall to House Committee on Military
Affairs, February 7, 1840, J. W. Allen, "Report on the Petition of John H. Hall,"
April 25, 1840, Coffin to Stanley, June 6, 1842, Petition of John H. Hall, HR.

ited Harpers Ferry. Among the most prominent were Simeon North, Lemuel Pomeroy, Asa Waters, Eli Whitney, and Marine T. Wickham. Springfield's Roswell Lee, having frequently inspected the Rifle Works and served as acting superintendent at Harpers Ferry on two separate occasions, also became intimately acquainted with the inventor and his machinery. In 1826 he had not only hosted the Carrington committee but had helped select its members. Since Hall's contracts implicitly guaranteed the free use of his inventions at both national armories, Lee had ready access to his associate's drawings and patterns. Finally, since the Ordnance Department encouraged the use of Hall's machinery by others and possessed a complete set of specifications, it also served as a transmission center, a point for the diffusion of technical information.[37]

There were other equally effective means by which Hall's improvements could reach the Connecticut Valley. One likely agent was Carrington. As chairman of the special committee investigating the Rifle Works in 1826 he possessed firsthand knowledge of the subject and, judging from the enthusiastic language of his report, had more than passing interest in Hall's cutting machinery. Moreover, his role as a transmitter is all the more credible in light of recent research on Eli Whitney and the milling machine.

In challenging the great-man theory of history, Edwin A. Battison not only refutes the legend that Whitney invented the modern mode of milling iron but also demonstrates that a machine used at Whitneyville "must have been made after Whitney's death in 1825." Battison's further research indicates that the so-called single "Whitney" machine possesses several important features of Hall's cutting machines. Of particular interest is a self-acting carriage which disengages its screw feed by a spring-actuated slide once the work traverses the cutter (Figure 12). It would be presumptuous to assume that the extant "Whitney" machine represents a duplicate copy of Hall's straight-cutting engine. Nevertheless, it is possible that this machine was built from sketches and information supplied

37. Talcott to Poinsett, March 17, 1840, Letters Sent to the Secretary of War, OCO.

Figure 12. The famous "Whitney" milling machine (ca. 1827) may reflect Hall's influence. The spring-actuated slide release, only partially visible, is located behind and to the right of the worm and worm wheel (*foreground*). Photograph courtesy of the New Haven Colony Historical Society.

by Carrington to his former employers upon returning to Connecticut in January 1827.[38]

Several of Hall's armorers probably also carried technical information to the Connecticut Valley. In June 1827 two of his three principal workmen departed Harpers Ferry for New England. Soon afterwards, Hall reported, "Three others upon whom I relied most for the correct execution of the arms, left me also, as I would not comply with their unreasonable demands for an increase of wages, which they would never have made but for the departure of the others." The loss of five experienced workmen—nearly one-third of his skilled labor force—within the short space of two or three months upset Hall. Their knowledge of his business and machinery had lightened his labors considerably and allowed him to devote more time to administrative affairs. "But now," he lamented to the paymaster at Harpers Ferry, "I am again caught short-handed without much hope of filling their places."[39]

Hall's payrolls reveal that William Robinson, Jacob M. Mong, and Samuel Brantree, full-time machinists; Lenox Compton, a stocker; and Elizur B. Cogsill, a screw maker, left the Rifle Works during the summer of 1827. Although their exact destinations are unknown, presumably one or more of them accompanied Lee to Springfield when the latter ended his duties as acting superintendent at Harpers Ferry in June 1827. In any event, Lee's correspondence indicates that he employed a number of former Hall workmen. In 1830, for instance, he acquired the services of an able machinist named John Norman. By 1832 he had added two more armorers to the Springfield roster.[40]

By far the most skilled and most likely person to transmit Hall's

38. Battison, p. 33; Battison, "A New Look at the 'Whitney' Milling Machine," *Technology and Culture* 14 (1973):592–598. Although Carrington had resigned as Whitney's foreman in 1825, he continued to frequent the Whitneyville factory as a government inspector of contract arms until his retirement in 1830.

39. Hall to Bomford, June 25, December 10, 1827, Letters Received, OCO; Hall to William Broadus, June 27, 1827, payrolls and accounts, Harpers Ferry armory, 1827, GAO.

40. Payrolls and accounts, Harpers Ferry armory, 1827, GAO; Stubblefield to Lee, June 15, 1827, Letters Received, SAR; Hall to Bomford, September 20, October 12, 1830, March 24, 1831, Letters Received, OCO; Bomford to Lee, October 29, 1830, to Norman, October 29, 1830, Letters Sent, OCO.

techniques to New England was Nathaniel French. French's career exemplifies the roving tendencies of early American mechanics. Of unknown origin and background, he first appeared on Hall's payroll in January 1821. The man undoubtedly possessed prior experience and ability because he received the relatively high wage of $1.75 a day, a sum second only to the earning power of Hall's shop foreman, Robert Blanchard, and the forger Timothy Herrington. Employed as a pattern maker and machinist, French remained at the Rifle Works through the summer of 1827. Evidently the promise of higher wages lured him away, for he accepted the position of master machinist at North's works in Middletown, Connecticut. After spending two years with North, French again packed his belongings and returned to Harpers Ferry for another sojourn at Hall's factory. Finally, at the summons of Lee, he took a permanent position at the Springfield armory in May 1831.[41] The implication is clear: in less than five years Nathaniel French had plied his trade at three leading arms-making establishments. The mechanical information he carried from one armory to another cannot be dismissed lightly, for here was fresh know-how and the ability to execute such work embodied in one and the same person.

French undoubtedly familiarized North with Hall's methods and assisted him in introducing a number of mechanical improvements at Middletown. This is not to say, however, that North was unacquainted with the practice of milling iron. Around 1816 he had constructed the earliest known milling machine in America, a light, ingeniously conceived device equipped with a hand-cranked rack-and-pinion feed capable of producing small, flat surfaces with a cutter shaped like a rotary file (Figure 13). During the 1820s and 1830s he continued to elaborate upon milling designs, so much so that his Staddle Hill shops became the center of Middletown's flourishing machine trade.[42] That North used the talents of others to update and improve his manufacturing techniques does not

41. Payrolls and accounts, Harpers Ferry armory, 1821–1830, GAO; French to Lee, March 7, 22, 1831, Letters Received, SAR; Lee to French, March 14, April 5, 1831, Letters Sent, SAR; Fitch, "Rise of a Mechanical Ideal," p. 526.

42. For more extended analyses of North's mechanical contributions, see Smith, "Hall, North, and the Milling Machine," pp. 574–577, 587–590; and Battison, "Eli Whitney and the Milling Machine," pp. 27–28.

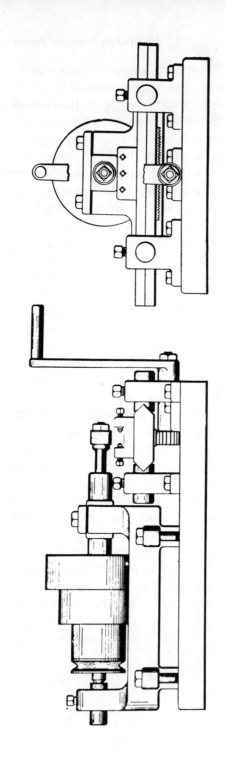

Figure 13. The "North" milling machine, ca. 1816. Reprinted from the *American Machinist* 23 (March 8, 1900):217.

lessen his reputation as a highly creative inventor-entrepreneur. Quite to the contrary, his concern for keeping abreast of the latest mechanical improvements evinces qualities of open-mindedness and flexibility common among the most successful innovators of the nineteenth century. No better example can be cited than Roswell Lee, who, through watchful attentiveness to techniques being developed elsewhere, by the 1830s had fashioned the Springfield armory into one of the largest and most progressive arms-making establishments in the United States. Even Hall, who had visited the Connecticut Valley in 1818 and examined machinery at North's factory and at Springfield, was not averse to borrowing promising ideas. There is good reason to believe that his patented cutting machinery was directly influenced by North's earlier design. Such practices were common among factory masters and mechanics at the time, a central factor that helps to explain the highly integrated character of the firearms industry as well as the relative speed with which arms makers assimilated the new technology during the 1840s and 1850s. Out of this ongoing process of cooperation, transfer, and convergence emerged the technical watershed known as the American System.

While Hall's influence was most deeply felt in New England, former employees at the rifle factory carried his technological ideas to other sections of the country as well. By 1837, for example, several had taken positions at government arsenals in Pittsburgh, St. Louis, Washington, Augusta, Georgia, and Mount Vernon, Alabama. A number of his most skilled workmen migrated to Cincinnati, Ohio. Among them was Otis Dudley, a New Englander who served as Hall's most trusted subordinate prior to moving west in 1832, and John Griffiths, an Englishman who left Harpers Ferry in 1841 to set up shop in Cincinnati as a private arms maker and government contractor. Whether Dudley or Griffiths engaged in the manufacture of machinery for commercial sale is not known. Both men possibly associated with Miles Greenwood, a leading foundry owner and early pioneer of the now-famous Cincinnati machine-tool industry.

Finally, Hall trained a number of young men who went on to become able mechanics in their own right. Among those who spent their formative years at the Rifle Works were John H. King, Philip Burkart, Armistead M. Ball, Jerome B. Young, and Daniel J. Young. "Commencing at the bottom of the ladder and ending second in command," King succeeded William A. Hall as director of the rifle factory in 1842 and held the position nearly eight years before going into private business at Martinsburg, Virginia. During this period he not only supervised the retooling and renovation of the Shenandoah shops, but also found time in 1844 to devise an ingenious machine for letting in all component fittings on stocks for the new Model 1841 Harpers Ferry rifle. Burkart started out as a common laborer and by 1844 had worked his way up to foreman of tools and machinery at the Rifle Works. The following year he was promoted to assistant inspector and in 1849 became foreman of the barrel shop at the neighboring Musket Factory. A strict taskmaster and thoroughly competent, Burkart joined his son-in-law, James H. Burton, at the Virginia state armory in Richmond with the outbreak of the Civil War. Like King and Burkart, Ball began his career in 1830 and moved through the ranks to succeed Burton as master machinist at Harpers Ferry in 1849 and Benjamin Mills as master armorer in 1859. Considered one of the best machinists in the United States, Ball proved "uncommonly clever" at designing self-acting machinery for drilling and rifling cast-steel gun barrels. Although the Ames Manufacturing Company and several other New England firms sought his services, Ball stayed on at Harpers Ferry, evidently professing Confederate sympathies until his untimely death in June 1861. Except that they adhered to the Union cause and worked at the Washington arsenal during the war years, Daniel and Jerome Young followed a similar development, rising from machine tenders to the ranks of master machinist and foreman of turners respectively at the Rifle Works and Musket Factory by 1841. Although none of these men garnered the prestige associated with Cyrus Buckland, Thomas Warner, and other New England contemporaries, they nonetheless represented Hall's contribution to a new generation of gifted inventors and mechanics that emerged in the arms business during the 1830s, 1840s, and 1850s.[43]

43. Fitch, "Rise of a Mechanical Ideal," pp. 522–523; Ames to Burton, October 18, 1858, James T. Ames Papers, Bingham Collection.

All told, Hall emerges as a pivotal figure in the annals of American industry. Although he conducted his operations in a small pilot plant and failed to achieve significant economies of scale, he did produce the first fully interchangeable weapons in the United States. In pursuing this elusive goal, he was proceeded by numerous harbingers who had popularized the concept of interchangeability, promoted and supported such work, and made important technical contributions to the machine production of standardized parts. What was so startlingly new about the Harpers Ferry experiment was the extent to which Hall had mechanized his operations and the impressive results he had actually achieved. No one at the time of the Carrington committee report had been able to master the problem of attaining complete interchangeability in firearms. Much of the excitement generated by the special investigations of 1826 can be traced directly to Hall's success in combining men, machines, and precision-measurement methods into a *practical* system of production. Until then, no one—not even the chief of ordnance—was absolutely certain that mechanized processes could be applied successfully to every aspect of gun making. Hall established the efficacy of such methods and, in doing so, bolstered confidence among arms makers that one day they would achieve in a larger, more efficient manner what he had done on a limited scale. In this sense, Hall's work represented an important extension of the industrial revolution in America, a mechanical synthesis so different in degree as to constitute a difference in kind.

As much as contemporaries acknowledged Hall's achievement, some felt that his methods had limited applicability to the manufacture of smooth-bore muskets, the staple weapon of American troops. Eli Whitney Blake, a zealous guardian of his uncle's reputation, voiced similar sentiments in 1835 when he stated that Hall had purposely designed his rifle for interchangeable production, adding that, whenever insurmountable technical difficulties arose, the inventor eliminated them by changing his model accordingly. Such liberties, Blake averred, contrasted sharply with the experiences of private contractors who were provided with a pattern musket and expressly forbidden to deviate from the model unless sanctioned by the Ordnance Department.[44]

Implicit in Blake's argument was the assumption that the common musket was more complex and therefore more difficult to manufacture than Hall's patent rifle. Herein lay the weakness of his argument, for even the most cursory examination of existent specimens reveals that the operating parts of the Hall rifle are more numerous and complex in design than those of the common military musket. Indeed, this very point became one of the most frequently cited reasons why field commanders objected to arming troops with Hall's breechloaders and why the War Department eventually discontinued their production at Harpers Ferry in 1844. Hall, to be sure, enjoyed considerable leeway in altering his model as he saw fit. Between 1823, when the first patent rifles were delivered, and 1841, the year of Hall's death, eleven alterations were made. Except for the substitution of band pins for band springs on the gunstock in 1828, however, none of these changes was aimed at circumventing technical production problems. If anything, they demanded even greater machining capacity. The irony is that each change slightly altered the original pattern with the result that not all 19,680 rifles produced under Hall's supervision at Harpers Ferry are completely interchangeable. This is particularly true of the first parcel delivered in 1824 which contained significant differences in the contour of the receiver, the shape of the cock, and the size of the lower jaw. These components were made before Hall had completed his tools and machinery and do not match those made at later dates. By 1828 Hall had regularized his methods to such an extent that all the major parts of his rifles could thereafter be exchanged. All subsequent alterations were minor in character and did not affect the interchangeability of his rifles.[45]

44. Blake to the editor, *Daily National Intelligencer* (Washington), July 8, 1835. Gene S. Cesari advances a similar argument in "American Arms-Making Machine Tool Development," pp. 237, 252.

45. In addition to the original Model 1819 flintlock rifle, Hall also produced 2,020 smooth-bore percussion carbines between 1837 and 1840. Designated as the Model 1836 carbine, it closely resembled the older rifle and, except for a simplified guard bow and the addition of an implement box in the butt of the stock, its parts could be exchanged with those of the rifle. Two other models—a percussion rifle introduced in 1841 and a brass-mounted carbine introduced in 1842—were briefly

There is a saying in chess that a threat is often stronger than its execution. This phrase, though something of a truism, aptly characterizes Hall's situation at Harpers Ferry. Throughout the 1820s and 1830s the civilian superintendents who managed the Musket Factory and arsenal viewed Hall as a baneful outsider who threatened to usurp their authority and disrupt traditional practices at the government works. While these fears never materialized, the intense animosity they aroused go far to explain the uneven and uneasy history of the Harpers Ferry armory during the antebellum period.

Subtle influences molded social relations at Harpers Ferry. Provincial attitudes, family cliques, vested interests, and personal jealousies combined in a curious fashion to keep outspoken outsiders such as Hall from social acceptance and respectability. Frequently the director of the Rifle Works found himself ostracized by well-to-do community leaders and excluded from their gatherings. Being a New Englander with different attitudes, values, and beliefs, he with his family remained isolated from the mainstream of life at Harpers Ferry. Unless one appreciates how positions of authority within the armory were tied to a larger network of economic and political control within the community, how these linkages in turn fostered managerial behavior that held the machine suspect and hampered inventive activity, and how cultural patterns—especially extended family bonds, personal friendships, and deeply embedded craft traditions—reinforced these attitudes, it is difficult, perhaps even impossible, to comprehend fully the trying circumstances under which Hall worked. It is even more difficult to understand the cool disdain which most people in Harpers Ferry reserved for novelties that threatened to upset their accustomed lifestyles and accelerate the pace of change.

---

manufactured at the Rifle Works after Hall's death, but these changes were instituted by a special board of Ordnance officers sitting in Washington. For further information, see Huntington, *Hall's Breechloaders*, pp. 189–212.

CHAPTER 9

Politics and Technology, 1829–1859

I n the exciting and highly partisan ferment of Jacksonian democracy a public institution such as a federal armory could not escape intimate and, as it proved, almost disastrous involvement in politics. Since jobs, contracts, and general economic well-being weighed in the balance, aloofness from politics was out of the question. The local situation at Harpers Ferry, for example, did not permit a neutral political position. Where certain persons might wish to stand apart, the pressure of local opinion demanded some practical commitment. As it was, the pressure more often came from civilian administrators who by their control of hiring and firing policies were scarcely loath to politicize even armaments technology. The result was unfortunate both for labor relations at Harpers Ferry and for the efficiency, productiveness, and reputation of the government works between 1829 and the outbreak of the Civil War.

Compared especially with John Hall, those who presided over the main armory and arsenal at Harpers Ferry sorely lacked administrative ability. In Inspector General John E. Wool's judgment, Hall managed the Rifle Works "with great regularity, order and system, whilst that of the Armory was the reverse."[1] What is more, the situation seemed to worsen with the passage of time.

Months before James Stubblefield had submitted his resignation in 1829, spoils-minded supporters of "Jackson and Reform" besieged the War Department with applications for the offices of

1. Wool to General Alexander Macomb, October 31, 1829, Inspection Reports, OIG.

superintendent and paymaster at Harpers Ferry. Between April and August 1829, the newly appointed secretary, John H. Eaton, had received at least fifteen petitions for the latter office alone. Accompanied by numerous letters of recommendation, these documents testified to the widely varying backgrounds of the solicitors, who ranged from a newspaper editor of Harrisburg, Pennsylvania, to a physician from Culpeper, Virginia. While he did not present himself as a candidate for superintendent, Hall nonetheless held strong views about the position and did not hesitate to express them to the secretary of war. Convinced that many serious reforms were needed at Harpers Ferry, he urged that someone be selected for the office who would divorce armory interests from those of the local community. "In order to indicate and maintain the public rights and to introduce and preserve a system of order and economy," he emphasized, "we need, at the head of the establishment a man of great personal integrity, activity and firmness of purpose."

Such a man, possessed also of the necessary practical and scientific knowledge and habits of steady application to business would here find a field which would reward his best efforts, and the benefits anticipated from the present administration be made to be seen and felt in their best form. Situated as I am in regard to this Armory & knowing how perfectly incompetent to the situation are several of the individuals who are about to present themselves, or will be presented by their friends, as candidates for the office I trust you will excuse the liberty.

Hall meant these remarks to apply to Fontaine Beckham and Henry Strider, the two candidates most favored by the Junto to be Stubblefield's successor. Although both men had actively campaigned against Andrew Jackson in the election of 1828, they were now feigning neutrality and camping at the new administration's doorstep. Even more appalling, in Hall's opinion, neither possessed the slightest experience as arms makers.[2]

Evidently Eaton heeded this advice because, in selecting the new superintendent, he passed over several politically appealing candidates and conferred the appointment on Thomas B. Dunn of Antietam, Maryland. The choice, a judicious one, reflected concern for

2. Hall to Eaton, May 25, 1829, Letters Received, OCO.

a number of practical considerations. As a faithful Democrat and a strong supporter of the party's local congressman, Michael C. Sprigg, Dunn easily met the political test. As a life-long resident of the Northern Neck, he gratified the wishes of those who felt that someone with local roots and connections should govern the armory. As superintendent of the Antietam Iron Works, he also satisfied Hall's criteria for "integrity, firmness, industry, and experience in a kindred business." Most members of the arms-making community agreed with James Baker's assessment of the new superintendent. "He is a very gentlemanly man," the Philadelphian wrote Roswell Lee, "well calculated to restore peace to the society and correctness to the Establishment."[3]

Upon entering office early in August 1829, Dunn proceeded to "clean house" at the armory by firing a number of inept and undesirable workmen. At the same time, he rehired Thomas Copeland and Charles Staley, both of whom had been dismissed several years earlier by the Stubblefield administration. In his first official letter to the Ordnance Department, Dunn also called for the appointment of a new master armorer. Certain that Armistead Beckham stood behind many illicit practices at Harpers Ferry, he recommended a former inspector named Joseph Hoffman as a replacement. "However," he added, "I . . . do not wish to be considered as urging *his* appointment—all I want is some competent man and the sooner the better." Interestingly enough, Beckham refused to resign, and no one in Washington seemed willing to force the issue. Dunn therefore solved the problem by relieving Beckham of his most important duties and reassigning them to armorers of his own choosing. Although this action was of questionable legality, it effectively eliminated Beckham's decision-making influence at the factory. Thereafter he served as master armorer in name only.[4]

Dunn next turned his attention to administrative reform, placing particular emphasis on the restoration of labor discipline. Several years earlier during the special investigations of 1827, Lee had

3. Sprigg to Bomford, June 29, 1829, Letters Received, OCO; Baker to Lee, October 14, 1829, Letters Received, SAR.
4. Dunn to Bomford, August 21, 26, September 2, 1829, Letters Received, OCO.

introduced a series of work regulations at Harpers Ferry. After Stubblefield had returned to office, however, no attempt had been made to enforce them. In 1829 the new superintendent not only revived these regulations but also insisted that they be rigorously observed by supervisors as well as regular armory personnel. Among other things, the rules forbid loitering, gambling, and consuming alcoholic beverages on armory premises, made unexcused absences punishable by immediate dismissal, and held each armorer personally responsible for the damage or destruction of tools consigned to his use. To improve standards of workmanship, Dunn also issued a special set of instructions to his inspectors. Known as the "Yellow Book," it meticulously stated their respective duties and provided for much more stringent inspections than had ever before been followed at Harpers Ferry.[5]

From the beginning Lee made a special effort to assist the superintendent and ease the transition. In a letter dated August 18, 1829, he congratulated Dunn on his recent preferment and extended an invitation to visit Springfield to become better acquainted and to discuss questions of mutual interest to both national armories. Sickness and pressing business commitments prevented Dunn from making this trip, but a brisk correspondence ensued. In addition to routine subjects, he consulted the Springfield superintendent on the establishment of labor regulations, the inspection of muskets, and, in an effort to improve standards of uniformity, the exchange and comparison of finished work. He also requested and received aid in recruiting armorers and purchasing raw materials. Finally, with Lee's full cooperation, he had plans in the offing for the adoption of standardized bookkeeping methods and the installation of triphammers for welding gun barrels at the Musket Factory. Before any of these projects materialized, however, a shocking incident ended Dunn's brief but earnest career.[6]

In spite of the favorable impression he made upon fellow factory

5. Dunn to Bomford, October 12, 1829, Letters Received, OCO; Bomford to Lee, October 15, 1829, Letters Sent, OCO; Dunn to Lee, October 21, 1829, Letters Received, SAR; Barry, *Strange Story of Harper's Ferry*, p. 25.

6. See Dunn-Lee correspondence, August 18, 1829–January 16, 1830, Letters Sent-Letters Received, SAR.

masters and the high regard he enjoyed in Washington, Dunn was never popular with the labor force at Harpers Ferry. To workmen long accustomed to minimal supervision and even less discipline, he bore the trappings of an unconscionable martinet. Moreover, the severity of his rules not only became a source of bitter complaint but also a cause for violence. On more than one occasion angry workers, gathering at local taverns, had denounced the nature of his caustic reprimands and had even accosted the superintendent outside the armory gate. The controversy came to a bloody climax on January 29, 1830, when an armorer named Ebenezer Cox entered the superintendent's office, pulled out a concealed gun, and shot Dunn at point-blank range. Since the assassin had been dismissed from the public works during the previous summer and refused employment several times since, his motive seemed abundantly clear to members of the Ordnance Department.

Although tried, convicted, and executed for the crime, Cox became something of a folk hero among the workers at Harpers Ferry. In later years disgruntled armorers never tired of citing Dunn's fate as a blunt reminder to superintendents of what could be expected if they became overzealous in executing their duties and impinged on the traditional freedoms of employees. Such attitudes contrasted sharply with the indignant commentaries which followed Cox's action among arms makers in other sections of the country. Lemuel Pomeroy, the private contractor from Pittsfield, Massachusetts, best summarized their feelings when he wrote Lee, "We ought to be thankful to that Kind Providence which has cast our lot in a part of the country where the moral sense of the people is our security . . . from such terrible outrages." Dunn's assassination only reinforced Pomeroy's opinion that the residents of Harpers Ferry were an incorrigibly vicious lot and that the federal armory at that place was forever destined to be a strife-ridden, corruptible establishment.[7]

After Dunn's death, Colonel Bomford designated John Hall as acting superintendent until the War Department found a replace-

7. Pomeroy to Lee, February 9, 1830, Letters Received, SAR. Also see Major John Symington to Captain William Maynadier, July 12, 1849, Letters Received, OCO. For a sympathetic view of Cox and his co-workers, see Barry, pp. 25–26.

ment. Secretary Eaton had little difficulty reaching a decision because around February 1 he announced the appointment of George Rust, Jr., of Waterford, Loudoun County, Virginia. Characterized as a "Virginia Gentleman of great intelligence" and "sterling honor," Rust came from old Cavalier stock. In addition to high social standing, he possessed a large personal fortune which was substantially augmented by his wife's inheritance of two sprawling estates, "Rockland" and "Exeter", near Waterford. He had served as a member of the Virginia House of delegates from 1818 to 1823, held a brigadiership in the state militia, and was a successful candidate to the electoral college in 1828. Friends as well as adversaries acknowledged that he wielded more political influence than any other individual in the northern part of the state.

Rust undoubtedly possessed impressive political and social credentials, but as master of one of the largest manufacturing establishments in the country, he left much to be desired. Unlike his predecessors, he was an agriculturalist by avocation and admittedly knew nothing about arms making or any other remotely related field of metalworking. His appointment stemmed from a close personal friendship with President Jackson and from his ascendancy as "the shield and Banner of the Jackson party in Loudoun County." What was more serious, he considered the superintendency at Harpers Ferry a sinecure, requiring minimal personal attention and even less active service.[8]

Although Rust secured the removal of Armistead Beckham as master armorer, vigorously defended his prerogatives against the suspected intrusions of Hall, and repeatedly petitioned Congress for higher salaries and appropriations, he spent far more time attending to personal affairs than to those of the national armory. Pleading ill health and private business interests, he frequently absented himself from the works and delegated authority to his newly appointed master armorer, Benjamin Moor. In view of the

8. Sprigg to Bomford, June 29, 1829, L. Banks to Martin Van Buren, January 1, 1830, W. T. T. Mason to W. T. Barry, April 4, 1829, George Townes to Colonel Thomas Davenport, January 14, 1830, William Seldon to T. Bouldin, January 15, 1830, P. V. Daniel (letter), January 12, 1830, Letters Received, OCO; Rust to Roswell Lee, April 19, 1831, Letters Received, SAR; Ellsworth M. Rust, *Rust of Virginia* (Washington, D.C.: By the author, 1940), pp. 195, 198–201.

circumstances, Moor shouldered the extra burden remarkably well. Nevertheless, when Rust retired in March 1837, he left the office of superintendent in the same muddled condition as he found it in 1830. After seven years at the helm he completely failed to supply the entrepreneurial leadership the armory so desperately needed. As an administrator he proved honest and popular, but totally inept.

No sooner had Rust submitted his resignation than President-elect Martin Van Buren selected Edward Lucas, Jr., to fill the vacancy. Born near Shepherdstown, Jefferson County, Virginia, on October 20, 1780, Lucas had attended grammar schools in the community and in 1809 had been graduated from Dickinson College in Carlisle, Pennsylvania. After serving as a first lieutenant in the War of 1812, he studied law at Winchester, Virginia, gained admittance to the bar, and set up practice in Shepherdstown. Soon afterwards he also established a lucrative mercantile business at nearby Charlestown.

As a descendant of one of the oldest families in the Shenandoah Valley, an Episcopalian in religion, and a substantial land owner, Lucas ranked high among the gentry of Jefferson County. Yet, politically, he broke with tradition to express Republican sympathies in a predominantly Federalist stronghold. Elected to the state legislature for two successive terms beginning in 1819, he had become an avid supporter of Jackson and headed the "Jackson Committee" in Jefferson during the election of 1828. As a reward for loyal party service Lucas was offered the paymastership at Harpers Ferry, but declined the appointment to return to the Virginia House of Delegates in 1830. Two years later he ran for Congress and defeated his Whig opponent, J. R. Cooke, in a very close contest. Re-elected to a second term in 1834, he established close ties with Martin Van Buren, James K. Polk, Thomas H. Benton, Amos Kendall, and other high-ranking members of the Democratic party. Through these contacts as well as his own canny ability Lucas remained an extremely powerful force in Virginia politics until his death in 1858.

Evidently an understanding that Lucas would be appointed to the Harpers Ferry post had been reached between him and Van

Buren, because Lucas did not stand for re-election to the House of Representatives in 1836. Instead, he supported the successful candidacy of James M. Mason, a fellow Jacksonian from Winchester, Virginia, and graciously relinquished his seat on March 3, 1837. Exactly four weeks after returning to his home in Shepherdstown, he assumed command at Harpers Ferry.[9]

A politico by instinct and inclination, "Colonel Ed" Lucas was no more qualified to run the armory than his predecessor. Unlike Rust, however, he did establish a full-time residence at Harpers Ferry and attempted to educate himself to the duties of the office through on-the-job experience and by corresponding with John Robb, Lee's successor at Springfield. Although Moor continued to manage the works, Lucas considerably lightened the master armorer's labors by handling all business transactions related to the purchase of raw materials and the hiring of labor. This enabled Moor to devote more time to much-needed mechanical improvements as well as the preparation of a new model musket designed for interchangeable manufacture.

During his four years as superintendent Lucas proved particularly adept at securing and manipulating appropriations. Despite congressional efforts to prune military expenditures in response to the panic of 1837, he succeeded in obtaining the same annual allotments for the manufacture of arms plus a healthy infusion of special funds for the construction of buildings, millwork, and machinery. Whereas Harpers Ferry mustered over $170,000 in special appropriations between 1837 and 1841, Springfield received only $46,718. To all concerned, these advances seemed necessary to place the Potomac factory on an equal technical footing with its sister armory in New England. Unfortunately, however, Harpers Ferry had accumulated an operating deficit of $46,000 by 1839, most of which had accrued during Rust's tenure in office. To

9. *Biographical Directory of the American Congress* (Washington: GPO, 1961), p. 1240; Bushong, *History of Jefferson County*, p. 298; Lucas to Secretary of War, July 29, 1829, to P. G. Randolph, October 22, 1830, to Bomford, March 29, April 4, 1837, Rust to Van Buren, March 20, 1837, Letters Received, OCO; Bomford to Lucas, October 8, 1830, Letters Sent, OCO; P. G. Randolph to Lucas, October 8, 1830, to Daniel Bedinger, October 22, 1830, Letters Sent to the Secretary of War, OCO.

discharge this debt, Colonel Bomford instructed Lucas to reduce operations at the Musket Factory, specifying that the deficit had to be made up from funds earmarked for the "manufacture and repair of arms."

Lucas agreed that the armory's books had to be cleared and balanced, but he objected to the method of doing it. In effect, Bomford's order meant that numerous workers had to be discharged, something which the politically minded superintendent wanted very much to avoid. He therefore ignored his instructions and proceeded to discharge the debt from special funds, leaving the regular manufacturing appropriations intact. This action allowed him to keep the usual complement of workers, but it considerably retarded the construction of new buildings and machinery at the factory. It also brought a severe reprimand from Bomford, who refused to make further remittances to the paymaster until Lucas abandoned his stratagem. Confronted with the threat of a fiscal embargo, the wily superintendent had no choice but to scale down operations and reduce his labor force accordingly.

Retrenchment to Lucas was only a temporary expedient. His payrolls indicate that in 1839 the number of employees at the Musket Factory averaged 272 persons. The following year the roster dropped to 241, but by the first quarter of 1841 it had surged back to 267. These figures contrast sharply with those of the Springfield armory, which employed a smaller labor force while achieving a much more impressive production record at lower unit costs. For the most part, the Ordnance Department attributed this difference not to geographical price disparities but to overemployment and managerial chicanery at Harpers Ferry. As a result, one officer explained, when Lucas relinquished office in 1841 he left an unpaid deficit of $28,000 on the armory books.[10]

10. Lucas to Robb, April 24, 1838, January 16, 1839, Letters Received, SAR; William B. Calhoun to Bomford, March 28, December 29, 1838, January 31, February 12, 1839, Lucas to Colonel George Talcott, July 30, 1840, Talcott to Bomford, May 28, 1841, Major Henry K. Craig to Bomford, November 19, 1841, Colonel Benjamin Huger to Craig, June 18, 1853, Letters Received, OCO; Captain Alfred Mordecai to Secretary of War, May 29, 1841, Letters Sent to the Secretary of War, OCO; Joel R. Poinsett to Van Buren, November, 28, 1838, Letters Sent to the President, OSW; Benet, *Ordnance Reports*, 2:227–230; Payrolls and accounts, Harpers Ferry armory, 1839–1841, Second Auditor's Accounts, GAO; Deyrup, *Arms Makers*, pp. 233, 245.

Lucas' maneuvers as superintendent well illustrate the long-standing controversy that existed over the proper function of the Harpers Ferry armory. By ignoring Bomford's directive to reduce the number of operatives at the Musket Factory, he clearly sacrificed productive efficiency for political expedience. In doing so, Ordnance officers argued, he placed community convenience ahead of national need. Yet, considering that he held public office by virtue of politics, Lucas' position is understandable. He doubtlessly recognized the harmful effects that a large-scale turnout would have on the town and immediate vicinity. Such a measure would not only inflict hardship on the displaced workers and their families but also adversely affect the livelihood of local businessmen and shopkeepers who depended on their patronage. He also realized that the blame for these dismissals would be laid at his doorstep rather than at the Ordnance Department. In other words, the execution of Bomford's order meant jeopardizing his local popularity and possibly even undermining the very basis of his political strength in Jefferson County. This he refused to do.

As much as Lucas disliked reducing the labor force at Harpers Ferry, he did not hesitate to make selected removals. A master spoilsman, he had no compunction about dismissing perfectly competent armorers to make way for political sympathizers. Years later master armorer James Burton estimated that Lucas fired thirty-four workmen between May 1837 and the fall of 1840. Of this number twenty-eight were Whigs and six were Democrats. During the same period he hired thirty-three Democrats and six Whigs. From this Burton surmised that political allegiance rather than mechanical aptitude governed the selection of armorers at the Musket Factory.[11]

Since Burton did not arrive at Harpers Ferry until 1844, he could not speak from actual experience. But others could and did. While characterizing Lucas as "an uncompromising Politician" who "knew no man by any other merit," a machinist named William Chambers found him totally unqualified to manage the Musket

11. Symington to Charles M. Conrad, May 21, 1851, Lucas to Talcott, July 30, 1840, Letters Received, OCO; Lucas to Robb, January 16, February 22, 1839, Letters Received, SAR; James H. Burton, "A list of names of operatives discharged from May 1837 till the fall of 1840," James H. Burton Papers, Yale University Archives.

Factory or any other arms-making establishment. Chambers further informed the secretary of war that Lucas "employed many hands who were of the same opinion with himself, to the great detriment of the Armoury; for the hands had to be curtailed in their work to make room for such. But this was not the worst; he was perpetually threatening to discharge those of a contrary sentiment and did actually carry out his threats as far as he could with any safety to the place—calling them by all the hard names and opprobedous epithets that a hot headed politician could invent. Thus a rod in terrorism was constantly hanging over our heads."[12]

Others agreed with Chambers. According to one inspector, Lucas wielded "the iron rod of Loco-Foco tyranny" with such vengeance that no one dared to speak out against these practices for fear of losing his job. Likewise Joseph Smith and nine other armorers accused the superintendent of being a zealous supporter of Martin Van Buren and of using "all the official influence in his power to control the sentiments & the votes of those employed in the Armory." Among other complaints Smith and his associates asserted that Lucas had not only established a partisan newspaper at Harpers Ferry but also provided its editors (Hayman and Smith) with free public housing at a time when a number of workmen were being denied similar privileges for lack of space. He had also sponsored several Democratic stump speakers yet prohibited Whig opponents from holding similar political rallies on armory grounds.

A far more serious charge concerned Lucas' interference in elections. "Just previous to every election," Smith declared, the superintendent initiated "the idea that there would be a discharge of hands soon, which had the effect to intimidate the Whigs & to urge Loco's to the polls." During the spring of 1839, for example, the superintendent's brother, William, was a candidate for Congress from the Harpers Ferry district. As the election approached, Smith recounted,

about sixty Armorers & workmen were notified that they were to be discharged, this number being about equally divided as it regards their politics. The reason assigned by the Supt. for this discharge was that there

12. Chambers to Secretary of War, December 13, 1841, Letters Received, OCO.

were too many hands employed: at the expiration of the notice however all the Van Buren men with two or three exceptions were *re-employed* & all the Whigs but two or three permanently discharged, & among them some of the most useful workmen in the establishment, to the great injury of the public service. To shew more plainly that the motive for discharging these men, was not the one assigned, we state what is notorious here, that within a short time afterwards as many hands were employed as had been discharged, & nearly all of them Van Buren men; the motive however may be fairly inferred from the fact that a number of Whigs were thus deterred from attending the polls, & his brother consequently elected, by *four votes.*

   Although universal white manhood suffrage did not exist in Virginia until 1851, the state constitution of 1830 extended the franchise to owners or tenants of a $25 freehold, $20 leaseholders, and heads of families paying taxes in an incorporated town. Since a majority of armorers either owned farms in outlying areas or paid taxes in the incorporated communities of Harpers Ferry, Bolivar, and Virginius, they qualified for the franchise and probably participated in local elections. State law did not provide for privacy, however, and since open voting still persisted from colonial times, each person's ballot became a matter of public notice and comment. Thus an ample opportunity was provided for Lucas to count votes and discover those who failed to support his candidates. During the presidential election of 1840, for instance, many armorers reported that Lucas hovered "near the polls nearly all day where his conduct was overbearing & insulting to the Whigs." If an armorer refused to vote the Democratic ticket, he had only two choices. He could stay home and hope to remain inconspicuous, or he could vote for the opposition and risk incurring the superintendent's displeasure. In either case, Lucas and his party gained the advantage.[13]
   As the sole dispenser of patronage at Harpers Ferry, Lucas carefully screened all candidates to ensure that jobs went to persons of sound political principles. After the death of his paymaster and in-law, Dr. Daniel Bedinger, in 1838, he blocked the appointment of

   13. John Avis to John Bell, March 20, August 22, 1841, Joseph Smith, et al., to Bell, [March 1841], Letters Received, OCO; "A Friend" to Burton, August 3, 1853, James H. Burton Papers, Yale University Archives; Bushong, p. 87.

Merriwether Thompson because, as he put it, the applicant "would not have my confidence or be acceptable to the friends of the Administration in this country." Instead he secured the appointment of Richard Parker, a young lawyer from Winchester, Virginia, whose father had been a member of the United States Senate and a loyal Jacksonian. Similarly in 1839 he refused to employ William N. Craighill as a clerk in the paymaster's office on the grounds that "he is a Federalist, dyed in the wool & opposed to the present Democratic Admn." Although he did not disparage the Federalists as a party, Lucas considered it suicidal to favor Whig applicants when equally qualified Democrats could be found for jobs. "On the contrary," he told Parker, "I have always been too liberal towards my political opponents, as my public course fully proves—and for which, I have deservingly incurred the censure of my political friends here & elsewhere—who say with truth, that the Federal Party have always been the most successful in obtaining office & place, because [they are] the most uncompromising to their opponents, & at the same time, the most hungry & importunate for office."[14] Lucas may have styled himself as too lenient a politician, but his intransigence as a spoilsman and guardian of Democracy at Harpers Ferry is incontrovertible. With the exception of two inspectorships, Jacksonian Democrats held every important position at the armory by 1840.[15]

If local biases and partisan attitudes governed the superintendent's actions, similar influences also operated at the paymaster's office. Since Samuel Annin's retirement in 1815, eight persons had held the paymastership at Harpers Ferry—none with distinction. As with those who managed the armory, influence rather than ability determined their selection, and many irregularities resulted. Absenteeism prevailed, books rarely balanced, and accounts were inaccurate and incomplete. Adding to the confusion, paymasters as well as clerks frequently used armory funds for private purposes. Some returned the money they supposedly borrowed

14. Lucas to Bomford, January 3, 1839, to Parker, January 25, 1839, to Poinsett, February 8, 1839, Letters Received, OCO.
15. Burton, "A list of names of operatives," James H. Burton Papers, Yale University Archives.

from the public coffers; others simply defaulted. In every case, the workers suffered when the paymaster subsequently failed to meet monthly payrolls on time.

One of the most mischievous practices consisted of paying armorers with depreciated currency. Since the War Department remitted money to the armory as treasury drafts on specie-paying banks in New York and Philadelphia, it became customary for paymasters at Harpers Ferry to exchange these notes for discounted currency and pocket the premiums. They then passed off this paper at face value in meeting payrolls and satisfying other armory obligations. While every incumbent from Lloyd Beall to Richard Parker practiced this ruse, it became especially prevalent during the 1830s. In November 1839, for instance, Parker exchanged two treasury drafts on New York banks for discounted bank notes and realized a profit of $1700. This sum exceeded his annual salary by more than four hundred dollars. In keeping with these activities, Parker also excelled at manipulating funds for political purposes. During the bitterly contested presidential race of 1840, for example, he borrowed money on the faith of his office and distributed it to various armorers as a means of inducing them to attend Van Buren rallies in Winchester, Virginia, and Frederick, Maryland. Although he later deducted the amount from their wages, he adamantly refused to extend similar favors to those who wished to attend Whig meetings.[16]

As an active member of the "Democratic Association" and a frequent speaker at local party gatherings, young Parker proved an invaluable asset to Lucas. "He denounced the Whigs in the roundest terms," several workers recalled, "attributing to them as a party an intention to resort to violence & calling upon his party to know if they were ready to meet them upon their own ground & fight them with their own weapons." In like manner, he urged fellow Democrats "to rally at the polls whether they had votes or not." Certainly Parker went to extreme lengths to ensure a majority for Van Buren in the election of 1840. With the financial aid of several friends, he purchased a plot of land in nearby Bolivar,

16. G. W. Cutshaw, et al. to Bell, April 12, 1841, John Strider to John Tyler, July 4, 1841, Poinsett to Symington, August 28, 1840, Letters Received, OCO.

divided it into sections, and "conveyed it to ten or twelve Loco Focos employed in the Armory so as to enable as many of them to vote upon it, as could by Law." To his chagrin, the scheme failed to produce the desired result. While Harpers Ferry and vicinity recorded a large Democratic majority, the rest of Jefferson County went Whig and subsequently elected William Henry Harrison by the slim margin of seventy-eight votes.[17]

The state of affairs at Harpers Ferry produced serious dissatisfaction at the Ordnance Department. By 1841, if not earlier, the armory had become a second-rate establishment and the source of considerable embarrassment to the military. Since the conduct of Stubblefield and Rust had already impaired its reputation, the activities of Lucas and Parker only served to strengthen opinion in Washington that reliance could not be placed on civilian appointees to act in the best interests of the government. For nearly twenty years Bomford and his staff had struggled to correct administrative abuses at the factory only to be ignored, scoffed at, and threatened with political reprisals. Lacking any effective means of controlling superintendents, the Ordnance office found it nearly impossible to enforce even the most salutary regulations. Despite constant prodding and scolding, Stubblefield, Rust, and Lucas repeatedly turned a deaf ear toward Washington and did exactly as they pleased. The War Department contributed to the growing disorder and abuse by selecting unqualified personnel to manage the armory. Unwilling to risk political disaffection, it had also refused to remove the incompetent. Whenever the chief of ordnance suggested means for remedying the situation, he was greeted with studied silence by high-ranking members of the Jackson and Van Buren administrations.

This posture changed, however, when Harrison and the Whig party won the election of 1840. With Harrison's inauguration, Bomford sensed that the new administration might be more amenable to instituting sweeping changes at the national armories. Since many complaints had been lodged against Lucas, the superintendent knew that it would be only a matter of time before the new chief executive demanded his resignation. Though less necessary, a sim-

17. Cutshaw, et al., to Bell, April 12, 1841, OCO; Bushong, p. 89.

ilar fate presumably awaited Robb of Springfield. At the same time Secretary of War John Bell made no secret of his intention to introduce a strict policy of retrenchment and reform, particularly at Harpers Ferry. Beset by scores of hungry office seekers and anxious to avoid the pitfalls of his predecessors, the solicitous Bell sought counsel from the chief of ordnance. Thus Bomford was given a golden opportunity to recommend scrapping the civilian superintendency system and placing both national armories under the immediate direction of Ordnance officers.

In presenting his case to the secretary of war, Bomford relied on his own knowledge as well as on copious information provided by the inspector of arsenals and armories, Colonel George Talcott, and the members of a special board of survey that had convened at Harpers Ferry in January 1841. Confidently predicting that the substitution of military for civilian superintendents would promote efficiency and eliminate waste, Bomford placed special emphasis on separating the armories from the malign political influences which, in Talcott's opinion, had tainted "every movement of the superintendent" and placed him "at the feet of a clique which sought only to fulfill their own base ends." Whereas a politician's first loyalty rested with his party, Bomford contended that "an officer may be expected to devote himself exclusively to the duties of his station and is beyond the reach or influence of extraneous matters that are found to control a citizen who has the supervision of hired workmen." To allay suspicion about potential abuses of military control, he assured Bell, "The authority of the officer is not more extensive or despotic than that of the citizen. Both are equally bound by the laws and regulations provided for the government of the armories and there is less possibility of the display of partiality or passion on the part of the officer. Besides—if one officer is found unsuited to the station, another can be readily substituted." For these reasons Bomford felt that the proposed plan should at least be given a trial. "If the just expectations of the Government are not fulfilled," he concluded, "it will be easy to return to the former mode of supervision."[18]

18. Bomford to Bell, April 6, 1841, Letters Sent to the Secretary of War, OCO. Also see Ordnance Board to Bomford, March 16, 1841, Bell to Bomford, March 23, 1841, Parker to Bomford, January 21, 1841, Talcott to A. M. Lea, September 23, 1841, Letters Received, OCO.

Bomford's brief evidently proved persuasive because on April 1, 1841, Bell announced his decision to place the national armories under the provisional supervision of Ordnance officers "until time shall be afforded to test the expediency of such an arrangement & until the sanction of Congress can be obtained." The same day he also notified Lucas and Robb that their services would be terminated as of April 15. Upon receiving instructions to carry out this order, Bomford immediately selected Majors Henry K. Craig and James W. Ripley to take command at Harpers Ferry and Springfield respectively. On April 16, 1841, both officers reported for duty.[19]

As expected, Bell's decision evoked a mixed response. Eager to see their Whig adversaries deprived of all-important patronage, Lucas and Parker endorsed the new system, expressing their conviction that it would succeed. In like manner, though for quite different reasons, members of the Ordnance Department praised the secretary for his willingness to remove the armories from partisan politics. None proved more sanguine in this expectation than Colonel Talcott. After visiting Harpers Ferry and Springfield in May 1841, he reported that there was every reason to believe that the plan would succeed and be very beneficial. A sampling of opinion at Springfield revealed that "not one man in ten, of those wholly disinterested," could be found opposed to having a military superintendent. Nevertheless, Talcott added, "it must not be forgotten . . . that the correction of errors and abuses will raise up opponents and those whose interests are affected will clamor and struggle to restore the old order of things."[20]

Truer words could not have been spoken. No sooner had Bell suspended the civilian superintendents than Congressman William B. Calhoun of Springfield, a Whig, objected vehemently to the plan as politically "full of mischief in *all respects*." "Instead of detaching the establishment from party influences," he declared, "the effect would be to awaken a fiery spirit of party." Worst of all, "the giving

19. Bell to Bomford, April 1, 1841, to Lucas and Robb, April 1, 1841, Craig to Bomford, April 16, 1841, Letters Received, OCO; Bomford to Craig, April 2, 1841, Letters Sent, OCO. For biographical information on Craig, see Military Service Histories of Ordnance Officers, vol. 1, OCO.

20. Talcott to Bomford, May 27, 28, 1841, Letters Received, OCO.

to the Armory any thing of a Military character would break down the spirit of the establishment."

The Workmen are among our best citizens; there is hardly a foreigner among them. They are nearly all married men, and as well settled as the great mass of our People. . . . They are respectable, intelligent, & substantial citizens, willing to submit to a wholesome control—but desirous, as free men, that the control should be steady, regular, and, as nearly as possible, in the same line of interest with themselves.[21]

Although Calhoun claimed a large following at Springfield, he received even more zealous support at Harpers Ferry. The most outspoken critic of the new system was John Strider, a local businessman and an intimate friend of the Beckham and Stubblefield families. An unbending Whig, Strider had worked for twelve years to defeat the forces of Democracy at the Ferry, only to be denied the spoils of victory by members of his own party. Embittered and frustrated, he addressed a long letter to President John Tyler, who had assumed office when Harrison died, in which he questioned the legality of Bell's decision to place the armories under military supervision. Specifically, he believed that the well-meaning secretary of war had been duped into approving a scheme intentionally devised by the Ordnance Department to increase its own appointive power. Not one to mince words, Strider reminded the president

that upon the broad basis of political rights, the members of a majority under a free government, are entitled to the enjoyment of the greater portion of the offices of profit and honour, in proportion to their relative number, when compared with the numerical strength of the minority. The reasons are many & obvious. The government is the property of the people, the offices are a part of that property. The greater part of that property belongs to the majority, and the municipal or constitutional claim upon the property itself guarantees the legal right of its possession and enjoyment.

Strider's definition of public property and its uses typified opinion at Harpers Ferry on the role of the armory in community affairs. It followed, then, that placing the office of superintendent in the

21. Calhoun to Bell, April 1, 1841, Letters Received, OCO.

hands of the military constituted an aberration of the democratic process. More important, it deprived Strider and his Whig friends of a key source of patronage at Harpers Ferry. Convinced that most voters placed their pocketbooks ahead of their principles, he warned that the Whig party could not expect to generate support and maintain a majority in Jefferson County without holding out tangible inducements to the electorate. Since armory jobs represented the sole inducement at Harpers Ferry, Strider pleaded with the president to restore the old system before the party became a victim of its own reform.[22]

Apart from purely political considerations, the new system, both Calhoun and Strider warned, would cause great disaffection among the workers. Much to the dismay of the Ordnance Department, their prediction came true. As early as August 1841, Talcott reported rumblings of discontent among the armorers at Springfield. Meanwhile, little did he know that a far more serious reaction was brewing at Harpers Ferry.[23]

For years the Potomac armory had suffered the reputation of being locally controlled, flagrantly mismanaged, and shamefully abused. It also claimed the dubious distinction of employing one of the most troublesome and disorganized labor forces in the country. According to one officer, "workmen came and went at any hour they pleased, the machinery being in operation whether there were 50 or 10 at work." Furthermore, he noted, "the shops were made places of business. *I* have seen four farmers at one time in one shop with paper and pencil in hand, surrounded by more than a dozen workmen, who were giving orders as to the number and weight of hogs they were to receive at killing time. All debts were done for in the Shops, it being a more convenient place of meeting than the houses of the workmen, and arrangements of all kinds, whether for politics, pleasure or business were concluded there." Along with these practices, armorers claimed the privileges of keeping frequent holidays, transferring jobs at will, drinking whiskey on the premises, and selling their tools "as sort of a fee simple inheritance." They also boasted that anyone who interfered with these rights

22. Strider to Tyler, July 4, 1841, Letters Received, OCO.
23. Talcott to Bell, August 6, 1841, Letters Received, OCO.

could expect the same fate as Thomas Dunn. "Instead of the operations of the Armory being under the control of the officers," Lucas exclaimed, they "were to some extent controlled by the workmen."[24]

Upon assuming command at Harpers Ferry, Major Craig immediately put a stop to such conduct by reviving and strictly enforcing Lee's regulations of 1827. Besides prohibiting unauthorized personnel in the workshops and the consumption of liquor during working hours, he installed a clock at the factory and insisted that a ten-hour day be observed by all workers. Since private contractors had long required their employees to labor twelve or more hours daily, Craig considered this mandate eminently reasonable. The labor force, however, thought otherwise.[25]

To armorers long accustomed to controlling the duration and pace of their work, the idea of a clocked day seemed not only repugnant but an outrageous insult to their self-respect and freedom. Since Craig was primarily responsible for its introduction, the clock reinforced already rife feelings about the pernicious influence of outsiders—particularly military men—at Harpers Ferry. Moreover its ineluctable cadence served to emphasize the rigorous discipline, regularity, and specialization so often associated with the coming of the machine. In this sense, the clock not only kept time but symbolically deprived armorers of the satisfaction of traditional craft labor. Every minute had to be accounted for and each accounting fostered further discontent.

By the fall of 1841 voices of opposition began to be heard when several letters and at least two long petitions reached the War Department denouncing Craig's actions as tyrannical and subversive of the rights of free men. Asked to investigate the question, Bomford replied that no one had intended to subject the armorers to

24. Symington to Maynadier, July 12, 1849, Lucas to Bomford, August 29, 1839, Letters Received, OCO. Also see Lucas to Talcott, February 7, July 30, 1840, to Bomford, March 23, 1841, Moor to Bomford, June 29, 1841, OCO; Bell to the President, May 31, 1841, Letters Sent to the President, OSW.

25. "Proceedings of a Board of Officers," February 1842, Reports of Inspections of Arsenals and Depots, OCO; Talcott to John C. Spencer, March 11, 1842, Letters Sent to the Secretary of War, OCO; Talcott to Spencer, May 17, 1842, Letters Received, OCO.

military discipline, nor had such a thing occurred at Harpers Ferry. Certainly, he declared, "no large private manufactory, nor any operations requiring numerous workmen, can be properly conducted without certain regulations and fixed hours for work, and the latter are still more essential when the operations are partly performed by machinery, as they are at the Armories." Having found no malfeasance on Craig's part, he adamantly reaffirmed that "the public interest will be best promoted by allowing the National Armories to remain under the superintendence of Officers of the Ordnance Department."[26]

Bomford's explanation evidently satisfied the War Department, but it totally failed to appease the armorers at Harpers Ferry. As the months passed, opposition grew stronger. Although Bomford attempted to allay antagonism, the issue was joined on March 21, 1842, when the entire labor force, led by pieceworkers, walked off the job in protest and selected spokesmen to present their grievances to President Tyler. Amid great fanfare, but without violence, the delegation chartered a boat on the Chesapeake & Ohio Canal and set off for Washington. In an interview with the president, they complained that Craig's petty regulations had degraded them to "mere machines of labor," rendering their situation little better than that of slaves. Such rules were dehumanizing, and they demanded an end to military control of the armory. Although Tyler promised to look into the matter and assured the armorers that no retaliatory measures would be taken against the "instigators and fomentors of the outbreak," he dashed hopes for immediate change by telling them to "go home and hammer out their own salvation." Go home they did, but they felt no better about the military system under which they worked.[27]

The "clock strike" of 1842 represented an emotion-laden response to the consequences of rationalization. Not bread-and-butter issues but an assault on traditional customs and freedoms precipitated the crisis. Under civilian superintendents the armorers had

26. Bomford to Spencer, December 20, 1841, Letters Sent to the Secretary of War, OCO.

27. Craig to Talcott, March 21, 22, April 1, 1842, to Spencer, March 28, 1842, Letters Received, OCO; Barry, p. 32.

submitted to successive divisions of labor as well as the limited
introduction of machine processes. None of these innovations had
been popular, however. As much as the new technology helped to
lighten physical labor, increase output, and yield a more uniform
product, older artisans continually complained about the devalu-
ation of old skills and the loss of artistry in their work. In addition
to the repetitive and less fulfilling nature of piecework, the incessant din of machinery wore on nerves, shortened attention spans,
and acerbated tempers. Yet, all these changes had been made with
the implicit understanding that the armorers would continue to
control the tempo and duration of their labor. When the Ordnance
Department instituted rigid work rules demanding daily attendance, punctuality, and steady intensity, this tradition was suddenly broken. For mutual rights and obligations, military superintendents substituted unilateral controls. In place of individual
freedom and self-determination, they imposed collective standards
of behavior. There was no choice—an alien factory discipline characterized by time-oriented goals and impersonal bureaucratic constraints *had* to be absorbed. Workers were expected to abandon
irregular habits and become more dutiful, efficient, and obedient.
Those who resisted faced harassment, fines, and, ultimately, dismissal. Such an agenda was intolerable. It not only struck at the
craft ethos but also threatened basic norms within the community.
These factors awakened the armorers to a spirited defense of the old
order.

Although the strikers failed to wring any concessions from the
government, the episode did not pass unnoticed. By redirecting
public attention to the armory question, the walkout at Harpers
Ferry brought increased pressure on the Ordnance Department to
justify its policy. Since Bomford had relinquished the duties of
chief of ordnance on February 1, 1842, the burden of proof fell
upon his successor, Talcott. Well acquainted with armory affairs,
Talcott met the challenge by addressing a long letter to the secretary of war in which he reviewed the department's position point
by point. Unlike his predecessor, he did not hesitate to speak his
mind sharply and forcefully.

First and foremost, the new chief of ordnance emphasized, "the

Regulations governing the workmen are *not changed*, they are *merely enforced*, every man being required to commence work at a fixed hour and labor during working hours."

The old practice of coming & going to suit the pleasure of each and working or playing—*in* hours or *out* of hours was an abuse formerly tolerated, but never sanctioned by regulations, & the pretext that *because men work by the piece*, they should be allowed to run machinery when they please & be absent whenever it suits their whim, finds no favor at private workshops nor can it be allowed where the work of one man depends on that done by another, for carrying on and keeping up all branches to a proper standard. The Master Armorer cannot keep all branches in a suitable state of advancement, unless he can rely on the quantity of work to be done by each man.

Second, Talcott declared, "the *real ground of opposition* to the present mode of supervision is well known to be this":

The men have been paid high prices & were in the habit of working from 4 to 6 hours per day—& being absent whole days, or a week. At the end of a month their pay was generally the same in amount as if no absence had occurred. They are now required to work full time and during fixed hours (according to old regulations) and the master of the Shop keeps a time account showing the time *actually spent in labor*. Here is the *great oppression* complained of. At the end of a month the *quantity of labor performed*, or *product*, and the *time* which it is effected are seen by simple inspection of the Shop books. The degree of diligence used by each man is also known and hence results a knowledge of what is the *fair price* to be paid for piece work!!! The Armorers may attempt to disguise or hide the truth under a thousand clamors—but this is the *real cause* of their objections to a Military Superintendent. He enforces the Regulations which lay bare their secret practices (frauds—for I can use no better term). They can control a civil Supt. & have often done it! They have occasionally ousted one, and they have shot one.

In effect, the Colonel concluded, "We say to the Armorers—here are our Regulations; if you will not abide by them—go elsewhere— for we know that as many good or better workmen can be had at any moment. They answer—no, we will not leave the armory. We insist on working for the United States and will fix our own terms!!!"[28] Having clearly delineated the sources of conflict and opposition, he called upon Congress to settle the matter once and for all.

Talcott's report proved controversial although effective. On the one hand, it earned him the lasting enmity of Calhoun, Strider, and all those who pressed for a return to the civilian system. On the other hand, it persuaded the secretary of war to sponsor legislation permanently establishing military superintendents at the national armories. Introduced in the Senate by the committee on military affairs, the bill met determined opposition in both houses. Nevertheless, after several conferences and some astute lobbying by the chief of ordnance, Congress sanctioned the measure and President Tyler signed it into law on August 23, 1842.[29]

With the superintendency question temporarily settled, the Ordnance Department turned to more pressing problems at Harpers Ferry. Although Craig had successfully implemented his controversial labor policies, much remained to be done to place the armory on an equal technical footing with Springfield. For one thing, manufacturing facilities at the Ferry presented a sharp contrast with the neat and orderly New England factory. Neglected for well over two decades, the armory was found by inspectors and visitors to be decidedly in bad, even wretched condition. "This whole establishment," one Ordnance inspector remarked, "is cramped for room, not having been constructed upon a plan arranged beforehand, but put up building after building as appropriations were obtained." To this officer as well as others, the armory shops looked like a string of dominoes, jutting in every direction and totally lacking in architectural or functional unity.[30]

28. Talcott to Spencer, May 17, 1842, Letters Received, OCO. Of related interest is Colonel Rufus L. Baker to Colonel Henry K. Craig, July 20, 1852, Reports of Inspections of Arsenals and Depots, OCO. For biographical information on Talcott, see Military Service Histories of Ordnance Officers, vol. 1, OCO.

29. U.S. *Statutes at Large*, 5:512; Charles Stearns, *The National Armories* (Springfield, Mass.: G. W. Wilson, 1852), pp. 4–6.

30. Maynadier to William L. Marcy, June 18, 1846, Letters Sent to the Secretary of War, OCO. Also see Talcott, Inspections of the Harpers Ferry armory, July 17–25, 1835, June 21–25, 1836, Reports of Inspections of Arsenals and

Confronted with a conglomeration of decrepit workshops, sheds, and shanties, Craig and Talcott saw no way to modernize operations without first renewing the entire physical plant. Yet, other than making piecemeal repairs, little could be done between 1842 and 1844 to improve the situation because a lingering depression severely restricted War Department finances. When Craig left Harpers Ferry in 1844, the workshops were in much the same dilapidated condition as he found them three years earlier. His successor, however, experienced better fortune. After assuming command in November 1844, Major John Symington, an eccentric but talented West Point engineering specialist, proceeded to draw up a master plan for the armory's complete renovation. Since economic conditions had improved and war with Mexico seemed imminent, he received permission to go ahead with the project from Secretary of War William L. Marcy in March 1845.[31]

Once under way, construction proceeded at a surprisingly rapid rate. With the completion of a sizable two-story boring mill at the Musket Factory in 1845, there followed a forging and smith's shop in 1847, a stocking and machine shop (called the "Bell Shop") in 1848, a tilt-hammer and barrel-welding shop in 1850, a grinding mill and sawmill in 1851, an annealing and brass foundry in 1851, and a rolling mill in 1854. During the same period the nearby Rifle Works received a finishing shop in 1847, a tilt-hammer shop in 1851, and a large machine shop in 1852. (See Map.)

Twenty-five structures were erected at Harpers Ferry between 1845 and 1854. Unlike the ones they replaced, the new buildings exhibited thoughtful and consistent planning. Constructed on heavy stone foundations with brick superstructures, cast-iron fram-

---

Depots, OCO; Bomford to Spencer, November 29, 1841, Letters Sent to the Secretary of War, OCO; Symington to Talcott, November 29, 1844, Letters Received, OCO; Adolphe Fourier de Bacourt, *Souvenirs of a Diplomat* (New York: Henry Holt, 1885), p. 225.

31. Symington to Talcott, November 9, 29, 1844, to Conrad, May 21, 1851, Talcott to William Wilkins, October 26, 1844, Craig to Talcott, November 12, 1844, Letters Received, OCO; Spencer to War Department, July 28, 1842, Letters Received, SAR. For biographical information on Symington, see Military Service Histories of Ordnance Officers, vol. 1, OCO; Cullum, *Biographical Register*, 1:131–132.

ing, sheet-iron or slated roofs, and large arched portals and windows, all conformed to the same general style of "factory Gothic" architecture. Inasmuch as the rolling mill adjoined the tilt-hammer shop, the forging shop the machine shop, and so on, they also formed a well-integrated functional whole where the flow of work from one stage of production to another was greatly facilitated. Beyond this construction, the renewal program included the enlargement of the armory canals, heavier millwork to handle the increased use of machinery, the installation of water hydrants and other fire-fighting equipment, and the construction of drainage ditches, cesspools, and cisterns for sanitation purposes. As a final touch, the grounds were walled off and landscaped to project a well-groomed appearance.[32]

If the physical plant stood in dire need of restoration by 1841, the same held true of the craft-oriented techniques which had been practiced at the Musket Factory since the days of Joseph Perkin. "In comparing this Armory with that at Springfield," Major Symington told the secretary of war, "it should be borne in mind that the mismanagement there had never brought that post into the condition of this." Springfield's location in the heartland of industrial America provided many opportunities for testing new ideas and advancing mechanical knowledge that Harpers Ferry had never enjoyed. Given this "great advantage," Symington observed, "new machinery was introduced there much earlier than here, their discarded machinery being generally far better than the wretched things we have been gradually getting rid of for five years."[33]

Certainly abundant testimony existed as to Harpers Ferry's technological backwardness. In 1832, as noted, Talcott had found the Musket Factory heavily dependent on manual labor and far behind manufacturing attainments elsewhere. While he detected some improvements by 1835, the works continued to operate without many useful machines well into the 1840s. Nonetheless, a slow but notice-

---

32. Annual reports of the Chief of Ordnance, 1845–1854, Letters Sent to the Secretary of War, OCO; Symington to Talcott, March 30, 1850, Letters Received, OCO.

33. Symington to Conrad, May 21, 1851, OCO.

able trend toward mechanization had begun to gather momentum during the late 1830s.

The person who presided over this change was the master armorer, Benjamin Moor. Born and raised in Springfield, Moor had entered the U.S. armory there as an apprentice in 1797 at the age of fourteen. After completing his training around 1804, he held successive positions at the Virginia state armory, Harpers Ferry, and a private armory in Canton, Massachusetts, before going to Philadelphia as an inspector of contract arms in December 1810. He returned to Springfield on a special assignment in 1812 and the following year accepted the post of master armorer at the establishment under Henry Lechler, a Pennsylvanian. Unfortunately, his superior proved totally incompetent and, although he prevented the works from falling into utter ruin, Moor became a victim of the purge that followed Lechler's removal as superintendent in 1815. Thanks to the good offices of Commissary General Callender Irvine, however, he remained in government service and received the appointment of master armorer at the Allegheny arsenal near Pittsburgh. There he remained until called to Harpers Ferry as Armistead Beckham's successor in May 1830.

Moor enjoyed a widespread reputation as a gifted armorer. Well traveled, he knew nearly every important arms maker north of the James River and east of the Ohio. Among his close friends were Marine T. Wickham, William Wade, Rufus L. Baker, Lemuel Pomeroy, John Clarke of the Virginia state armory, and, of course, Roswell Lee. Besides being "an ingenious mechanik, compleat workman & judge of every part of a fire arm," Moor was very likable. Everybody remarked on his easy-going manner and unquestionable integrity. Nevertheless, his childlike gullibility and reluctance to hurt anyone's feelings often allowed less scrupulous persons to take advantage of him. His greatest shortcoming was as a disciplinarian, and this caused problems that eventually led to a confrontation with the Ordnance Department in 1849.[34]

After spending fifteen trouble-free years at the Allegheny arse-

34. Irvine to Bomford, April 10, 1815, Lieutenant Daniel Tyler to Bomford, November 20, 1830, William Wade to Secretary of War, January 20, 1837, G. Z. to War Department, July 15, 1841, Letters Received, OCO.

nal, Moor was reluctant to break his ties in western Pennsylvania and move to the unsettled atmosphere of Harpers Ferry. Years later he informed a Senate committee that "my objection to it arose principally from the known difficulties of the station, as well as the onerousness of the duties and the meagre compensation of $600 per annum." Nevertheless, at the pressing solicitation of Bomford, he had accepted the appointment and reported for duty to Rust on June 1, 1830.

What the new master armorer witnessed upon his arrival at Harpers Ferry only served to increase his misgivings. For the most part he found the Musket Factory in a run-down condition, the environment disease-ridden, and the townspeople backward and inhospitable. "Every way considered," he confided to a friend, "there are customs and habits so interwoven with the very fibers of things as in some respects to be almost hopelessly remitless." Confronted with ramshackle facilities, inadequate water power, and outmoded techniques, Moor seriously doubted whether conditions could be improved even by numerous and expensive repairs and sweeping organizational changes. "If [the] Government wants [a] supply of good cheap guns," he suggested, it would be better to abandon the armory at Harpers Ferry and relocate it at Pittsburgh, "where the application of public funds for public supplies shall be in accordance with the spirit of economy." No one took the idea seriously.[35]

Encumbered with the added responsibility of serving as acting superintendent during the frequent absences of Rust, Moor considered his duties "multifarious, almost incessant, and somewhat perplexing." To be sure, innumerable problems existed at the armory. Labor discord and other difficulties rendered productive operations laborious, frustrating, and uncertain. Further complicating matters, Ordnance inspectors pronounced the arms manufactured at Harpers Ferry "40 per cent inferior" to those made at Springfield. The master armorer became convinced that something had to be done to reduce costs and to improve the quality of production at the Musket Factory. Most pressing of all was the need for

35. Moor to Major Rufus L. Baker, May 5, 1831, Letters Received, AAR; Memorial of Benjamin Moor [32A–H13.1], USS.

complete mechanical renovation. But with so many problems vying for attention, Moor found little time for these matters before 1837.[36]

Apart from his regular tasks at the armory, a particularly time-consuming assignment concerned the adoption of a new model army musket. Less than six months after his arrival at Harpers Ferry, Bomford had asked Moor to assist Lee and Lieutenant Daniel Tyler, the chief of contract arms inspectors at Springfield, in drawing up a report on the comparative merits of French and American firearms. Emphasizing the importance of selecting a design capable of manufacture with interchangeable parts, Bomford had asked the committee for specific recommendations. Although the members at first disagreed over the degree of standardization, they eventually settled their differences and set forth the basic specifications of a new model in 1832. With slight variations it represented a copy of a French musket made at Mutzig in 1822.

Moor had hoped that his part in the business would end at this point. However, with Lee's death in August 1833, he was charged with actually designing and making the patterns for the new musket. This required a great deal of time, but, after a number of lengthy delays, Moor finally submitted six specimens to the Ordnance Department in May 1836. Except for some minor alterations, Bomford approved the design and ordered him to proceed with twenty-four models and four sets of verifying instruments for eventual distribution to the Springfield armory and private contractors. Again delays occurred while Moor adapted John Hall's gauges to the new model musket. By December 22, 1838, however, he had completed the work and had taken the patterns and implements to Washington to exhibit to the members of a special Ordnance board. Highly pleased with the result, the board recommended their adoption to the secretary of war who, upon expressing his conviction that they would "challenge a comparison with any in the world," promptly sanctioned the measure. On December 31, 1838, Bomford notified the national armories to begin tooling up for their manufacture.[37]

36. Moor to Bomford, June 25, 1831, Letters Received, OCO.
37. Bomford to Robb, December 31, 1838, March 8, 1839, Letters Received, SAR; Secretary of War to Van Buren, November 30, 1839, Letters Sent to the

In contrast to the time and expense that had gone into preparing the new patterns and gauges, the production history of Moor's flintlock musket—officially designated the Model 1840—proved very short lived. Although the Springfield armory quickly retooled and manufactured over 30,000 arms of this type between 1840 and 1844, the Ordnance Department discontinued the model in 1842 in favor of a percussion musket devised by Thomas Warner, Moor's counterpart at Springfield (Figure 14). Ironically, Harpers Ferry, the place where the Model 1840 had originated, never produced the weapon. Instead the armory continued to make old Model 1816 flintlocks until the spring of 1845 when it switched to the Warner-designed percussion musket. This gun ultimately became the first fully interchangeable product to be made in large quantities at both national armories, an important technical watershed.[38]

Many of the component parts of the Model 1842 musket were copied directly from the earlier Harpers Ferry version, showing that Warner had profited from Moor's experience. Both men frequently corresponded with one another in 1839 and 1840 and also exchanged patterns and gauges. On two occasions they met to

---

President, OSW, For further information on what came to be known as U.S. Model 1840 (sometimes designated as the Model 1835) flintlock musket, see Bomford's correspondence with Lee, Tyler, Moor, and Lucas, 1831–1838, Letters Sent-Letters Received and Special File, OCO.

38. Talcott to Major James W. Ripley, June 1, 1842, Letters Received, SAR; Ripley to Craig, July 12, 1842, Warner to Craig, August 16, 1842, Letters Sent, SAR; Captain William A. Thornton to Bomford, March 15, 1841, Symington to Talcott, September 30, 1845, Huger to Craig, October 3, 1853, Letters Received, OCO; Fitch, "Rise of a Mechanical Ideal," p. 521; Benet, 2:228–230.

Also Moor had begun work on a new model percussion rifle before completing the Model 1840 musket. This weapon, a muzzle-loader, was intended to replace Hall's patent rifle in military service. The preparation of patterns and gauges took more than two years time, beginning in March 1839 and ending in November 1841. Alterations and improvements continued to be made as late as 1846, the year full-scale production began at the Rifle Works at Harpers Ferry. Considered one of the best-designed small arms of the antebellum period, the Model 1841, or "Mississippi" rifle as it came to be known, continued to be made at Harpers Ferry and by several private contractors—most notably Eli Whitney, Jr., of New Haven and Robbins, Kendall & Lawrence of Windsor, Vermont—until the mid-1850s. Six specimens displayed by Robbins & Lawrence at the London Crystal Palace Exhibition in 1851 won recognition for their excellence, thus paving the way for subsequent British investigations of the American System of manufacturing and large purchases of American machinery in 1854 and 1855.

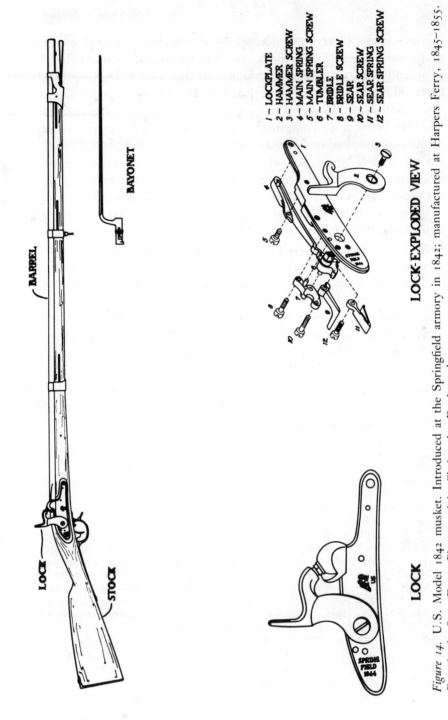

BARREL

BAYONET

LOCK

STOCK

LOCK

SPRING FIELD 1844

LOCK-EXPLODED VIEW

1 — LOCKPLATE
2 — HAMMER
3 — HAMMER SCREW
4 — MAIN SPRING
5 — MAIN SPRING SCREW
6 — TUMBLER
7 — BRIDLE
8 — BRIDLE SCREW
9 — SEAR
10 — SEAR SCREW
11 — SEAR SPRING
12 — SEAR SPRING SCREW

Figure 14. U.S. Model 1842 musket. Introduced at the Springfield armory in 1842; manufactured at Harpers Ferry, 1845–1855. Drawing by Steve Foutz. Photograph by Christopher Duckworth.

examine each other's work and to discuss problems related to the preparation of tools and machinery. Yet, Springfield took the lead in actually manufacturing common muskets with interchangeable parts. With Warner and such master machinists as Cyrus Buckland, Nathaniel French, and William Smith employed under the same roof, few if any manufacturing establishments in the United States equaled the New England armory in sheer mechanical talent during the late 1830s and early 1840s. The very presence of these men goes far to explain Springfield's international reputation during the antebellum period.[39]

Whereas Springfield concluded preparations for interchangeable production during the summer of 1840, Harpers Ferry had only partially accomplished the same objective by 1846. Even then, the practice of marking and hand filing lock components—unnecessary for truly standardized parts—continued at both works until the late 1840s. This indicates how extremely difficult the empirical process of coordinating and adjusting to new and unfamiliar techniques must have been.[40] It also reveals the extent to which Harpers Ferry, at least, required both retooling and internal reorganization during the 1840s. That these tasks became so great suggests that Moor could not have possibly accomplished the objective alone. He

39. Bomford to Robb, December 31, 1838, Letters Received, SAR; Talcott to Lucas, May 6, 1840, Letters Sent, OCO; Moor to Talcott, March 6, 1850, Correspondence Relating to Inventions, OCO; Fitch to Burton, November 11, 1882, January 13, 1883, Burton to Fitch, November 27, 1882, James H. Burton Papers, Yale University Archives; Fitch, "Rise of a Mechanical Ideal," pp. 521–523.

40. Robb to Talcott, December 1, 1840, Symington to Talcott, September 30, 1845, Letters Received, OCO; Talcott to Wilkins, January 6, 1845, Letters Sent to the Secretary of War, OCO; Payrolls and accounts, Harpers Ferry armory, 1846, Second Auditor's Accounts, GAO.

Paul J. Uselding, a practitioner of the new economic history, has attributed the lengthy time lag between the introduction of precision machinery at Springfield during the late 1830s and early 1840s and its full impact on productivity measures a decade later to "technological indigestion," that is, the inability of the productive system to assimilate the input of new and different embodied technologies—in this case machine tools—over a short period of time. A similar but more pronounced phenomenon can be observed at Harpers Ferry between 1841, the year reconstruction began, and 1854, the year that it ended. See Uselding, "Technological Change, Productivity Growth, and Linkages to Economic Development: 1820–1850" (Paper delivered at the Eleutherian Mills-Hagley Foundation Conference on Technology, Greenville, Delaware, October 31, 1969), p. 5.

necessarily depended on the ideas and assistance of others in working out the configurations of the new technology.

Moor had always been an ardent proponent of mechanized production. His willingness to adopt such a system at Harpers Ferry exasperated older, craft-trained artisans who branded him as "nothing more than a theorist, with his head crammed full of whims, yankey notions and useless machinery." Yet, amid growls of humbug and accusations of chicanery, around 1838 he began methodically to strip the workshops of obsolete tools and machinery and to replace them with more up-to-date equipment embracing the entire process of arms making. The work proceeded slowly—too slowly in the opinion of the engineering-oriented Ordnance Department—but the result was a truly impressive achievement.[41]

Most impressive was the strategy Moor used in completing the project. While he doubtlessly made some inventive contributions to the new stable of machinery erected at Harpers Ferry during the 1840s, his masterful ability to draw upon the innovations of others and mold them into a workable production system far overshadowed his aptitude as a machinist. Since Simeon North, John Hall, and Thomas Warner had already applied machinery to the manufacture of standardized parts, Moor had the good sense to see that much time, effort, and expense could be saved by taking advantage of their improvements. Hence, with the encouragement of the Ordnance Department and the approval of his superiors, he proceeded to tap these sources of information. From Hall, it will be recalled, he borrowed a sophisticated method of gauging components and adapted it to the manufacture of common muskets and rifles. In like manner, he began sending observers northward with instructions to scour the countryside for useful mechanical improvements and other knowledge relevant to the manufacture of interchangeable parts.

The effort paid handsome dividends. Provided with letters of

41. Anon. (Carey Thompson) to War Department, June 27, 1841, Lucas to Bomford, February 8, 1839, Letters Received, OCO; Bomford to Poinsett, July 5, 1838, Letters Sent to the Secretary of War, OCO; Burton to Fitch, November 27, 1882, James H. Burton Papers, Yale University Archives.

introduction by the secretary of war and chief of ordnance, Moor and his deputies gained access to virtually every important armory and machine shop in New England. At one time or another during the 1840s their itineraries included Middletown, Salisbury, and Whitneyville in Connecticut; Boston, Fall River, Fitchburg, Lawrence, Lowell, and North Chelmsford in eastern Massachusetts; Cabotville, Chicopee, and Pittsfield in western Massachusetts; Keene and Nashua, New Hampshire; and Providence, Rhode Island. Without doubt, however, their primary rendezvous was the Springfield armory. There they examined machinery, compared notes, and discussed common problems with such master mechanics as Buckland and Warner. After satisfying their curiosity and determining their needs, they then set out for other establishments to gain further information about new and interesting mechanical equipment.

In view of this strategy, mechanization at Harpers Ferry evinces two fundamental characteristics. First, Moor built a large number of machines from patterns and drawings obtained from Springfield and other New England factories. Covering a broad spectrum of mechanical types, they ranged in importance from relatively simple buff wheels and punch presses to more complex multi-spindled draw polishers and drill presses. In "getting up" these machines, Moor relied primarily on younger assistants—notably William Apsey, Henry W. Clowe, John H. King, Jerome B. Young, and James H. Burton—to work out the details.[42] Since the armory lacked foundry facilities, he procured the castings from private firms. Among those most frequently engaged were John Watchman of Baltimore, Hugh Gilleece of Harpers Ferry, and John Wernwag of Virginius. In every case, a standard procedure prevailed through-

---

42. Craig to Bomford, December 14, 1841, Letters Received, OCO; Fitch to Burton, September 5, 1882, Burton to Fitch, November 27, 1882, James H. Burton Papers, Yale University Archives.

Although nothing is known about his background, Apsey served as master machinist at Harpers Ferry from 1837 until his death in 1844. His successor was Burton, the young Virginian whom Major Craig had recruited from a Baltimore machine shop in April 1844. King and Young were former Hall workmen, and Clowe held the position of master millwright at the Musket Factory during the 1840s.

out the 1840s. Using patterns and specifications provided by the armory, the founder made the required castings and shipped them to Harpers Ferry to be machined, fitted, filed, and assembled into finished units. If the castings proved defective, the supplier technically forfeited payment. However, there is no evidence that this ever occurred. Apparently the government absorbed all losses for faulty castings.

Second, outfitting the armory involved outright purchases. Talcott had first suggested the idea in 1836 and the earliest acquisitions were made the following year. At that time Lucas contracted for five machines with three different firms—a wood planer from George Page of Keene, New Hampshire; a small screw lathe with extra gears from the Ames Manufacturing Company of Cabotville, Massachusetts; and an iron "slabbing" engine, turning lathe, and gear cutter from the Savage Manufacturing Company of Savage, Maryland. Evidently they all worked well, so well in fact that Lucas seriously considered discontinuing the office of master machinist and procuring all necessary machinery by special engagement.

Moor dissuaded the superintendent from implementing such a rash proposal, and limited purchases continued during the next three years. In 1839 the armory bought a screw-cutting and screw-tapping machine from Thomas W. Smith's Alexandria [Virginia] Foundry, a patented tenoning machine from J. W. Thompson of Boston, and several other woodworking machines from George Page. The following year saw the acquisition of a large engine lathe and an additional set of morticing and tenoning machinery from T. M. Edwards & Company of Keene, New Hampshire. Installed at Harpers Ferry by Edwards' principal machinist, Josiah Fay, the lathe and fixtures alone cost $2350. More important, the first recorded plain milling machine ever used at the Musket Factory arrived from the Springfield armory in the spring of 1840.[43]

While Lucas had favored purchases of machinery between 1837 and 1841, his military successors placed even greater emphasis on

---

43. Payrolls and accounts, Harpers Ferry armory, 1837–1840, GAO; Lucas to Talcott, May 25, 1839, to Bomford, May 31, August 20, 1839, Letters Received, OCO; Robb to Lucas, October 8, December 6, 1839, Letters Sent, SAR; Lucas to Robb, November 1, 1839, Letters Received, SAR.

the contract method. In 1842, for instance, Major Craig announced his intention to procure as many machines as possible from commercial builders because they could be gotten more quickly, economically, and efficiently than similar equipment could be made at Harpers Ferry. Craig carried out this measure, as is well shown by his correspondence with the chief of ordnance and Major Ripley, the military superintendent at Springfield. Unfortunately existent inventories, which are fragmentary and incomplete, do not record every purchase made between 1841 and 1844. It is known that Craig ordered a double-spindle machine for cutting wood screws from the Springfield armory in 1841. He also acquired three stocking machines, two drill presses, two barrel-turning lathes, one screw engine, one cone miller, one pin machine, and one "slabbing" machine from the Ames Company in 1843. Finally, in 1844 he bought a Daniels wood planer from J. H. Hitchcock of West Troy, New York, and a five-hundred-pound Nasmyth direct-action steam hammer from Merrick & Towne of Philadelphia.[44]

When serious illness forced Moor to take an extended leave of absence in 1845, his temporary replacement, Alexander Hitchcock of Troy, New York, continued his administrative strategy. So did James Burton, the talented Virginian who succeeded Moor as master armorer in 1849. The mechanization process, therefore, followed the same pattern during the superintendencies of Major John Symington (1844–1851) and his successor, Colonel Benjamin Huger (1851–1854). Of 121 new machines installed at the Musket Factory between 1845 and 1854, sixty were purchased at a total cost of $34,986. This sum exceeded the amount spent on the on-site construction of machinery by $3,686. Since Hall's machinery was easily adapted to the manufacture of muzzle-loading rifles, much less was spent at the Rifle Works. Thirty-three machines were installed at the Shenandoah shops between 1845 and 1854. Of this number, twenty came from private machine builders.[45]

44. Craig to Talcott, October 21, 1842, Hitchcock to Talcott, April 17, 1843, Letters Received, OCO; Payrolls and accounts, Harpers Ferry armory, 1843–1845, GAO; Craig to Ripley, June 7, 1841, Letters Received, SAR.

45. Craig to John B. Floyd, February 9, 1859, Letters Sent to the Secretary of War, OCO; Payrolls and accounts, Harpers Ferry armory, 1845–1851, GAO; Symington to Burton, December 17, 1849, November 25, 1850, Second Auditor's Accounts, GAO.

Under Craig, Symington, and Huger, Harpers Ferry purchased machinery from a number of sources. Eight firms have already been mentioned. In addition, the armory had transactions with Ralph Crooker of Boston, A. & W. Denmead & Son of Baltimore, Poole & Ferguson of the same city, Robbins & Lawrence of Windsor, Vermont, and the Lowell [Massachusetts] Machine Shop. However, none of these companies did the volume of work executed for the armory by the Ames Manufacturing Company and the American Machine Works of Springfield. Because of their location, both firms had ready access to patterns and drawings owned by the Springfield armory. Since most of Harpers Ferry's machinery was copied from Springfield's, Ames and American held a decided advantage when it came to bidding on contracts. Consequently they supplied four out of every five machines sent to Virginia between 1839 and 1854.

Though not known for innovations, the Ames Company deserves an important place in the annals of American industry. Founded in 1834 by Nathan P. and James T. Ames, it became one of the first firms in the United States to manufacture and market a standard line of machine tools to the general public. While the Ames brothers are deservedly famous for supplying gun-making equipment to Great Britain, Russia, and Spain in the middle of the nineteenth century, often overlooked is the fact that they also sold their wares to a number of pioneers of the American machine tool industry. In 1835, for example, they furnished Messrs. Merrick, Agnew & Tyler of Philadelphia with a milling engine and a screw-turning lathe. Later that year they sent an engine lathe to Coleman Sellers & Sons of the same city. They also equipped the Robbins, Kendall & Lawrence factory at Windsor, Vermont, when the latter company embarked on a short-lived but extremely significant career as a government arms contractor in 1845. Considering their varied engagements for textile machinery, mining equipment, millwork, cutlery, swords, small arms, cannon, bronze statuary, and other metal castings, the activities of the Ames brothers aptly illustrate how early tool builders served as key transmitters in the diffusion of new skills and techniques to technically related industries.[46]

46. Nathan P. Ames to Merrick, Agnew & Tyler, April 9, 1835, to Sellers & Sons, November 27, 1835, N. P. Ames Letterbook, James T. Ames Papers,

Although the Ames Company supplied Harpers Ferry with everything from iron planers to Boyden water turbines, two items dominated its sales lists: stocking and milling machinery. Between 1845 and 1854 the firm sold ten stocking machines to the armory. These were based on Cyrus Buckland's more streamlined versions of Thomas Blanchard's original equipment. Of all-metal construction and beautifully executed, they included units for spotting, facing, double-profiling, turning, and recessing the stocks of muskets and rifles. During the same period the Ames Company built at least twenty-six plain milling machines for the Musket Factory. The selling price for these was either $310 or $275, depending on whether they came equipped with double- or single-head supports. In addition, Symington purchased an index milling machine for $490 in 1846. Nothing is known about this particular design other than it had originated at the Springfield armory around 1844 and probably represented the prototype for a later "universal" indexed miller devised by Frederick W. Howe of Robbins & Lawrence in 1850.[47]

More reliable information exists on the plain milling machines produced by the Ames brothers. During the 1840s and 1850s workmen commonly referred to these machines as "M'Farland" millers. The term refers to Jacob Corey MacFarland, a mechanic who received his early training at the Springfield armory and later served as foreman of the Ames machine shop from 1845 to 1858. In view of his background, MacFarland had probably assisted Warner,

---

Bingham Collection; Robbins, Kendall & Lawrence to Talcott, March 10, 1845, August 25, 1846, Letters Received, OCO; Deyrup, pp. 125–126. For a brief history of the Robbins & Lawrence shops, see Merritt Roe Smith, "The American Precision Museum," *Technology and Culture*, 15 (1974):413–416.

47. Nathan P. Ames to Moor, January 24, 1845, N. P. Ames Letterbook, James T. Ames Papers; Talcott to Marcy, November 10, 1846, November 14, 1848, to George W. Crawford, November 1, 1849, to Conrad, November 4, 1850, Craig to Secretary of War, November 2, 1852, October 24, 1854, Letters Sent to the Secretary of War, OCO; Symington to Talcott, September 30, 1845, January 7, 1847, August 4, 1848, Letters Received, OCO; Payrolls and accounts, Harpers Ferry armory, 1851–1852, GAO. Historical treatments of the indexed machine are provided by G. Makers, "Old Machine Tools," *American Machinist* 32 (August 12, 1909):291; Robert S. Woodbury, *History of the Milling Machine* (Cambridge, Mass.: The M.I.T. Press, 1960), p. 38; and Guy Hubbard, "Development of Machine Tools in New England," *American Machinist* 60 (February 14, 1924):257–258.

French, and other machinists at Springfield during the late 1830s in working out the configurations of a fully evolved milling machine (Figure 15)—a synthesis of principles previously introduced by North and Hall. Once with the Ames Company, he no doubt added a few embellishments of his own. Nevertheless, the earliest plain millers made at the Ames shops were based on patterns loaned by the Springfield armory.[48]

Like the Ames Company, the American Machine Works also specialized in the manufacture of milling and stocking machines. Philos B. Tyler, who owned and operated the firm, relied exclusively on Springfield armory patterns in constructing machinery for Harpers Ferry. Although Tyler did not enter business until 1848, he nonetheless succeeded in drumming up orders by underbidding his neighbor and competitor, James T. Ames. Accordingly, between 1849 and 1854, he furnished four eccentric stock-turning lathes, twelve milling machines, and a number of other production units to the Potomac armory. Still, in terms of output, versatility, and quality of product, the record of the American Machine Works fell far short of that of the Ames Company.[49]

Whether constructed on the premises or purchased from private contractors, nearly every new machine installed at Harpers Ferry after 1839 revealed pronounced New England influences. To many

48. Edward G. Parkhurst, "More Early Milling Machines," *American Machinist* 23 (June 28, 1900):605–606; Martin Lovering, *History of the Town of Holland, Massachusetts* (Rutland, Vt.: By the author, 1915), pp. 604–607; Nathan P. Ames to Moor, January 24, 1845, James T. Ames Papers; James S. Whitney to Ames Manufacturing Company, November 28, 1855, Letters Sent, SAR; Craig to Whitney, November 21, 1855, Letters Sent to Ordnance Officers, OCO; Ripley to Secretary of War, June 11, 1861, Letters Received, SAR.
Since Robbins, Kendall & Lawrence initially equipped their plant with Ames-built machinery in 1845 and made abundant use of Springfield armory patterns thereafter, probably Frederick W. Howe's famous plain milling machine of 1848 was also based on the national armory's designs and specifications. See Robbins, Kendall & Lawrence to Talcott, March 10, 1845, OCO; Ripley to Robbins, Kendall & Lawrence, July 21, 1847, Letters Sent, SAR.
49. Whitney to Ames Manufacturing Company, November 28, 1855, January 9, 1856, to Tyler, January 9, 1856, J. W. Preston to Clowe, June 10, 1856, Letters Sent, SAR; Craig to Clowe, December 1, 4, 14, 1855, to Whitney, December 1, 1855, Letters Sent to Ordnance Officers, OCO; Payrolls and accounts, Harpers Ferry armory, 1849–1852, GAO.

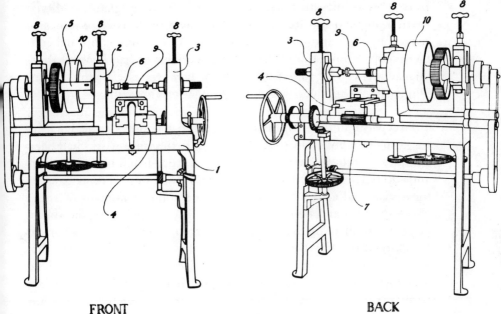

FRONT   BACK

1 BED
2 ADJUSTABLE HEADSTOCK
3 ADJUSTABLE TAILSTOCK
4 CARRAIGE
5 SPINDLE
6 MILL CUTTER
7 RACK & PINION FEED
8 ADJUSTING SCREWS
9 FIXTURE
10 DRIVE PULLEYS

Figure 15. This plain milling machine, designed by Frederick W. Howe and built by Robbins & Lawrence of Windsor, Vermont, reflects the engineering influence of the Springfield armory.

observers, the similarities between the techniques used at Springfield and those adopted at the Ferry were too obvious for comment. As one writer put it, everyone knew "where Mr. Moor obtained his machinery." The same held true for Moor's successor, Burton. Both men traveled extensively in New England, both visited Springfield as often as possible, both made copious notes of what they saw, and both made purchases. Through continual contact, cooperation, and exchange, Harpers Ferry had become thoroughly dependent on Yankee know-how by the 1840s. Technologically speaking, it was a Southern armory in name only.[50]

Except for minor additions, the reconstruction of the Harpers Ferry armory ended in 1854. By that date the waterworks had been completely rebuilt, new workshops erected, machinery installed, and interchangeable manufacture well under way. The new facilities presented a striking contrast with the physical plant of 1841. More spacious and better-lighted brick buildings had replaced earlier jerry-built structures. Neatly situated along paved streets and landscaped terraces, they completely altered the armory's appearance (Figures 16, 17, 18, and 19). Within these shops could be heard the clattering din of machinery. Men who formerly wielded hammers, cold chisels, and files now stood by animated mechanical devices monotonously putting in and taking out work, measuring dimensions with precision gauges, and occasionally making necessary adjustments. Slowly but perceptibly the world of the craftsman had yielded to the world of the machine.

Members of the Ordnance Department were delighted with the results. Although the inspector of arsenals and armories, Colonel Baker, always found room for improvement, he highly approved of the mechanical operations at Harpers Ferry and expressed satisfaction that "the System under which they are conducted is a very excellent one." At the same time, he commended the superintendent, Colonel Huger, and his staff for exercising diligence and efficiency in the discharge of their duties. All things considered,

---

50. Fitch to Burton, January 13, 1883, James H. Burton Papers, Yale University Archives; Symington to Burton, December 17, 1849, November 25, 1850, Huger to Burton, November 8, 1851, Second Auditor's Accounts, GAO.

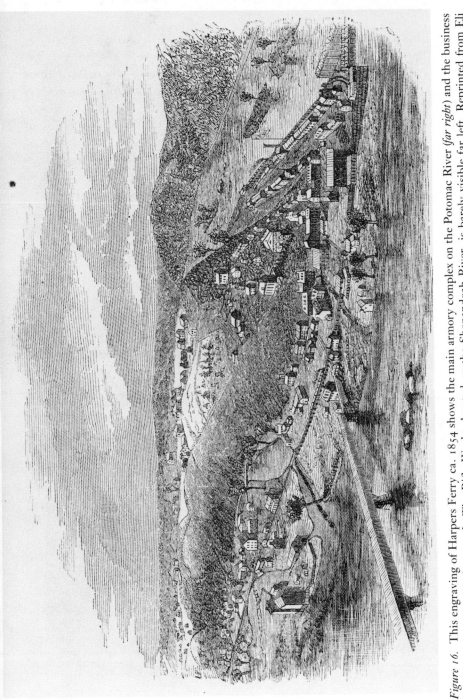

Figure 16. This engraving of Harpers Ferry ca. 1854 shows the main armory complex on the Potomac River (*far right*) and the business district (*right and center foreground*). The Rifle Works, located on the Shenandoah River, is barely visible far left. Reprinted from Eli Bowen, *Rambles in the Path of the Steam Horse* (Philadelphia: Bromwell & Smith, 1855), p. 191.

Figure 17. Lithograph of the U.S. Musket Factory, Harpers Ferry, 1857. Photograph from the National Archives, No. 127-G-523492.

Figure 18. View of Virginius Island on the Shenandoah River at Harpers Ferry, 1857. The U.S. Rifle Works (*far right*) is adjacent to the Shenandoah Canal. Photograph courtesy of the Harpers Ferry National Historical Park, No. 490.

Figure 19. Sketch of the U.S. Rifle Works at Harpers Ferry, ca. 1851, by Lieutenant James G. Benton. From the United States Military Academy, Library Manuscript Collection, West Point, N.Y. Photograph courtesy of the Harpers Ferry National Historical Park, No. 794.

Baker attributed the armory's remarkable resurgence as a progressive manufacturing establishment to the managerial skill of its military supervisors.[51]

Even though the armory had been almost entirely rebuilt and stocked with new and improved machinery, certain questions arose during the early 1850s over what had actually been accomplished. When it came to expenditures, the cost of producing a musket in 1854 showed no appreciable reduction over the cost prior to 1842. This point became a source of bitter political contention when the House committee on military affairs, a body that favored a return to civilian superintendents at the national armories, seized upon it as a means of criticizing the "waste and extravagance" fostered by the Ordnance Department. Colonel Huger responded by pointing out that certain price disparities were bound to exist since the additional amount of labor and raw materials required to manufacture a percussion musket exceeded that required for an old model flintlock by eighty-three cents. Besides, he argued, the real benefit of the new system accrued not from cost efficiencies but from the quality of the arms produced. Specifically, he considered the Model 1842 musket so *"much superior"* to the Model 1816 that no fair comparisons between the two could be made. Unfortunately, however, the quality of product remained an intangible factor that could not be demonstrated statistically to a cost-conscious Congress.[52]

Despite problems of economy, members of the Ordnance Department emphasized that the Harpers Ferry armory had made unprecedented strides under military supervision. One of the most outspoken proponents of the system was Craig, who had been the first military superintendent at Harpers Ferry. Having succeeded Talcott as chief of ordnance in 1851, he repeatedly declared that the change from civilian to military superintendents "produced a great, if not entire reformation of the abuses formerly existing" at the

51. Baker to Craig, July 20, 1852, July 19, 1854, Reports of Inspections of Arsenals and Depots, OCO; Craig to Secretary of War, November 11, 1853, Letters Sent to the Secretary of War, OCO; Major William Bell to Craig, May 18, 1854, Letters Received, OCO.

52. Stearns, pp. 8–9; Huger to Craig, June 18, 1853, Letters Received, OCO. Also see Symington to Talcott, October 23, 1846, OCO.

national armories. In 1852, for instance, he wrote the secretary of war that "my experience in the Affairs of these establishments for many years convinces me, fully and entirely, that since they were placed under the present system, many and important improvements have taken place, which are, in my judgement, due to the System." Nevertheless, he added, "like all other reformations it has met with opposition, and the reformers have had to encounter the ill will and hostility of those, who had profited by the abuses and are interested in restoring the former state of affairs."[53]

When Craig made these remarks, he did so amid renewed debate over the so-called armory question. Actually the civilian-military controversy had never really ended at Harpers Ferry. While supposedly settled by Congressional legislation in 1842, much dissatisfaction and dissension continued throughout the 1840s. Time and again pieceworkers and inspectors complained about having to keep regular hours. Older artisans bemoaned the mechanization of production that lowered their wages and, in some instances, put them out of work. Others grumbled about the rigor with which military superintendents enforced armory regulations. Still others opposed the system out of mere cantankerousness and personal animosity toward their superiors. To make matters worse, the Striders, Beckhams, and other local Whig stalwarts fanned discontent by publicly attacking the "despotism and oppression" of military rule.[54]

Antagonism and ill-feeling became especially visible after a falling-out occurred between Symington and Moor. The two men had never liked each other, but the situation worsened during the spring of 1845 when Moor, hemorrhaging from the lungs and thought to be dying, took a leave of absence from the armory. Assuming that the master armorer's days were numbered, Symington replaced him with Alexander Hitchcock, who had been the

53. Craig to Secretary of War, October 28, 1851, November 2, 1852, November 11, 1853, Letters Sent to the Secretary of War, OCO.

54. Symington to Talcott, June 29, 1846, August 4, 12, 1848, May 12, 1849, to Bedinger, June 29, 1846, to Maynadier, July 12, 1849, to Conrad, May 21, 1851, Statement made by John H. Strider to Henry Bedinger and presented to the Secretary of War, July 26, 1848, S. E. Cohen to "Squire," August 8, 1849, Anon. to Millard Fillmore, October 30, 1850, Letters Received, OCO.

chief workman at the Watervliet arsenal near Troy, New York. He did so, however, unbeknownst to Moor who, after an almost miraculous recovery, complained bitterly about the superintendent's conduct to members of Congress and the secretary of war.[55]

Thereafter relations between the two men steadily deteriorated only to be pushed to the breaking point in July 1849 when Symington summarily dismissed three inspectors—Zadoc Butt, Joseph Ott, and Moor's eldest son, William—for troublemaking and insubordination. Three months later he also asked for and secured the removal of the master armorer himself, using the pretext that "old Moor" had outlived his usefulness. "The antiquated notions of the Master Armorer," Symington wrote the secretary of war, "his want of forethought for the necessities of the work, his dislike to improvements and incapacity to perceive their advantages, and the absence of resource in any unforseen contingency, continually interfered with my plans." Since Symington's remarks were both harsh and, for the most part, grossly unfair, Talcott attempted to make it clear that no one questioned Moor's integrity or conduct as master armorer. Rather, he stated, "The introduction of new machinery and improved processes of manufacturing have changed the duties, and consequently the requisite qualifications for that office, and one who might have been eminently competent formerly, would now, if he had not kept up with the improvements of the time, be altogether unfit. Such was the case with Mr. Moor, he had been so long habituated to a certain routine, to which he was, as it were, bred, that it was next to impossible to introduce any other, under his management." Like the techniques he had helped to replace during the 1840s, Moor had become out-dated.[56]

Moor's dismissal boded ill for the military system at Harpers Ferry because it prompted a number of disgruntled Democrats to

55. Moor to Talcott, May 20, 1845, Symington to Talcott, July 30, 1845, Letters Received, OCO; Talcott to Moor, May 21, 1845, to Symington, May 29, July 31, 1845, Letters Sent, OCO.

56. Symington to Conrad, May 21, 1851, OCO; Talcott to Conrad, March 7, 1851, Letters Sent to the Secretary of War, OCO. Also see Symington to Maynadier, July 12, 1849, to Talcott, October 23, 25, 1849, William H. Moor and Zadoc Butt to Talcott, August 9, 1849, Lieutenant James G. Benton to Maynadier, September 1849, Moor to Talcott, October 31, 1849, Letters Received, OCO.

put aside political differences temporarily and join with Whig adversaries in a concerted effort to restore local civilian control over the armory. Among the most prominent of these was the former paymaster, Richard Parker. As a member of the 31st Congress, Parker not only insisted on the re-appointment of Moor but also demanded the removal of Symington. Still seething about his forced retirement, Moor even entered politics for the sole purpose of destroying the military system. Elected to the Virginia House of Delegates in 1850, he spent two years in Richmond canvassing support and making antimilitaristic speeches warning of a "return to the Savage state" in which republican institutions and individual liberties would wither away under the foot of military dictators. At the same time, his friend and fellow Democrat, Charles James Faulkner of Martinsburg, succeeded Parker as the Valley's representative to Congress in 1851. An astute politician, Faulkner played an extremely important role in deciding the armory question. As chairman of the powerful House committee on military affairs, he deftly coordinated the forces of opposition at Springfield and Harpers Ferry, called for a full-scale investigation of armory management, and spearheaded an attack in Congress that eventually ended the military system.[57]

While the civilian-military controversy had many different ramifications, Colonel Huger, the superintendent at Harpers Ferry, came closest to the truth when he wrote the chief of ordnance that "the question here is between the *INNS & OUTS*, and as a great many outs desire *each* office it would only tend to make Mr. F. [Faulkner] a favorite, to find that he tried to open the door for them."[58] The Faulkner-Moor coalition undoubtedly commanded a large following at Harpers Ferry. However, not everyone supported their efforts to wrest control from the military. For example, Lucas remained aloof from the proceedings and, in so doing, helped split local Democrats into two opposing factions. At the same time, a number of armorers such as Philip Burkart,

57. Parker to Conrad, March 3, 24, 1851, Letters Received, OCO; Clowe to Burton, November 1, 1851, Moor, "Objections to the Military Superintendencies of the National Armories," James H. Burton Papers, Yale University Archives. On Faulkner's activities, see Erskine S. Allin to Burton, January 21, May 26, August 2, 1852, August 5, 1853, Burton Papers.

58. Huger to Craig, June 8, 1853, Letters Received, OCO.

George Mauzy, and A. J. Wright openly opposed a return to civilian superintendents as destructive of good order and progress at the armory. Since one of the most hotly debated issues concerned the legitimacy of James Burton's appointment as Moor's successor as acting master armorer in 1849, he naturally sided with his military benefactors. Burton became so involved in the controversy that Congressman Faulkner accused him of interfering in the presidential election of 1852 and demanded his immediate removal from office.

Apart from events at Harpers Ferry, the most telling clashes between the antimilitary faction and the Ordnance Department occurred in Washington. Beginning in 1851 both sides exchanged accusations, insults, and rebuttals in an acrimonious debate lasting over three years. On several occasions the controversy seemed to have run its course with the Ordnance Department maintaining the advantage. Each time, however, the persistent Faulkner revived the question at opening sessions of Congress until he finally succeeded in attaching a rider to an army appropriations bill of 1854 that provided for the restoration of civilian superintendents at the national armories. Although Secretary of War Jefferson Davis opposed the measure, the exigencies of interest politics prevailed. Beset by Governor John H. Steele of New Hampshire, former superintendent Robb of Springfield, Faulkner, and many Democratic friends who favored the bill, President Franklin Pierce signed it into law on August 5, 1854.[59]

With the demise of the military system the curtain of reform quickly fell at Harpers Ferry. No sooner had Faulkner secured the appointment of Moor's former assistant, Henry W. Clowe, as superintendent, than the armory reverted to the same routines practiced by Stubblefield, Rust, and Lucas years earlier. With "enemies to punish and friends to reward," Clowe dismissed more men in four years than his military predecessors had in thirteen. Most of them possessed skills that the superintendent could ill afford to lose.

59. Symington to Burton, July 5, November 4, 1853, Burton to Major W. W. Mackall, June 21, 1853, to Symington, October 24, 1853, Faulkner to Pierce, n.d., J. W. Preston to Burton, May 17, 1852, James H. Burton Papers, Yale University Archives; Petition presented to the Secretary of War, n.d. [1851], Letters Received, OCO; *Springfield (Massachusetts) Republican*, February 23, 1853; Benet, 2:543–545; U.S. *Statutes at Large*, 10:578.

Nevertheless, as those who survived the purges well knew, politics and its influences controlled the destiny of the armory. According to one machinist, Clowe "used the whole power and patronage of his office to cajole and intimidate the votes of the Armorers and others for the benefit & support of his candidate Mr. Faulkner." Fortunately for him, Burton escaped proscription by accepting a position with the Ames Manufacturing Company before Clowe took office.[60] His departure proved to be a serious loss, as order and discipline lapsed under his well-intentioned but unassertive successor, Samuel Byington. At the same time overemployment and unreasonably high wages were instituted, expenditures exceeded appropriations, favoritism prevailed in letting jobs and contracts, production fell off drastically, and the quality of workmanship steadily deteriorated. By December 1858 the situation at Harpers Ferry had grown so bad that Secretary of War John B. Floyd had no choice but to remove Clowe, a fellow Democrat, from office. Soon afterwards Colonel Ripley, formerly of Springfield, reported that conditions at the armory were "most unsatisfactory."

A much larger force was borne upon the rolls, than was warranted either by the means on hand, or the pressure of work; and I was given to understand that much of this labor was of that expensive and unprofitable class whose merit lies not so much in mechanical skill, as in other qualifications less difficult to acquire.

The mal-administration of the Armory affairs has by no means been confined to carelessness in watching over its financial interests, or in affording in its shops an asylum for ignorant and indifferent work men. The general regulations of the Department governing its operations, have in many ways been violated, and in every instance which has come under my notice, the Government has invariably been the looser.[61]

60. Thomas K. Laley to Craig, August 13, 1857, Letters Received, OCO; Craig to Jefferson Davis, March 10, 1854, Letters Sent to the Secretary of War, OCO; Burton to Major William H. Bell, May 22, 1854, Huger to Burton, May 24, 1854, James H. Burton Papers, Yale University Archives.

Actually Burton remained with the Ames Company only a year. In June 1855 he accepted a five-year appointment as chief engineer at the Enfield armory in England.

61. Ripley to Craig, April 14, 1859, Reports of Inspections of Arsenals and Depots, OCO.

On December 24, 1858, the War Department announced the appointment of Alfred M. Barbour as Clowe's successor. Barbour, a lawyer from nearby Charlestown and scion of one of Virginia's oldest families, faced a severe test. The armorers had reverted to their old habits and political adversaries were creating quite a stir because the new superintendent was neither a businessman nor a practical mechanic. The only reason he received the post, an editorialist in the *Springfield Republican* commented, was that he either stood "in somebody's way for Congress, or could help somebody else [namely Charles J. Faulkner] to stay in Congress."[62] Sarcastic remarks such as this, however, were the least of Barbour's worries. Production costs had skyrocketed since 1854 and output per worker had fallen to its lowest point since 1845. Barbour's task was to lift the armory out of its current state of depression. The need was urgent, and the Ordnance Department pressed for a quick solution.

No sooner had Barbour entered office than Congress, in a surly mood after several stormy debates on sectional issues, cut the armory's operating budget by 38 percent. This action placed the new superintendent in the unenviable position of having to release more than one hundred armorers and institute a 10 percent wage reduction for those who remained. Much to everyone's surprise, he implemented these measures in a firm and nonpartisan manner. "I have endeavoured to eschew politics," he told the Chief of Ordnance. "I consider that the interest of the Armory imperatively demands this course—and that one of the most serious obstacles to its success in the past has been caused by mixing it up with the politics of the country. I have made Mechanical skill and fidelity to duty the sole tests for promotion in the Armory. It is very unfortunate to have politicans meddling with our affairs and I hope they will meet with no countenance with the Department."[63] Barbour's determination to dissociate politics from arms making contrasted sharply with the practices of his civilian predecessors. No superintendent since Thomas Dunn had even attempted such a policy. Yet Barbour managed to thwart tradition without causing serious disaffection

62. *Springfield Republican*, January 17, 1859.
63. Barbour to Craig, May 25, 1859, Letters Received, OCO.

among the armorers and strife within the councils of the local Democratic party.

During the spring and summer of 1859 the new superintendent worked tirelessly to restore order and discipline at the factory. Old work rules were reinstated, stricter time-keeping procedures were introduced, and efforts doubled to improve the quality and precision of manufacturing operations. While none of these measures endeared Barbour to the workers, his spry manner, positive attitude, and penchant for fair treatment did earn their respect. Concern for the improvement of working conditions and little things such as the provision of fresh drinking water in the workshops helped to make regularized labor more tolerable. Even members of the Ordnance Department, usually outspokenly critical of civilian superintendents and their many follies, acknowledged that Barbour had the makings of a first-class manager, the likes of which had not been seen since Roswell Lee. "He deserves much credit for the reforms he has already introduced," one inspector reported in 1859. Although Harpers Ferry still fell short of Springfield's production record, output per worker had begun to rise and unit costs were gradually declining. By October the situation had improved so much that Barbour traveled to New England to visit the Springfield armory and consult with the Ames Manufacturing Company about building a mill for rolling gun barrels at Harpers Ferry. It was during his absence that a small band of abolitionists led by John Brown seized the armory. Unfortunately Brown's raid and the emotional tensions it evoked denied Barbour the opportunity of demonstrating his full potential as an administrator. Technical improvements continued and productivity increased slightly, but the armory never regained the operational efficiency it had enjoyed in 1854.[64]

64. Ripley to Craig, April 14, 1859, May 12, 1860, Reports of Inspections of Arsenals and Depots, OCO; Barbour to Craig, July 30, September 10, 1860, Letters Received, OCO.

CHAPTER 10

The Community in Crisis, 1859–1861

John Brown's raid of 1859 sent a tremor through the entire South, but nowhere was its impact more traumatic than at Harpers Ferry and other settlements along Virginia's northern border. The insurgents had taken the town completely by surprise during the early morning hours of Monday, October 17. At first the residents reacted with shock and disbelief. Then panic set in; terror-stricken families bolted their doors and cowered in fear as rumors circulated that slaves were committing "rapine and murder" in the streets. Gunshots and hysterical cries could be heard from the vicinity of the armory yard while the sexton of the Lutheran church tolled the alarm and messengers scurried to Charlestown, Shepherdstown, and other nearby villages spreading the word: insurrection at Harpers Ferry! "I was fearful of a bloody civil war," one slave owner later recalled; "I was under the impression that, unless they were there in great numbers, they would not be foolish enough to make an attack on the borders of two slaveholding States." Such reasoning convinced nearly everyone that thousands of abolitionists were storming the arsenal and pillaging the Ferry.[1]

Once it became apparent that the dreaded invasion force was no

1. U.S., Congress, Senate, *Report of the Select Committee of the Senate Appointed to Inquire into the Late Invasion and Seizure of the Public Property at Harper's Ferry*, 36th Cong., 1st sess., 1860, Rep. Com, No. 278, pt. 2, p. 17. The background, action, and aftermath of the Brown raid are thoroughly documented by Stephen B. Oates, *To Purge This Land with Blood: A Biography of John Brown* (New York: Harper & Row, 1970); Benjamin Quarles, *Allies for Freedom: Blacks and John Brown* (New York: Oxford University Press, 1974); and Oswald G. Villard, *John Brown, 1800–1859: A Biography of Fifty Years After* (Boston: Houghton Mifflin, 1911).

more than a handful of fanatics, mob psychology gained ascendancy. Shock turned to anger and anger to rage as hundreds of half-crazed citizens encircled the armory brandishing weapons, shouting epithets, and demanding revenge. By Monday forenoon uncontrollable passions had erupted. Drunken vigilantes roamed the streets, tense militiamen fired recklessly at anything that moved, and curious onlookers edged along fences and buildings to catch a glimpse of the desperadoes. Brown "sowed to the wind and reaped the logical harvest," one eyewitness, a minister, reported.[2] To be sure, Brown's band had initiated the attack and drawn first blood. During the fighting they had killed five persons—four whites and one free black—and wounded a number of others. At least two of the dead victims were unarmed. Yet none of these acts was as wantonly brutal as the atrocities committed by the Virginians themselves.

The first raider to fall was Dangerfield Newby, a forty-eight-year-old mulatto who was cut down around 11 A.M. on Monday by a sniper's bullet which entered the lower part of his neck and ripped his throat from ear to ear. Although Newby died almost instantly, rowdies dragged his mangled corpse into a culvert, beat it with sticks, and cut off the ears as souvenirs. As a final gesture of contempt residents allowed swine to root at the body. Other depredations followed. By early Monday afternoon Brown realized that the situation was hopeless. Twice he attempted to open negotiations under a flag of truce only to have one emissary seized and two others gunned down. Later that day the captured messenger, William Thompson, was dragged into the street by a howling mob and shot in cold blood in retaliation for the death of the town's elderly mayor. The executioners then dumped the body into a watery grave and used it for target practice. William Leeman, the youngest raider, met a similar fate when in a fit of panic he attempted to escape by swimming to the Maryland side of the Potomac. He got as far as midstream before a hail of bullets forced him to seek refuge on an exposed rock which jutted out of the river. There he

2. Samuel V. Leech, *The Raid of John Brown at Harper's Ferry as I Saw It* (Washington, D.C.: The Desoto, 1909), p. 19. Leech's reference is to Hosea 8:7, "for they have sown the wind, and they shall reap the whirlwind."

crouched, exhausted and terrified, as a party of avengers swam toward him. Soon after his capture one of the men put a pistol to his head and shot him. Other members of the guerrilla band were taunted, cursed, and spat upon as they lay mortally wounded in the armory yard. Such senseless acts of brutality evinced the unmitigated feeling of outrage and insecurity that pervaded the community.

Tensions remained high during the six-week interval between Brown's capture on October 18 and his execution on December 2. Alfred Barbour first received word of the "Harper's Ferry outrages" while visiting Samuel Colt in Hartford, Connecticut, and rushed back to Virginia full of apprehension and dread. He arrived on October 21, two days after Brown and four other prisoners had been transferred to the county jail at Charlestown. The village looked like an armed camp, the citizens were still anxious and on edge. Rumors persisted that armed bands of fugitive slaves were roaming the mountains and that a large army of Yankee abolitionists was poised on the Pennsylvania border, ready to invade Virginia and free Brown. The national armory was in disarray as troops billeted in the workshops amid strewn tools and broken window panes. Acting superintendent Archibald Kitzmiller, master armorer Benjamin Mills, and master machinist Armistead Ball, all of whom had been held hostage by Brown, were physically exhausted and emotionally drained. Although Kitzmiller "rang the bells at the usual hours" and attempted to resume operations on the morning after Brown's capture, the armorers were unable to work. All of the men had been deeply shaken by the experience and needed time to calm their nerves and collect their thoughts. Adding to the turmoil, Mills, whose resignation as master armorer had already been accepted by the secretary of war, hastened his departure. The task of restoring order and initiating production therefore fell primarily upon Barbour. More than a week elapsed before necessary repairs were made and operations began to get back to normal. Even then a great deal of anxiety existed among the workmen.[3]

3. Barbour to Colonel Henry K. Craig, October 18, 1859, January 3, 1860, Kitzmiller to Craig, October 19, 1859, Letters Received, OCO.

The sensational disclosures that emerged during Brown's trial and subsequent government investigations appalled Southerners. Soon after the capture authorities discovered a carpetbag full of incriminating documents at Brown's secret headquarters in nearby Maryland. In addition to the maps of seven Southern states marked with slave statistics, the cache contained letters addressed to Brown by Frederick Douglass, Samuel Gridley Howe, Gerrit Smith, and many other Northern abolitionists. Portions of this correspondence were obtained and subsequently published by New York and Baltimore newspapers, thereby generating much publicity about the covert nature of the undertaking. The proud and unrepentant Brown increased suspicion and antagonism by refusing to implicate his benefactors or discuss the sources of his support. At the same time sympathetic writers such as Henry David Thoreau extolled Brown's humanitarian principles and hailed the Harpers Ferry raid as "the best news that America has ever heard." Speaking at Boston and Concord, Ralph Waldo Emerson expressed a common sentiment among abolitionists when he characterized Brown as a new American saint whose suffering and death "will make the gallows glorious as the cross." The analogy with Christ and the idea of martyrdom met with Brown's wholehearted approval. He had always felt that God had called him to a special destiny and now it was close to being fulfilled. As he went to the scaffold on December 2, he issued one last portentous statement: "I John Brown am now quite *certain* that the crimes of this *guilty land: will* never be purged *away*; but with Blood."[4]

Brown's eloquence during the trial, his spirited "Vindication of the Invasion," and the ringing eulogies that followed his execution sent a paroxysm of wrath throughout the South. Not to be outdone by abolitionist rhetoric, Southern writers from Richmond to Mobile responded with venomous diatribes branding Brown as a "monomaniac," a "lawless ruffian," a liar, murderer, and thief. Despite a large body of Northern opinion that bitterly castigated the raid and vilified Brown and his "gang of runaway niggars," secession-minded fire-eaters encouraged the notion that all Northerners were

4. Henry David Thoreau, *A Plea for Captain John Brown* (Boston: David R. Godine, 1969), p. 28; James E. Cabot, *A Memoir of Ralph Waldo Emerson*, 2 vols. (Boston: Houghton Mifflin, 1888), 2:597; Villard, p. 554.

"Black Republicans" and that all Republicans were bloodthirsty abolitionists bent on inciting a slave war against the South and destroying its cherished institutions. A joint committee of the Virginia legislature gave widespread currency to this sentiment when, on January 26, 1860, it reported "the existence, in a number of Northern States, of a widespread conspiracy, not merely against Virginia, but against the peace and security of all the Southern States." Even at the highest political levels hyperemotionalism was quickly displacing reason.[5]

As sectional tensions increased, Harpers Ferry remained in a state of perpetual alarm. Warnings that "resolute bands of desperadoes" were planning "to fire the principal towns and cities of Virginia" along with actual acts of incendiarism in the immediate vicinity convinced people that gangs of runaway slaves and Northern infiltrators were lurking in the neighborhood, intent upon gaining revenge. Once again, panic spread through the community, and armed citizens formed night patrols to enforce curfews, question strangers, and picket the town. In addition to requesting a garrison of federal soldiers at the armory, local residents raised four militia companies which, fully armed and uniformed, ceremoniously paraded through the streets on the first Saturday of each month. By January 1860, nearly every able-bodied man at Harpers Ferry was a member of the local guard.[6]

Even though these actions were intended to inspire public confidence and bolster morale, deep-seated fears continued to plague the townspeople throughout 1860 and 1861. Older, established residents, particularly slave owners, shuddered at the thought that John Brown had actually lived undetected in their midst for more than three months. How many other abolitionists or abolitionist sympathizers were residing in the community? Were they plotting another attack? Would the slaves join them this time? Were their families safe? Could Northerners who lived in the town be trusted? No one knew the answers to these questions—indeed,

---

5. Barton H. Wise, *The Life of Henry A. Wise of Virginia, 1806–1876* (New York:Macmillan, 1899), p. 256.

6. Wise, p. 257; Alfred Barbour to Craig, January 3, February 14, April 4, 1860, Letters Received, OCO; J. A. Russell to G. Dwight, July 23, 1861, Letters Received, SAR; Barry, *Strange Story of Harper's Ferry*, p. 67.

no one knew whether such questions were even justified—but the gnawing insecurity and suspicion they generated seriously eroded many personal relationships. Yankees had never been very popular at Harpers Ferry; now they were watched more closely than ever.

The events that had thrust Harpers Ferry into the national limelight contained the seeds of social disintegration and physical destruction. As the nation girded itself for a bitter presidential contest, the Virginia assembly responded to the Brown raid by passing on January 21, 1860, a bill "For the better defence of the State." In an attempt to free the commonwealth from an almost absolute dependence on Northern-controlled arms outlets, the act reactivated the old Virginia Manufactory of Arms, a publicly-owned enterprise closed since 1822. The products of the factory would, in turn, be distributed among a growing number of militia regiments in an effort to discourage further abolitionist disruptions and to protect the state "from unjust National legislation threatening interests vital to the entire South." Of $500,000 appropriated for the purpose, $320,000 was designated for rebuilding the state armory at Richmond and $180,000 for the outright purchase of munitions and heavy ordnance. The act also empowered the executive to appoint a three-man "Armory Commission" to award contracts and oversee both operations. Governor John Letcher quickly selected George W. Randolph, Colonel Francis N. Smith, and Philip St. George Cocke—all prominent Virginians with extensive military experience—to fill the posts. Randolph, "an estimable well informed & capable gentleman" and the most active member of the group, served as informal chairman of the commission until 1862.[7]

The Virginia Manufactory, renamed the Richmond armory in 1861, was the first state-owned arms factory in the United States. Constructed between 1798 and 1808, the plant consisted of a foundry, a boring mill, and an imposing main building which faced northward on the James River & Kanawha Canal. The latter, a two-story masonry structure, was magnificently designed with a large vaulted belfry in the center, a wide arched entrance to an open

7. James H. Burton, "History of the Richmond Armory," 1893, James H. Burton Papers, Yale University Archives.

quadrangle in the interior, and projecting wings at the sides. Together these buildings afforded over 37,000 square feet of floor space and represented one of the largest and most functionally integrated manufacturing establishments of the early nineteenth century. Thirty-eight years of neglect, however, had taken a severe physical toll, and by 1860 the complex stood in almost complete disrepair. Before extensive manufacturing operations could begin, the shops had to be renovated and equipped with new and up-to-date machinery.[8]

Although Randolph, Smith and Cocke began their duties in February 1860, the actual job of outfitting the armory moved slowly. The first item on the agenda concerned personnel. Since the factory was starting up anew and none of the armory commissioners possessed adequate mechanical expertise, a master armorer had to be found to advise them on technical matters and to supervise the works once manufacturing operations got underway. The commissioners first approached Harpers Ferry's talented machinist, Armistead Ball, about taking the job. When Ball, who had just been promoted to master armorer at Harpers Ferry, declined the post, they turned to Salmon Adams of the Springfield armory. Adams began working as a consultant to the commission in May 1860, even though his appointment as master armorer was not made final until the following fall.

Adams' first assignment was to determine the armory's potential production capacity and to prepare a list of tools, fixtures, and machines needed to manufacture 5,000 arms annually. Once this job was completed, the commissioners invited proposals for supplying the equipment from several private machine builders. At first it appeared that the Ames Manufacturing Company would get the lucrative contract. But when the xenophobic Richmond *Enquirer* vehemently protested that such action would "not only involve mortification to our state pride, but very great injustice to the enterprise and skill of our own contractors and workmen," the commissioners quickly reconsidered. After reopening the bidding, they awarded the $156,590 contract on August 23 to J. R. Ander-

8. Ibid.; Giles Cromwell, *The Virginia Manufactory of Arms* (Charlottesville: University Press of Virginia, 1975), pp. 24–30, 150–152.

son & Company, owners of the Tredegar Iron Works in Richmond. Supplementary orders increased the Tredegar contract to $172,364. Later that fall part of the work was subcontracted to the Ames Company and several other New England firms, including Samuel Colt's Patent Fire Arms Manufacturing Company of Hartford, Connecticut. Most of these purchases consisted of stocking and forging machinery.[9]

Although the contract negotiations consumed a great deal of time, preparing the machinery took even longer. Since the Tredegar Works had never equipped a small-arms factory, the owners and state commissioners agreed that much time and expense would be saved if they could avail themselves of the large collection of machine patterns and specifications already on hand at Springfield. In the past the Ordnance Department had followed a very liberal policy in making this information available to private contractors and representatives of foreign governments. But with the approach of the election of 1860 and the worsening of the sectional crisis, Colonel Henry Craig and other Ordnance officials had become extremely chary about opening up the armories to anyone, especially Southerners.

As early as May 14, Randolph had written to the War Department for permission to have a model rifle-musket and some extra components made at the Springfield armory at the commission's expense. Although Secretary of War John Floyd, a Virginian, acceded to the request, the project languished for more than five months. By November J. R. Anderson had engaged James Burton, the former master armorer at Harpers Ferry who had recently returned from the Enfield armory in England, to engineer the Richmond armory contract. Both men were eager to proceed with the machinery. Evidently believing that Floyd's earlier order also sanctioned the Tredegar Works' use of drawings and patterns at Springfield, Burton accompanied Adams to New England as Anderson's representative. Upon their arrival around November 22, superintendent Isaac H. Wright agreed to prepare the model rifle-

9. Randolph to John B. Floyd, May 14, 1860, Letters Received, OCO; Burton, "History of Richmond Armory"; Charles B. Dew, *Ironmaker to the Confederacy: Joseph R. Anderson and the Tredegar Iron Works* (New Haven: Yale University Press, 1966), pp. 53–55, 64–65.

musket but refused to assist Burton in securing the necessary mechanical drawings and patterns without specific authorization from the secretary of war. Another exchange of letters resulted in which Adams and Randolph asked for "free access" to the engineering files and pattern rooms at both national armories. "As this privilege was accorded to the British Government [in 1854]," Randolph wrote Floyd on December 1, "I respectfully ask that it may be granted to the Agents of the State of Virginia." In the meantime, Burton had returned to Richmond, upset and disappointed by the cool reception at Springfield. When Floyd ultimately issued the directive on December 4, the master machinist hurried to Harpers Ferry—a more hospitable environment—and returned several weeks later with a large portfolio of drawings. Even then, the Virginia officials waited with baited breath as "the heated opposition of the republicans" at Springfield delayed further the completion of the model rifle-musket and pattern parts for the Richmond armory.[10]

By the time the Armory Commission completed these arrangements, Abraham Lincoln had been elected president of the United States and a cluster of Southern states stood on the brink of secession. A majority of Virginians, opposing in principle any extremism and favoring compromise, had spurned traditional partisan affiliations and voted for John Bell and Edward Everett of the Constitutional Union party. At Harpers Ferry, a Democratic stronghold since the 1830s, the election outcome was different. There Stephen A. Douglas and the advocates of popular sovereignty won the largest number of ballots, although every other precinct in Jefferson County returned majorities for Bell. The "Southern Rights" ticket of John C. Breckinridge and Joseph Lane ran a distant second in the county with less than 500 votes. Lincoln, the despised "Black Republican," registered no votes at all.[11]

Virginia walked a political tightrope as the nation awaited the inauguration. An aura of dread and suspicion surrounded the new administration, relations with the Federal government were rapidly

10. Ibid.; Benet, *Ordnance Reports*, 3:1–2; Randolph to Floyd, December 1, 1860, February 10, 1861, Adams to Wright, January 26, 1861, Letters Received, OCO.

11. Bushong, *History of Jefferson County*, p. 100.

deteriorating, and the events of each passing day brought increased feelings of tension and uncertainty. The state's political and economic future swung in the balance, and public opinion was seriously divided over what course to pursue. By the time South Carolina passed an ordinance of secession on December 20, a growing number of Virginians felt that the commonwealth should sever its ties with the Union and form a confederacy of Southern states. Another faction, headed by former governor Henry A. Wise, argued for a policy of armed neutrality. Many others counseled restraint, maintaining that Virginia was the eldest of the original thirteen colonies and, further, that considerations of sentiment and self-interest demanded some form of reconciliation between the sections. As the psychology of crisis intensified, a special session of the legislature, meeting on January 14 in Richmond, called for a state convention to settle Virginia's status within the Union once and for all.

The election of delegates to the state convention provoked a spirited contest in Jefferson County. "I am nominated," Alfred Barbour wrote to Congressman Alexander R. Boteler on January 24, 1861, "they are moving heaven and earth to beat me." Barbour, an avowed Unionist, faced stiff opposition. William Lucas, a former Democratic congressman from the Northern Neck and the brother of the late superintendent at Harpers Ferry, and Andrew Hunter, a prominent Whig and prosecuting attorney during John Brown's trial, had joined forces in a determined effort to swing the county into the secessionist camp. Despite rumors that scores of voters intended to abandon the Union party and support the states rights ticket, Barbour stubbornly refused to believe that his party, "the Bell men," would be defeated and conducted a vigorous stump-speaking campaign throughout the county. When the results came in on February 4, the superintendent and his Unionist running mate, Logan Osburn, won a convincing victory over Hunter and Lucas. Barbour only needed now a leave of absence from the armory to attend the convention in Richmond. Eager to see Union sentiments represented at the assembly, his superiors in Washington quickly acceded to the request.[12]

Barbour arrived at Richmond on February 13, just in time for the

12. Bushong, pp. 100–101; Barbour to Boteler, January 24, 1861, Boteler to Joseph Holt, January 26, 1861, Letters Received, OCO.

opening session of the convention. He was in an optimistic frame of mind, very much aware of the historical significance of the occasion and hopeful that the secessionist impulse might be blunted. The distinguished membership and setting of the convention also inspired confidence. A more impressive array of political dignitaries had not assembled at the capital since the constitutional convention of 1829. Most promising of all, a large majority of delegates (120 out of 152) were committed to maintaining peace and keeping the state within the Union.

Once the preliminary fanfare ended and the convention settled down to the knotty problem of deliberating Virginia's future relationship with the Federal government, Barbour's enthusiasm dampened. Initially he had expected that the proceedings would last about two weeks and that he could return to duty at Harpers Ferry no later than March 1. But the convention dragged on, seriously divided over ends and means. When three weeks elapsed and a key report submitted on March 9 by the Committee on Federal Relations failed to gain sufficient support, Barbour began to have serious doubts about ever effecting a compromise. Great pressure was being exerted upon him to change his position and side with the secessionists. Fire-eaters had branded him as a "submissionist," trusted friends and relatives urged him to abandon the Union, and agents from South Carolina, Georgia, and Mississippi worked tirelessly to reinforce these views. By March 21 the endless barrage of pleas and exhortations began to have an effect. Barbour, without any word of explanation, wrote the Ordnance Department requesting the removal of Federal troops stationed at Harpers Ferry. The following day he submitted a letter of resignation to the chief of ordnance. Although Craig delayed acceptance of the resignation "to afford Mr. Barbour time to reconsider and recall it," certain suspicions had been aroused. There was no mistaking a new and unfamiliar note of formality and distant feeling in the superintendent's correspondence. On April 4 Barbour voted one last time against a motion to report an ordinance of secession. Thereafter he moved with the drift of events toward breaking the bonds of Union.[13]

13. Henry T. Shanks, *The Secession Movement in Virginia, 1847–1861* (Richmond, Va.: Garrett & Massie, 1934), pp. 158–178; Wise, pp. 269–273; Barbour to Craig, March 21, 22, 1861, Letters Received, OCO.

The fall of Fort Sumter and President Lincoln's subsequent call for 75,000 volunteers to suppress the "rebellion" precipitated Virginia's withdrawal from the Union. News of the surrender reached Richmond on Saturday, April 13, and the city buzzed with excitement. "Great rejoicing," Burton noted hastily in his diary, "100 guns fired, and [a] torchlight procession." Confederate flags could be seen atop a number of buildings as crowds gathered in the streets to celebrate the occasion. Spontaneous public demonstrations continued throughout the weekend. Secessionist feeling was definitely gaining momentum when the state convention reconvened on Monday, April 15. As soon as reports of Lincoln's decision to use coercion circulated throughout the city the following day, Unionist strength weakened further. Even so, the outcome still seemed far from certain. When William Ballard Preston of Montgomery County introduced a motion for secession on the afternoon of the 16th, the convention adjourned without taking a vote.[14]

One of the most aggressive members of the convention was Henry Wise, governor from 1856 to 1860 and a militant states righter. At the outset he had argued that Virginians should "fight in the Union" by declaring neutrality, seizing Federal installations, and defending their borders against all intruders. Lacking support, however, he eventually abandoned the strategy and opted for outright secession. By April 16, Wise was tired of debate and ready for action. He had already organized a "Spontaneous Southern Rights assembly" to prod the convention but was convinced that Unionist delegates would continue to drag their feet and obstruct passage of the secession ordinance unless confronted with a crisis. Drastic measures were needed, and Wise was willing to take them. That evening he called a secret meeting at the Exchange Hotel to discuss the capture of the Harpers Ferry armory.

When the group assembled around 7 P.M., Alfred Barbour, Turner Ashby, John A. Harman, and Captain John D. Imboden were among those present. Wise outlined his plan, expressing concern that if they did not act quickly Federal troops would reinforce the armory and make the task even more difficult at a later date. All agreed that not a moment should be lost, and a committee

14. Burton, "History of Richmond Armory."

of three immediately left the meeting to see Governor Letcher and enlist his support. When Letcher refused to sanction the scheme, the collaborators decided to act on their own. It was a bold, even treasonous move. Wise, acting as commander, issued a set of written orders and the meeting adjourned around 11 P.M. Soon afterwards Barbour left for Harpers Ferry while Ashby, Harman, and Imboden hastened to Augusta and Fauquier counties to raise troops for the impending invasion.

The next morning Wise went to the convention and, in an impassioned speech during which he drew a pistol and laid it on his desk, announced that 2,000 "volunteers" under Colonel Ashby and General Kenton Harper were pressing toward Harpers Ferry. Why had he initiated the movement? Because, he explained, war was inevitable, Virginia stood in peril, and the governor was vacillating. Unionists sat in stunned silence as Wise harangued the assembly and called for decisive action. Clearly general sentiment had shifted from moderation to secession, and little could be done to restore the balance. Robert Y. Conrad, John B. Baldwin, and several other "cooperationists" vehemently protested against the Harpers Ferry invasion as "unauthorized and illegal," and Robert E. Scott, leader of the moderates, attempted to amend Preston's ordinance by calling for a conference of border states still in the Union. But the cause of compromise was lost. When news arrived that Governor Letcher had succumbed to Wise's pressure and had issued orders to reinforce the troops marching on Harpers Ferry, the way was clear for secession. That afternoon the final vote was taken. Of 144 delegates present, 88 supported the ordinance, 55 opposed it, and one member abstained. The die was cast.[15]

While clerks were counting and recording ballots at the secession convention on April 17, Barbour was on a westbound passenger train somewhere between Richmond and Strasburg, Virginia. After more than two months' absence he was returning home, charged with the responsibility of rallying public support at Harpers Ferry and preparing the community for the seizure of the national armory and the confiscation of its valuable stores of arms,

15. Wise, pp. 274–281; Shanks, pp. 202–205.

munitions, and equipment. It was not a pleasant task, nor were there many assurances that his mission would succeed.

Barbour arrived at Harpers Ferry at 9 A.M. the following morning and immediately proceeded to address a crowd gathered outside the armory gate. The superintendent appeared tired and disheveled. He had been traveling over twenty-four hours, his face was drawn, and there were signs—or at least some people thought they saw signs—that he had been drinking. The atmosphere was tense. A momentary hush fell over the crowd as Barbour announced that the convention had signed an ordinance of secession the previous day and that he supported the resolution. It was a difficult decision, he admitted, but the choice was either the Union or his native state and he loved the latter more. At that very moment, he continued, Virginia troops were marching on Harpers Ferry and before the day ended they would take the town and control the national armory. He urged the audience to accept "the new order of things" and to pledge allegiance to the Old Dominion. If they resisted, he warned, there would be unnecessary fighting and bloodshed.

Barbour's remarks incited a near riot. Although a majority of onlookers applauded the superintendent, others shook their heads in utter disbelief. Only a few months earlier they had voted for him because of his "unconditional" stand on the Union. Now he was advocating secession. "Treason," someone shouted. Amid cheers and angry denunciations a fistfight erupted. No sooner had the melee been broken up than a young man dressed in a Confederate uniform appeared in the street to stand guard at the local telegraph office. This action provoked another outburst of name calling and scuffling. Tumult prevailed as the town reeled with excitement. Workmen streamed out of the armory shops to join in the frenzy. Except for a flourishing barroom trade, all business activity ceased.

In a desperate effort to protect the government property until reinforcements arrived from Washington, Lieutenant Roger Jones, in charge of Federal troops stationed at the armory, attempted to obtain the assistance of the local militia companies at Harpers Ferry. Not surprisingly, his call for help went unanswered since almost all the officers and men in these units supported secession.

By nightfall Jones had fifty regulars and fifteen volunteers to defend the works. With such a meager force at his command, he could only establish sentries at various entrances to the town and make preparations to destroy the armory if help failed to arrive in time. About nine o'clock he received a message that an advance guard of 360 men and four artillery pieces had been sighted about a mile and a half west of the town and that a larger invasion force was rapidly approaching from Charlestown. Convinced that the situation was hopeless, he ordered his men to set fire to the factory and retreat northward toward Chambersburg, Pennsylvania. A loud explosion and the sight of flames and smoke billowing into the sky caused the Virginia troops to quicken their pace. By the time they reached Harpers Ferry a carpenters shop, the main arsenal, and some 15,000 arms were lost, though citizens, fearful that the entire town would be destroyed, had acted quickly to extinguish the fires in all the other armory buildings. At midnight the main contingent under Colonel Ashby marched unmolested into Harpers Ferry. The town had been taken without firing a shot.[16]

The April 18 seizure of the Harpers Ferry armory placed nearly two complete sets of machinery and tons of valuable stock at the disposal of the Confederate government. The Virginia militia also confiscated 4,287 finished firearms and enough components to assemble between 7,000 and 10,000 weapons of the latest design. For a short time after the raid the Rifle Works on lower Hall Island continued in operation while workmen at the larger Musket Factory disassembled and packed over 300 machines, thousands of feet of belting and shafting, and 57,000 assorted tools for shipment to the Richmond armory. Here the most vexing problems occurred in transporting the spoils to Richmond. Since the Baltimore & Ohio Railroad ran through enemy territory, the only accessible route southward was a spur line which terminated at Winchester, Virginia. There hundreds of crates had to be transferred to wagons, teamed over twenty miles to Strasburg, and then reloaded on the

16. Russell to Dwight, July 23, 1861, SAR; Anon. to Simon Cameron, May 3, 1861, Letters Received, OCO; Barry, pp. 96–99; Claud E. Fuller and Richard D. Steuart, *Firearms of the Confederacy* (Huntington, W.Va.: Standard Publications, 1944), pp. 25–26.

Manassas Gap Railroad for the final leg of the journey to Richmond. This was not only expensive but also time consuming. As a result, the task of stripping the Rifle Works did not begin until the first week of June. By then Colonel Joseph E. Johnston, the Confederate commander at Harpers Ferry, expressed concern that the Union army would retake the town before the equipment could be removed. In a last great burst of activity, crews working under the direction of Burton, the newly appointed superintendent of the Richmond armory, completed the job in less than two weeks. When the last consignment left for Richmond on June 13, well over $300,000 worth of equipment (including 132 additional machines and thousands of spare parts and tools from the Rifle Works) had been evacuated. So complete was the inventory that Burton recommended canceling the Richmond armory's extensive machinery contract with J. R. Anderson & Company and filling all subsequent needs by special arrangement. This, in turn, freed the Tredegar Works to fill other pressing orders for the Confederacy.[17]

During the early morning hours of June 14, Johnston's troops blew up the Baltimore & Ohio Railroad trestle at Harpers Ferry, burned the main armory buildings, and made a hasty withdrawal towards Winchester. Exactly two weeks later Confederate raiders returned to destroy the Rifle Works and a wagon bridge over the Shenandoah river. Both actions, along with the threat of further attacks, triggered a series of evacuations which sent hundreds of residents scurrying to safer surroundings. Many of these refugees were unemployed armorers who left Harpers Ferry and never returned—a severe economic loss to the community. Their destinations varied. While a handful of Union sympathizers found positions at Springfield and other Northern factories, a majority of mechanics migrated south and went to work at the Richmond armory. There the superintendent as well as four out of five shop foremen were former Harpers Ferry employees. Still others went on to the Fayetteville arsenal in North Carolina where the Confederate government eventually installed the machinery taken from the

17. Burton, "History of Richmond Armory"; Inventory of Musket [and Rifle] Machinery Taken at Harpers Ferry, *Messages from the Governor of Virginia*, Doc. No. 25 (Richmond: William Ritchie, 1861), 5:103–137; Dew, p. 124.

Figure 20. The main armory at Harpers Ferry, ca. August 1862. In the background are Loudoun Heights (*upper right*) and Maryland Heights (*upper left*). Photograph courtesy of the Smithsonian Institution.

Rifle Works. Richmond and Fayetteville formed the backbone of Confederate ordnance manufacture during the Civil War. Neither factory could have played this role without the rich human and technical resources of the Harpers Ferry armory.[18]

Four years of warfare exacted a devastating toll at Harpers Ferry. Between April 1861 and General Robert E. Lee's surrender at Appomattox court house, the town shifted from Confederate to Union control and back again at least eleven times. Equally indica- tive of the settlement's strategic importance and vulnerability was that the Baltimore & Ohio bridge which spanned the Potomac at the Ferry was destroyed and reconstructed nine times during the same period. By 1865 the local landscape presented a dismal sight. What had once epitomized Thomas Jefferson's pastoral ideal had become a garden despoiled, a victim of the powerful technologies it had helped to create. It seemed as if some supernatural force had swept over the gorge with fire. Houses, churches, and shops had been burned and looted. The local population had fallen sharply. Facing bankruptcy and impoverishment, many businessmen had boarded up and abandoned their stores to seek a new life elsewhere. The national armory, once the mainstay of the town's economy, stood in ruins (Figure 20). By an ironic twist of fate one of the few government buildings left intact was the little enginehouse where John Brown had been captured in 1859. To some people it was a monument to righteous aspirations, to others a grim symbol of hatred and fear; to everyone it served as a haunting reminder that Harpers Ferry had passed its prime. Lost fortunes, broken tradi- tions, and thwarted hopes—this was the legacy wrought by civil war.

18. Burton, "History of Richmond Armory"; Barry, pp. 105–106.

Cultural Conditions and Technological Change: In Retrospect

From its establishment in 1798 to its fiery destruction in 1861, the armory at Harpers Ferry remained a chronic trouble spot in the government's arsenal program. Tradition-bound and often recalcitrant, the civilian managers and labor force accommodated themselves to industrial civilization most uneasily. The situation at the Springfield armory presented a marked contrast. There factory masters as well as mechanics seemed to embrace the new technology without any of the hesitancy or trepidation of their contemporaries in Virginia.

Compared with Springfield, the main branch of the armory at Harpers Ferry never matured into the progressive, trend-setting institution originally envisioned by Benjamin Stoddert in 1799. For more than forty-five years the works had on its rosters over two hundred armorers at much higher wages than those normally paid in New England. Generally more funds were appropriated to Harpers Ferry, yet fewer arms were produced at unit costs 9 to 22 percent greater than those at Springfield.[1] Some of these disparities arose from the high price of labor and raw materials along the inner Potomac Valley. Part of the difference is also attributable to the products made at the two armories. While Springfield concentrated

1. The 9 to 22 percent figures are rough averages based on adjusted estimates of annual production costs at the two national armories between 1810 and 1854. These estimates include adjustments for interest, depreciation, and insurance. The averages also vary over selected periods of time. Adjusted cost estimates and production statistics for the Harpers Ferry armory are presented in Table 1. For the Springfield armory, see Deyrup, *Arms Makers*, pp. 229–231.

primarily on the manufacture of smooth-bore muskets and musket appendages, Harpers Ferry produced a more diversified aggregate of muskets, rifles, pistols, and specially executed ordnance equipment. In addition to higher manufacturing outlays, other charges such as repeated repairs necessitated by flood damage and lengthy closures brought about by frequent epidemics entered the cost accounting equation. Given these circumstances, it is remarkable that Harpers Ferry's production costs did not exceed those at Springfield by more than they actually did. In terms of economic efficiency, therefore, the former's essentially labor-intensive system remained reasonably competitive with the latter's increasingly mechanized operations through the mid-1830s, a much longer period than might be expected.

Initially new machining processes induced only marginal savings at Springfield.[2] In the long run, of course, these savings steadily accrued and played an important part in persuading other arms makers to adopt labor-saving machinery. But during the developmental period between 1815 and 1845, the economic wisdom of investing in such equipment was by no means evident to most manufacturers. General acknowledgment of the cost efficiencies resulting from the increased use of machinery did not become widespread until the middle of the century, and then only after many design problems and organizational bottlenecks associated with integrating men with machines had been resolved at a few federally owned or subsidized establishments. This was indeed one of the most enduring contributions made by government to industry during the antebellum period. Under the auspices of the Ordnance Department, successive administrations encouraged, supported, and rewarded arms makers in developing a large stable of ingenious woodworking and metalworking machinery. These devices formed the keystone of modern interchangeable manufacture, one of America's greatest contributions to world technology in the nineteenth century. Once the economic advantages of mechanized production over traditional labor-intensive methods became manifest, the know-how developed by arms makers found numerous applications

2. Cf. Deyrup, *Arms Makers*, pp. 49–52, 131–132, 229–230; and Uselding, "Technical Progress at the Springfield Armory," pp. 303–307.

elsewhere in the economy. A direct offshoot was the rise of commercial machine-tool building, an industry that appeared around 1850 and played a pivotal role in disseminating the new technology to other metalworking businesses, most notably manufacturers of sewing machines, hardware, agricultural machinery, and railway equipment.

As much as questions of economy enter into discussions of the long-term growth of the arms trade, they do not sufficiently explain why mechanization first appeared in the industry when and where it did. In fact, purely economic considerations had little to do with the initial development and application of machinery in arms making. Admittedly, a certain type of demand did exist, but it was decidedly noneconomic. The primary stimulus here came from the Army, particularly its desire for more uniform and precisely made weapons whose components could be exchanged in the field for new ones whenever the need arose. Until the mid-1840s the military's foremost concern was not economy but whether the Ordnance Department, through its supervision of the national armories and private contracting system, could persuade manufacturers to depart from customary practices, mechanize their operations, and deliver arms of consistently higher quality. Time and again Ordnance officers emphasized these points in their daily correspondence and reiterated them in private conversation. While they attempted to justify this strategy by appealing to the potential economic benefits that would flow from the adoption of mechanized techniques, the argument was primarily a rhetorical device to woo Congress and gain financial support. Only after John H. Hall, Simeon North, and several other government-supported innovators had demonstrated the efficacy of machine-made firearms with interchangeable parts did others generally begin to contemplate the savings that might arise from the widespread application of precision methods.

It is no coincidence that the earliest arms makers to mechanize their operations—North, Nathan Starr, R. and J. D. Johnson, Asa Waters, and others—did so while engaged on Ordnance Department contracts. Because of the considerable investment involved, even these pioneering firms moved toward mechanization haltingly and, then, mainly because of pressures put upon them by the chief of ordnance.[3] A good example is Thomas Blanchard's patented gunstocking machinery, one of the most versatile and important series of inventions to originate in the antebellum firearms industry. Like most innovations, Blanchard's methods contained flaws and produced only marginal savings during the first few years of use. Royalty fees and construction costs made their adoption expensive—so expensive, in fact, that Waters, who first tested Blanchard's eccentric lathe in 1819, could not afford to install one until 1825 or 1826. Yet, even at the most primitive stage of development, Blanchard's machines produced much more uniform work than could be accomplished by artisans with hand tools. Consequently, arms makers, no matter how hard-pressed they were for funds, almost had to install Blanchard's equipment if they wished to meet Ordnance Department specifications and continue on government contracts. Not until the 1840s, with the rise of a number of new, more heavily capitalized companies such as Colt, Remington, and Robbins & Lawrence, did highly mechanized operations become commonplace among the largest arms producers. By then almost all the elements of the new technology—machine movements as well as gauging devices—had reached a level of refinement sufficient to allow their users some leeway in calculating the economic niceties of high-volume production and the profitability of nonmilitary markets.

Throughout the period from 1815 to 1861 the Harpers Ferry armory maintained an aloofness from the forces shaping industrial civilization. While the armory's work force never completely repudiated the new technology, they did not eagerly embrace it either. Instead they vacillated between the old and the new, unwilling to break bonds that tied them to the past yet unable to dissociate themselves from the gathering momentum of technical creativity. There were, to be sure, those at the armory who endorsed change and whose thoughts and actions served to further the new technology. But they constituted a small minority who, like John Hall,

3. See, for example, Nathan Starr to John H. Eaton, July 10, 1829, OCO; Statement of Simeon North, April 1, 1852, Letters Received, OCO.

Benjamin Moor, and James Burton, stood apart from the rest of the community and were frequently belittled for their efforts.

Disenchantment with the new and unfamiliar found many emotional outlets at Harpers Ferry. While the armory experienced a certain amount of physical violence and sabotage during its sixty-three-year history, most workers confined the expression of their feelings to vituperative language and obstinacy on the job. Not surprisingly, they directed their most scornful criticism not at the machines themselves, but at persons who encouraged their use. Over the years no individual or group became more closely identified with this issue and aroused greater ire than the Ordnance Department. Ever since assuming supervision of the national armories in 1815, the chief of ordnance and his staff had unflaggingly promoted the uniformity system and its adjunct, mechanization. The realization of this goal demanded a greater degree of discipline, organization, specialization, and—most odious of all—central control from Washington than most armorers were willing to accept. Such things, in the minds of many Harpers Ferryians, not only smacked of military despotism but also struck at freedoms that a majority of workers had come to consider proprietary rights.

This controversy dated back to the 1790s when George Washington, Tobias Lear, and their mercantile friends had insisted on building an arsenal at Harpers Ferry. Since Washington's commitment was prompted primarily by regional considerations—particularly his desire to make the new Federal City the economic hub of America—his decision to proceed with the project at all costs encouraged the belief that the arsenal existed more to serve local needs than the national interest. In addition to enhancing land values and promoting economic growth along the Potomac Valley, the armory was envisioned by residents of the area as a fountainhead of lucrative contracts, jobs, and other politically related emoluments. While this view clashed with what federal officials considered the armory's primary function—the well-regulated production of military firearms—little could be done to alter the situation because the War Department possessed neither the staff nor the leadership to exercise firm control over armory affairs until after the War of 1812. In the meantime, the armorers, left on their own for the most part, developed a spirit of autonomy that strengthened their determination to keep the works under local control and hardened their attitudes toward outside interference. Consequently when the Ordnance Department sought to redirect the armory toward more nationally oriented goals after 1814, it encountered determined opposition not only from the artisans whose working habits they wished to alter, but also from a majority of townspeople who identified their well-being with that of several families who literally owned the town of Harpers Ferry and, through strong political influence, controlled all key offices at the armory. This, in turn, hampered innovation at the factory because neither James Stubblefield nor his civilian successors exhibited much interest in introducing novel techniques if as a result wages became lower or the labor force under their command was reduced. To have done so would have jeopardized the very basis of their local power and support. The bloody fate of Thomas Dunn—the one superintendent who had attempted to cooperate with the Ordnance Department, reorganize the factory, and update its methods of production—served as a blunt reminder to anyone who contemplated changing the status quo. Even John Hall, whose mechanical contributions constituted the most important innovation at the works, did not escape the intimidating force of local opinion.

The transition from craft to machine followed a very long and circuitous path at Harpers Ferry. Except for the thirteen-year period from 1841 to 1854, when Ordnance officers replaced civilian superintendents, a coalition of workers, businessmen, and local politicians—all united in one way or another by bonds of friendship, blood relations, and mutual interest—dominated every aspect of armory affairs. To a large extent the primitive technology that prevailed at Harpers Ferry reflected their parochial concerns. Not until the 1840s, when Benjamin Moor and a younger group of mechanics successfully retooled the armory for interchangeable manufacture, did the basic stages of production begin to assume a modern character. For a time it appeared as if the armory would close the technological gap that for thirty years had separated it from Springfield and the highly proficient New England arms industry. Yet, even then proponents of the new system encountered

resistance, not only from workers who complained of their sub-
jugation to the discipline of the machine but also from other members of the displaced coalition who, through the unceasing efforts of
their congressional representatives, succeeded in regaining control
over the works in 1854. From that time until Alfred Barbour's
appointment as superintendent in December 1858, the armory reverted to the traditions that had more or less characterized its operations before 1841. Barbour improved the situation, but John
Brown's raid and the resulting sectional crisis soon diverted his
attention to the larger issue of "irrepressible conflict."

Few societies relish change. It taxes customs, unsettles social
relations, intensifies anxieties, and upsets the general rhythm of life
in a community. Change is even less welcome when it challenges
people who are securely rooted in a familiar environment and
satisfied with the way things are. So it was with Harpers Ferry
between 1815 and 1861. Even though the town had its slaves,
indigents, and other blatant socioeconomic inequalities, these shortcomings were more than offset by the freedom, contentment, and
stability which the community's white majority enjoyed.

If one factor conditioned Harpers Ferry's attitude toward
change, it was lack of contact with the outside world. Unlike
Springfield which was part of a rapidly growing industrial region,
Harpers Ferry remained secluded in a sparsely populated agrarian
hinterland. The nearest settlement of any size was Frederick, Maryland, a farm community some twenty miles away. Hagerstown, to
the north, and Washington, to the southeast, were farther away,
and neither possessed significant industry. Not until 1834, when
the Baltimore & Ohio railroad reached the junction of the Potomac
and Shenandoah rivers, was Harpers Ferry placed within reasonable access of Baltimore and the rising manufacturing centers of the
East. This cloistered condition fostered a cultural milieu much
more akin to the elemental and primitive folkways of the back-
country than to the increasingly specialized urban life of the East.

As part of a larger rural culture, the armorers and townspeople of
Harpers Ferry were deeply influenced by the intimacy and sense of
involvement that flowed from a relatively unfettered, insular existence. Like so many other Southerners, they found it extremely

difficult "to conceptualize social unity in terms other than personal relationships."[4] Day in and day out, friends, family duties, and other narrow localistic concerns absorbed their foremost thoughts and interests. In this scheme of things, larger societal goals acquired little meaning. Since local allegiances placed a premium on the individualistic components of everyday life, attempts to organize society for more rigorous and impersonal ends invariably met with derision and scorn. At Harpers Ferry existed doubtlessly a sense of community, but contrasted with the expansive "go-ahead" vision of Springfield and other industrial settlements, it was myopic in quality—at once restricted in scope, static in attitude, and provincial in its processes.

The predominance of personal over institutional arrangements in the social spectrum at Harpers Ferry can be seen in religion and education. For years the town lacked churches and schools, two agencies of acculturation which in communities such as Springfield had served to define common values and goals, reconcile them with national ends, and instill a conviction that such norms were both necessary and good. Through formal organization and an expansive network of preachers and teachers, both church and school brought a broader vision of American society to people who otherwise would have remained untouched by events outside their local environments. Both idealized political, economic, and social order. By emphasizing the harmony of godliness, progress, and democracy and inveighing against strikes, idleness, and other forms of deviant behavior, they implanted a sense of discipline, conformity, and adaptability among populations which helped ease the transition to an urban-industrial age. In an era of acceleration, neither the power of the pulpit nor the molding influence of education was lost to the architects of progress who wished to dictate right conduct and discourage dissent while erecting the new order.[5]

4. C. Vann Woodward, "The Southern Ethic in a Puritan World," *William and Mary Quarterly* 25 (1968):347–348.

5. The relationship between social control and industrial progress has been largely overlooked by business and technological historians. Here the author has been particularly influenced by the writings of Donald G. Mathews, "The Second Great Awakening as an Organizing Process, 1780–1830: An Hypothesis," *American Quarterly* 31 (1969):23–43; Hugo A. Meier, "Technology and Democracy, 1800–1860," *Mississippi Valley Historical Review* 43 (1957):618–640; Clifford S. Grif-

While other communities were becoming acclimated to social control and consolidation, Harpers Ferry remained isolated from the forces of change. For a long time after 1798 the settlement was little more than a missionary outpost, visited infrequently by evangelists and even less often by itinerant tutors. Although the armory set aside facilities in 1810 for ecumenical worship, a formal ministry was not established at Harpers Ferry until 1825. This was the Methodist-Episcopal church, an amalgam of Wesleyanism and the old Anglican establishment in Virginia. Between 1830 and 1851, five other denominations—Catholic (1830), Presbyterian (1841), Methodist Protestant (1843), Lutheran (1850), and Episcopalian (1851)—erected churches in the town. Owing to fluctuations in membership, however, all faced chronic financial difficulties and had to struggle to survive. This collective institutional weakness perhaps explains why the temperance movement, a powerful force for moral rectitude and discipline in other communities, failed to gain a foothold at Harpers Ferry until the late 1840s. Even then, the Sons of Temperance, whose most active organizer and vigorous proponent was Major Symington, did not command a sizable following. By and large the townspeople remained unaffiliated with specific religious sects and reform movements during the antebellum period. Whatever moral instruction their children received was imparted primarily at home.[6]

Similar trends could be seen in education. For a short time after

---

fin, "Religious Benevolence as Social Control, 1815–1860," *Mississippi Valley Historical Review* 44 (1957):423–444; and Lois W. Banner, "Religious Benevolence as Social Control: A Critique of an Interpretation," *Journal of American History* 60 (1973):23–41. While varying in emphasis, John Higham's "From Boundlessness to Consolidation: The Transformation of American Culture, 1848–1860," *William L. Clements Library* (1969; Bobbs-Merrill Reprint Series in American History, H-414), highlights powerful trends toward consolidation during the antebellum period.

6. Colonel George Talcott to William L. Marcy, February 16, 1849, Letters Sent to the Secretary of War, OCO; Philip R. Smith, Jr., "History of the Methodist Episcopal Church, 1818–1868, and the Free Church" (Research paper, HFNP, 1958); Smith, "St. Peter's Roman Catholic Church" (Research paper, HFNP, 1959); Smith, "The Methodist Protestant Church and Odd Fellows Hall" (Research paper, HFNP, 1958); Smith, "History of the Evangelical Lutheran Christ Church, Camp Hill, 1850–1868" (Research paper, HFNP, 1959); Smith, "Protestant Episcopal Church" (Research paper, HFNP, 1959); Charles W. Snell, "The Presbyterian Church" (Research paper, HFNP, 1959).

the War of 1812, the armory conducted a totally inadequate school in which the children of workmen were taught how to raise plants and shoot firearms. Thereafter sporadic attempts to establish bonafide grammar schools began in controversy and ended in failure. The most notable effort to improve elementary instruction occurred in 1822 when Hall and a group of concerned parents organized a Lancastrian school at Harpers Ferry; but after a harried fifteen-year existence, it too floundered for lack of public support. Since the town's leading families sent their children either to local academies or to boarding schools in the East, they saw little need for public schooling. Their indifference was, in turn, mirrored by the rest of the community. This meant that, until the first permanent free school system was established in the 1850s, a child's educational development depended largely on parental initiative. Such a meager cultural endowment no doubt strengthened the corporate role of the family, but it did little to soften pervasive attitudes which viewed strangers as well as strange ideas with great suspicion.[7]

Outsiders such as George Bomford could never quite fathom why the people of Harpers Ferry seemed so well satisfied with their condition. Obviously a few privileged families controlled the economic, political, and social life of the community. Clearly a great majority of townspeople willingly acquiesced to their paternalistic leadership. Yet, what Bomford and others failed to perceive, much less appreciate, was that the blade of paternalism cut two ways. To be sure, Harpers Ferryians showed a measure of deference to their social superiors, but their sense of dependency was not absolute. Indeed, they frequently spoke their minds on issues. If something displeased the populace, the ruling families could count on being told about it, and often in a boisterous and ill-mannered way. Even the lowliest armory workers could and often did voice their opinions on subjects that in other communities would have been considered ill-suited and out of place. And well they could, because the

7. "Description of Harpers ferry in the state of Va.," June 24, 1821, typescript, West Virginia Collection, West Virginia University Library; Hall to Bomford, November 1, 1822, Letters Received, OCO; Edward Lucas, Jr., to Talcott, September 7, 1849, Second Auditor's Accounts, GAO.

local elite knew that their power and well-being depended on maintaining the trust and support of less-distinguished neighbors. Since much of their local power emanated from controlling key armory appointments and since this, in turn, rested with their ability to exercise political influence, for them the votes of the armorers (who constituted a majority of the town's free white electorate) assumed critical importance. This is why civilian superintendents rarely flaunted their authority or did anything that seriously jeopardized their relationship with the workmen. Like good politicians, they catered to the interests of their constituents while enjoying the power, prestige, and emoluments of office.

If political exigencies tended to blur class distinctions and induce a spirit of like-mindedness among Harpers Ferry's white majority, the institution of slavery further strengthened their attitude of cohesiveness. No one knows exactly how much the existence of slavery held white society together, but the "relative completeness" of black bondage doubtlessly served as the "utmost affirmation" of white freedom.[8] The preservation of this condition loomed ever present to the white population of Harpers Ferry.

Numerous writers have noted slavery's effect on the meaning of work in the antebellum South, pointing out that the "peculiar institution" degraded physical labor and enshrined the leisure-laziness myth as part of the Southern way of life. To a degree this syndrome existed among the workmen at Harpers Ferry, but far more important was the extent to which black servitude made them status-conscious. Only a handful of slaves were employed at the national armory, and then only at menial tasks such as shop sweeping. Yet, because slave holding heightened one's social standing in the community, the very possession of slaves by few armorers made them extremely jealous of their rights and sensitive about the honor and dignity accorded their jobs. Any organizational or technical change that even slightly threatened to undermine their status was therefore firmly resisted. For instance, the most frequently repeated protests during the "clock strike" of 1842 accused the Ordnance Department of subverting the workers' freedoms and

8. Earl E. Thorpe treats this theme in *Eros and Freedom in Southern Life and Thought* (Durham, N.C.: Seeman Printery, 1967), pp. 10, 64.

making them "slaves of machines." In a patriarchal society both terms conveyed deeply felt meaning.

Many cultural strains fortified the "cake of custom" at Harpers Ferry. Not least of these was the lingering influence of craft traditions. For more than forty years after 1798 craft-trained artisans, through seniority and a continued monopoly of supervisory posts, exerted a powerful influence on their co-workers as well as on armory affairs. All had served long apprenticeships in the gun trade and were highly skilled in a wide variety of tasks. The emphasis the "mysteries" of gunsmithing placed on manual dexterity made their methods more artistic than mechanical, more individualistic than organized. To many, the possession of skill represented something other than a means of earning a living. It was a calling, a way of life. Above all, the craft ethos instilled a sense of workmanship which served as a source of deep personal pride, creativity, and satisfaction. Because the element of individuality loomed so large in their lives, they proved to be a fiercely independent breed who balked at any attempt to regulate, systematize, or depersonalize accustomed work procedures at the armory.

By a curious twist of fate tradition moved the Springfield armory in a different direction. There, owing primarily to a scarcity of trained gunsmiths, the early labor force had come from diversified backgrounds. Many first employees were unskilled recruits who came to the factory from farms and villages that dotted the countryside of western Massachusetts. The need for specialization, therefore, existed from the very beginning and since few if any workers felt threatened by the adoption of machinery and divisions of labor, the armory quickly became involved in the development and assimilation of the new technology.

Springfield's dynamism and devotion to progress issued from a number of sources. An enviable location, excellent transportation facilities, willing workers, and extraordinarily gifted managers provided advantages enjoyed by few other antebellum communities outside New England. Behind these resources stood a carefully marshaled heritage which, steeped in religion and accentuated by social control, extended back to the mid-seventeenth century. While the underpinnings of Congregational orthodoxy had been

334

diluted by continuous infusions of new people and secular ideas, many distinct shades of the Puritan ethic could still be seen in widely shared norms that encouraged industriousness, sanctioned discipline, and viewed change as a positive force for social betterment. Such deeply internalized values provided excellent incentives at Springfield, and invariably visitors noted them while commenting upon the community's vitality and flare for innovation.

Early on there were signs that Harpers Ferry had little in common with Springfield. By 1818 attempts at cooperation had become tainted by rivalries, petty jealousies, and fears. Springfield's concern with getting ahead, its commitment to the uniformity system, and the well-meaning but unmistakeable one-upmanship of Roswell Lee irked many Harpers Ferryians. Moreover, Ordnance reports that repeatedly acknowledged Springfield's technical superiority over Harpers Ferry and the increasing presence of New Englanders sent to update and instruct their more backward Virginia contemporaries bred further resentment. By 1825 if not earlier Springfielders and Yankees in general had acquired an unpopular reputation at the Potomac works.

Yankeeism meant many things to the residents of Harpers Ferry. On the one hand, the term had broad sectional connotations which became associated with Northern radicalism, especially antislavery agitation. More narrowly construed, the word epitomized the restless ambition, aggressiveness, and sharp practices of outsiders like John Hall. Equally significant was that it symbolized New England's compulsive fascination with change, specialization, and centralization of power. To people who believed that men were more important than institutions and held that, instead of trying to alter conditions, one should find his place in society and live according to its conventions, Yankeeism assumed repressive qualities which were viewed as a threat to white freedom and individuality. That Harpers Ferry's citizenry succeeded in stunting the growth of change and consolidation for more than forty years is a measure of their attachment to a preindustrial way of life. In the end, the stamina of local culture is paramount in explaining why the Harpers Ferry armory never really flourished as a center of technological innovation.

Map and Tables

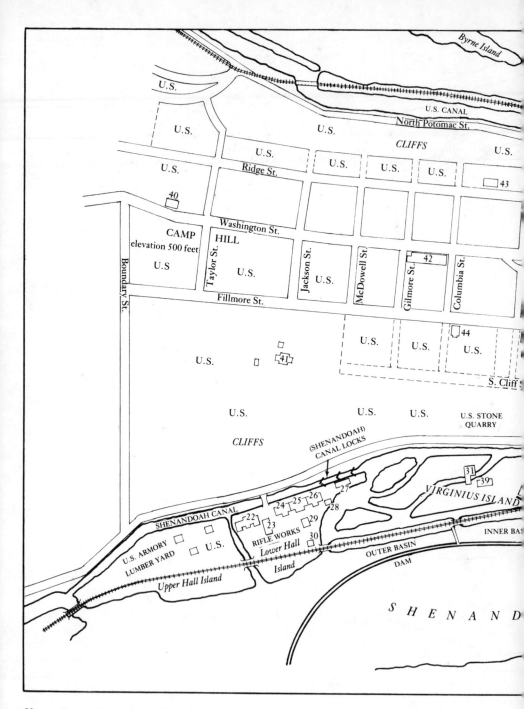

Harpers Ferry, 1859. Adapted from 1859 base map, Harpers Ferry National Historical Park.

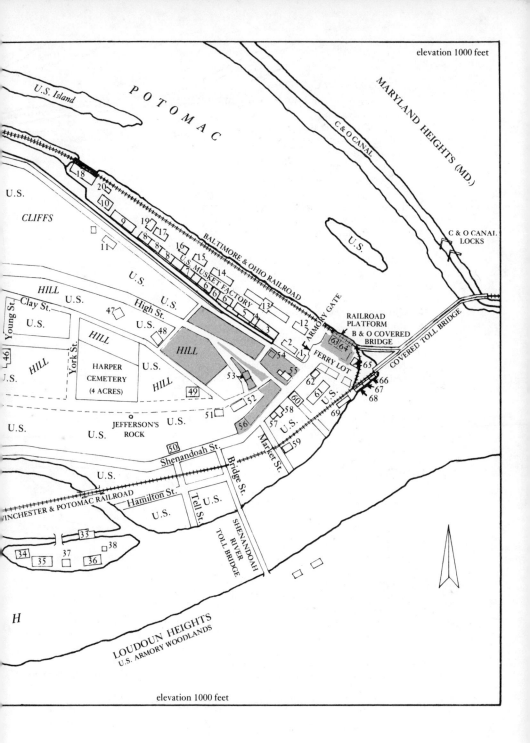

elevation 1000 feet

POTOMAC

U.S. Island

MARYLAND HEIGHTS (MD.)

C & O CANAL

U.S.

CLIFFS

C & O CANAL LOCKS

18

20

10

9

19

17

8

8

8

8

16

15

14

BALTIMORE & OHIO RAILROAD

U.S. MUSKET FACTORY

11

7

6

6

6

5

4

13

3

12

ARMORY GATE

RAILROAD PLATFORM

B & O COVERED BRIDGE

63/64

COVERED TOLL BRIDGE

HILL

Young St.

Clay St.

U.S.

U.S.

U.S.

High St.

U.S.

47

48

U.S.

HILL

York St.

HILL

HILL

HILL

U.S.

HARPER CEMETERY (4 ACRES)

HILL

U.S.

49

53

52

54

55

2

FERRY LOT

65

66

67

68

62

61

U.S.

51

50

JEFFERSON'S ROCK

U.S.

U.S.

U.S.

56

57

58

60

69

U.S.

46

U.S.

Shenandoah St.

Bridge St.

Market St.

59

U.S.

WINCHESTER & POTOMAC RAILROAD

Hamilton St.

Tell St.

U.S.

U.S.

33

34

35

37

36

38

SHENANDOAH RIVER TOLL BRIDGE

H

LOUDOUN HEIGHTS

U.S. ARMORY WOODLANDS

elevation 1000 feet

U.S. Musket Factory, nos. 1–21.

1. Enginehouse (site of John Brown's capture)
2. Armory offices
3. "Bell" or finishing shop
4. Polishing shop and washhouse
5. Boring mill
6. Stocking and machine shop
7. Millwright shop (stock storehouse prior to 1859)
8. Grinding mill, sawmill, and carpenters shop
9. Tilt-hammer and barrel-forging shop
10. Lumber house and coal bin
11. New stock and storehouse
12. Warehouse
13. Smith and forging shop
14. Annealing shop and brass foundry
15. Proof house
16. Charcoal house
17. Old stock storehouse
18. Rolling mill
19. Limehouse (exact location unknown)
20. Icehouse (exact location unknown)
21. Armory magazine (on Magazine Hill)
—The U.S. Canal enters the armory workshop area from the northwest.

Rifle Works on Lower Hall Island, nos. 22–30.

22. Finishing and machine shop
23. Filing shop
24. Machine shop
25. Barrel-drilling and finishing shop
26. Tilt-hammer and forging shop
27. Annealing furnace and proof house
28. Coal house
29. Stockhouse
30. Proof house

Selected buildings on Virginius Island, nos. 31–39.

31. Tannery and oil mill built in 1824 by Townshend Beckham. Converted into an iron foundry in 1835 by Hugh Gilleece.
32. Island Flour Mill built in 1840 on approximately the same site as Fontaine Beckham's earlier mill. Owned and operated in 1859 by James S. Welch and Abraham H. Herr.
33. Sawmill and lumberyard established in 1824 by Lewis Wernwag.
34. Machine shop constructed by Lewis Wernwag ca. 1832. Operated by his son, John, in 1859.
35. Site of a cotton mill built in 1849; destroyed by flood in 1852.

36. Cotton factory erected by the Harpers Ferry & Shenandoah Manufacturing Company in 1847; operated by Abraham H. Herr in 1859.
37. Smith shop dating from 1834.
38. Office adjoining Wernwag's sawmill and lumberyard.
39. Chopping mill built in 1840 by Hugh Gilleece.

Miscellaneous buildings at Harpers Ferry, nos. 40–69.
40. Lutheran church (Camp Hill)
41. Armory superintendent quarters and grounds
42. Fire house and public square (Camp Hill)
43. Female seminary (Camp Hill)
44. Armory paymaster's clerk house
45. Armory superintendent's clerk house
46. Armory paymaster quarters and grounds
47. Methodist Protestant church
48. Methodist-Episcopal church
49. St. John's Protestant Episcopal church
50. Presbyterian church
51. Catholic school
52. St. Peter's Roman Catholic church
53. Harper-Wager house, built between 1775 and 1782; used as a tavern ca. 1803.
54. Tavern and store
55. Dry goods store (former Harpers Ferry Hotel, 1835)
56. Small shops (former Stagecoach Inn operated by James Stephenson)
57. Old master armorer's house (1818–1858)
58. New master armorer's house (1859)
59. Market house
60. Small arsenal (U.S.)
61. Large arsenal (U.S.)
62. Old superintendent's office
63. Potomac Restaurant, located on original Ferry Lot
64. Wager House Hotel, located on original Ferry Lot
65. Baltimore & Ohio Railroad depot, located on original Ferry Lot
66. Toll house for covered bridge over Potomac River
67. Baltimore & Ohio Railroad office
68. Gault House Saloon
69. Winchester & Potomac Railroad depot

Approximate location of Wager owned or leased commercial and residential properties between about 1796 and 1836 (shaded area)

Table 1. Output, size of labor force, and unit costs at the Harpers Ferry armory, 1801-1860

Year	Flintlock muskets 1	Percussion muskets (M1842) 2	Rifled muskets (M1855) 3	Flintlock rifles (M1803/14) 4	Percussion rifles (M1841) (M1855) 5	Pistols 6	Pattern arms 7	Verifying instruments (sets) 8
1801	293							
1802	1,472							
1803	1,048							
1804	156			772			4 (muskets) 4 (rifles)	
1805	0			1,716				
1806	136			1,381			8 (pistols)	
1807	50			146		2,880		
1808	3,051					1,208		
1809	7,348							
1810	9,400 8,600*							
1811	10,000							
1812	10,200						4 (pistols)	
1813	9,000						5 (muskets) 4 (rifles)	
1814	10,400			1,600				
1815	5,340			1,508				
1816	6,416			2,052			5 (muskets)	
1817	8,513			2,726			6 (muskets)	
1818	9,892			2,700			8 (muskets) 9 (rifles)	
1819	7,020			3,324			2 (muskets) 2 (rifles) 6 (pistols)	
1820	9,856			1,793				
1821	10,320							32
1822	10,000 11,500*						32 (muskets)	
1823	12,200							
1824	10,559							
1825	14,000							
1826	8,720							
1827	12,020							
1828	10,000							
1829	8,915 8,895*							
1830	10,130							
1831	11,160							
1832	12,000							
1833	12,000							
1834	12,000							
1835	10,000							
1836	9,150						6 (muskets)	
1837	8,200							
1838	12,000						24 (muskets)	4
1839	5,850							
1840	8,304							
1841	8,650						2 (rifles)	
1842	6,575						1 (rifle)	

iscellaneous oducts	Accoutrements, appendages, and extra parts 10	Alterations and repairs 11	Output in musket equiva-lents (M.E.) 12	Number of production workers 13	Output (M.E.) per production worker 14	Ordnance Department estimate of cost per musket ($) at H.F. 15	Writer's estimate of cost per musket ($) at H.F. 16
			293			15.78†	84.56
			1,472	25	58.9	13.44	19.38
			1,048	17	61.6	13.44	26.59
	3,088		1,309	34	38.5	13.44	23.53
	6,864		2,386	48	49.7	13.44	13.84
	5,524		2,119	63	33.6		16.70
(wall piece)	584	5,500	4,022	67	60.0		9.67
(wall pieces)			4,078	82	49.7		18.10
		590	7,496	148	50.6		17.87
		691	9,573	197	48.6		15.10
(harpoon guns)		1,392	10,390	209	49.7		13.90
		3,225	11,308	210	53.8		14.66
(torpedo locks)		843	9,347	208	44.9		20.32
(torpedo lock boxes)							
	6,400	564	12,765	210	60.8		12.42
	6,032	25	7,442	202	36.8		19.08
	8,200		9,318	214	43.5		25.73
(cannon locks)	10,904		12,365	224	55.2	14.25	14.73
	17,550		13,883	227	61.2	14.33 15.00*	14.43
	25,796		11,853	232	51.1		15.76
	20,940		12,486	243	51.4		14.23
	10,766		10,780	224	48.1	12.97	15.64
	25,875		10,579	234	45.2	11.76	15.12
	6,094*						
	11,264		12,313	253	48.7	11.62	14.05
	102,760		11,587	268	43.2	14.26	14.95
	43,774		14,438	262	55.1	11.56	12.88
	91,154		9,632	219	44.0	12.48 11.75*	17.83
	46,103		12,481	230	54.3	11.63 11.97*	14.22
	79,383		10,794	207	52.1	12.40	14.89
	60,873		9,524	197	48.3	15.13 13.42*	16.76
	42,139		10,551	197	53.6	11.25 11.04*	14.19
	7,728		11,237	242	46.4	11.25 11.14*	14.41
	766		12,008	272	44.1	11.62 12.44*	15.60
	32,638		12,326	249	49.5	11.80 10.76*	13.77
	59,391		12,594	255	49.4	11.66 11.17*	14.16
	2,226		10,022	229	43.8	10.78	15.89
	10,447		9,314	243	38.3	11.73	17.19
	106,424		9,264	262	35.4	13.14	17.73
	66,274		13,003	294	44.2	11.61	15.64
	57,524		6,425	272	23.6		28.33
	23,564		8,540	241	35.4	12.87 15.48*	21.34
	24,850		8,919	267	33.4	12.00 14.75*	24.38
	29,651		6,882	246	28.0	15.87	21.46

Table 1 (continued).

Year	Flintlock muskets 1	Percussion muskets (M1842) 2	Rifled muskets (M1855) 3	Flintlock rifles (M1803/14) 4	Percussion rifles (M1841) (M1855) 5	Pistols 6	Pattern arms 7	Verifying instruments (sets) 8
1843	3,105						1 (rifle)	
1844	608						1 (musketoon) 1 (rifle)	
1845		2,225					1 (rifle)	2
1846		12,203 12,209*			700		2 (rifles)	9
1847		12,000 (400 w/o bayonets)			3,054			6
1848		11,000			2,802			
1849		8,300			1,925			1
1850	82	9,600			2,676			
1851		11,100			3,050			1
1852		13,400			3,227			
1853		10,101			2,762			
1854		9,000			2,761			
1855		7,700			2,339			
1856								
1857			1		10		1 (rifle)	1
1858			8,581		1,719			
1859			6,489		2,466			
1860		190	7,349		2,701			1

The data for this table have been gathered primarily from three sources: *American State Papers: Military Affairs* (2:481; 5:913-923), Benet, *Ordnance Reports* (2:230), and *Annual Reports of the Secretary of War* in the congressional serial set. Where possible, these figures have been checked and verified with the appropriate archival records the Office of the Chief of Ordnance (OCO, entries 5, 21, and 1003).

Columns 1 through 11 show the range of products made annually at Harpers Ferry. Since each class requires different inputs of capital and labor during the manufacturing process, an estimate of aggregate output was obtained by converting all items into "musket equivalents" (col. 12). The following values were assigned in making computations: rifles and rifle-muskets were counted as 1.35 musket equivalents (M.E.); carbines and torpedo locks as 1 M.E.; musketoons and pistols as .83 M.E.; all pattern and experimental long arms as 10 M.E.; pattern pistols as 8 M.E.; rampart and wall pieces as 4 M.E.; harpoon guns as 1.5 M.E.; cannon locks as .5 M.E.; repairs, alterations, and torpedo locks as .25 M.E.; all appendages as .01 M.E.; extra barrels as .24 M.E.; extra bayonets as .08 M.E.; cannons as 25 M.E.; and sets of verification gauges as 11 M.E. (if made before 1836) and __ M.E. (if made after 1836). These values are based on appraisals made by armory and ordnance officers during the antebellum period.

Miscellaneous products	Accoutrements, appendages, and extra parts 10	Alterations and repairs 11	Output in musket equivalents (M.E.) 12	Number of production workers 13	Output (M.E.) per production worker 14	Ordnance Department estimate of cost per musket ($) at H.F. 15	Writer's estimate of cost per musket ($) at H.F. 16
	112		3,116	180	17.3		48.23
	22,813		856	199	4.3		186.76
	8,820 2,056 (barrels) 296 (bayonets)		2,890	267	10.8	12.24	71.54
	36,128 800 (barrels) 405 (bayonets)		13,978	296	46.6	11.87	17.46
	52,201		16,795	301	55.8		16.78
(rampart guns)	58,953 600 (barrels) 600 (bayonets)		15,573				16.98
	102,310	278	12,006	260	46.2	12.72	21.03
2 (experimental rifles)	153,197 330 (barrels) 75 (barrels)	10,836	17,574	264	66.6	11.16 11.73*	13.83
	210,622		17,349	279	62.2	9.21	15.29
	113.672		18,893	245	77.1	9.99	15.38
) (experimental rifles)	109,513		15,015	240	62.6		15.23
3 (experimental rifles)	49,916		14,916	221	67.5	11.98	14.10
000 (musketoons) 3 (experimental rifles)	57,958 3,325 (bayonets) 326 (barrels)	2,241	12,372	142	87.1		18.02
	83,597 3,179 (bayonets)	4,132	2,123	225	9.4		105.93
	51,511 4 (bayonets)	10,077	3,070	309	9.9		78.44
	60,685 2,251 (bayonets) 6 (barrels)	3,097	15,464	400	38.7	16.75	19.32
2 (experimental rifles)	54,242 1,433 (barrels) 4,512 (bayonets)	3,060	13,121	254	51.7	16.15	
3 (experimental rifles) 1 (breechloading cannon)	44,303 1,447 (barrels) 3,585 (bayonets)	2,270	15,468	265	58.3		

Column 13 represents the average number of production workers as derived from contemporary payrolls and inspection reports. Output per worker (col. 14), probably the most reliable index of productivity growth, is obtained by dividing column 12 by column 13.

Periodically members of the armory and Ordnance Department staffs estimated the average cost of a musket at Harpers Ferry. These are entered in column 15. Since estimates frequently differed, alternative figures are indicated by an asterisk (*). A dagger (†) appears after $15.78 in column 15 because this figure represents a general average for all arms made at Harpers Ferry between 1801 and 1822. Column 16 represents the author's calculation of unit costs at the armory. For the sake of uniformity, Deyrup's accounting method was followed (see Arms Makers, pp. 50-51, 229-232), allowing 2 percent for depreciation, ½ percent for insurance, and 6 percent for interest on capital and renewing the calculation for each year. This method admits certain biases—no attempt was made to convert annual expenditures into constant dollars, for example—but they do not vitiate the basic purpose of this column: to provide a means of gauging cost differences between the Harpers Ferry and Springfield armories.

Table 2. Arms and accessories manufactured at the Hall Rifle Works, 1819-1844

Year	Flintlock rifles and bayonets 1	Percussion rifles and bayonets 2	Flintlock carbines 3	Percussion carbines 4	Wipe 5
1819					
1822					
1823	22				
1824	980				1,000
1825					
1826					
1827	1,000				
1828					8,998
1829					
1830					
1831					
1832	4,360				4,360
1833	3,670				3,660
1834	970				940
1835	1,714				1,714
	1,000*				
1836	1,809				
1837	1,200		1,017		1,770
	800*		1,018*		2,217
1838	2,934				
	3,334*				2,934
1839					
1840	1,023				
1841	190		1,003		2,026
1842					190
1843		300			
1844		2,700		1,001	1,999

rew vers	Bullet moulds 7	Charges and flasks 8	Spring vises 9	Ordnance Department estimate of cost per rifle without appendages 10	Ordnance Department estimate of cost per rifle with appendages 11
)00	1,000	1,000	100	$18.83+	$20.59+ $21.57+*
)00	1,000	1,000	100	$13.93	$15.93 $17.82*
360	436	20	436		$14.50+
670	367	20	367		
940	47	4,734	47		$21.13
714	171		171	$15.70 $14.50*	$17.81
770	177	3,000	177		
217	301	881	200		
934	395		1,334		$16.32†
026	55		2,026		$19.19
190	25		49		$21.73
950	131		501		$19.17

Sources: Bomford to Barbour, January 31, 1827, Letters Sent to the Secretary of War, OCO; Bomford to Hall, ovember 17, 1828, Letters Sent, OCO; Craig to Bomford, November 19, 1841, Moor to Talcott, February 11, 343, Huger to Craig, June 18, 1853, Letters Received, OCO; "Proceedings of a Board of Officers," February 342, Reports of Inspections of Arsenals and Depots, OCO; U.S., Congress, House, Committee on Military .ffairs, *Report on the Petition of John H. Hall*, 24th Cong., 1st sess., 1836, H. Rept. No. 375, pp. 11-12; *American ate Papers: Military Affairs* (5:915-917; 6:805); Benet, *Ordnance Reports* (1:305; 2:230).
Symbols: Asterisks (*) denote alternative estimates; daggers (†), a general estimate for the years 1832 through 338; and plus signs (+), cost estimates that do not include adjustments for interest and depreciation. Estimates ade by the Ordnance Department after 1832 generally included charges for interest, but not depreciation.

Bibliography

PRIMARY SOURCES
Manuscript Materials

The Adams Papers. Massachusetts Historical Society. Boston, Mass. (Examined on microfilm.)

James T. Ames Papers. Private collection of Harold J. Bingham. Cromwell, Conn.

James H. Burton Papers. Yale University Archives. New Haven, Conn.

Document File. Office of the Superintendent, Harpers Ferry National Historical Park. Harpers Ferry, W. Va.

Hall-Marmion Letters. Office of the Superintendent, Harpers Ferry National Historical Park. Harpers Ferry, W. Va.

Portland Scrapbook. Maine Historical Society. Portland, Me.

Records of the Adjutant General's Office (AGO). Record Group 94. National Archives. Washington, D.C.

Records of the Allegheny Arsenal (AAR). Record Group 156. National Archives. Washington, D.C.

Records of the Office of the Judge Advocate General (JAG). Record Group 153. National Archives. Washington, D.C.

Records of the National Park Service (NPS). Record Group 79. National Archives. Washington, D.C.

Records of the Office of the Chief of Ordnance (OCO). Record Group 156. National Archives. Washington, D.C.

Records of the Office of the Inspector General (OIG). Record Group 159. National Archives. Washington, D.C.

Records of the Office of the Quartermaster General (OQG). Record Group 92. National Archives. Washington, D.C.

Records of the Office of the Secretary of War (OSW). Record Group 107. National Archives. Washington, D.C.

Records of the Patent Office (RPO). Record Group 241. National Archives. Washington, D.C.

Records of the Springfield Armory (SAR). Record Group 156. National Archives. Washington, D.C.

Records of the United States General Accounting Office (GAO). Record Group 217. National Archives. Washington, D.C.

Records of the United States House of Representatives (HR). Record Group 233. National Archives. Washington, D.C.

Records of the United States Senate (USS). Record Group 46. National Archives. Washington, D.C.

Francis O. J. Smith Papers. Maine Historical Society. Portland, Me.

Papers of George Washington. Library of Congress. Washington, D.C. (Examined on microfilm.)

West Virginia Collection. West Virginia University Library. Morgantown, W. Va.

Public Documents

American State Papers, Class V, *Military Affairs*. Vols. 1–7. Washington: Gales and Seaton, 1832–1861.

Annals of Congress. Vols. 3–6. Washington: Gales and Seaton, 1849.

Benet, Stephen V., ed. *A Collection of Annual Reports and Other Important Papers, Relating to the Ordnance Department*. Vols. 1–3. Washington: Government Printing Office, 1878.

Fitch, Charles H. "Report on the Manufactures of Interchangeable Mechanism." In *Tenth Census of the United States (1880): Manufactures*. Vol. 2. Washington: Government Printing Office, 1883.

State of Virginia. Inventory of Musket [and Rifle] Machinery Taken at Harpers Ferry. *Messages from the Governor of Virginia*, Doc. No. 25. Richmond: William Ritchie, 1861.

U.S., Congress, House. *Report on the Petition of John H. Hall*, 24th Cong., 1st sess., 1835–1836, H. Rept. No. 375.

——, Congress, Senate. *Report on Improvements in Fire-Arms*, 25th Cong., 1st sess., 1837, Senate Ex. Doc. No. 15.

——, Congress, Senate. *Military Commission to Europe in 1855 and 1856: Report of Major Alfred Mordecai*, 36th Cong., 1st sess., 1860, Senate Ex. Doc. No. 60.

——, Congress, Senate. *Report of the Select Committee of the Senate Appointed to Inquire into the Late Invasion and Seizure of the Public Property at Harper's Ferry*, 36th Cong., 1st sess., 1860, Rep. Com. No. 278.

——, Ordnance Department, *Ordnance Notes, No. 25*. Washington: Ordnance Office, 1874.

——, *Statutes at Large*. Vols. 1–5. Boston: Little and Brown, 1848.

Printed Sources: Letters, Memoirs, Travel Accounts

Adams, Henry. *The Education of Henry Adams*. Boston: Houghton Mifflin, 1918.

Bernhard, Karl, Duke of Saxe-Weimar Eisenach. *Travels through North America during the Years 1825 and 1826*. Vol. 1. Philadelphia: Carey, Lea & Carey, 1828.

Bradley, Milton. *History of the United States Armory, Springfield, Massachusetts.* Springfield: Milton Bradley, 1865.

Cabot, James E. *A Memoir of Ralph Waldo Emerson.* Vol. 2. Boston: Houghton Mifflin, 1888.

Caldwell, John E. *A Tour through Part of Virginia in the Summer of 1808.* Edited by William M. E. Rachal. Richmond, Va.: The Dietz Press, 1951.

Calhoun, John C. *The Papers of John C. Calhoun.* Edited by W. E. Hemphill. Vol. 3. Columbia: University of South Carolina Press, 1967.

Coke, Edward T. *A Subaltern's Furlough.* New York: J. & J. Harper, 1833.

de Bacourt, Adolphe Fourier. *Souvenirs of a Diplomat.* New York: Henry Holt, 1885.

Fairfax, Thomas, Lord. *Journey from Virginia to Salem, Massachusetts, 1799.* London: Privately printed, 1936.

Jefferson, Thomas. *Notes on the State of Virginia.* Philadelphia: R. T. Rawle, 1801.

Leech, Samuel V. *The Raid of John Brown at Harper's Ferry as I Saw It.* Washington, D.C.: The Desoto, 1909.

Martin, Joseph. *A New and Comprehensive Gazetteer of Virginia, and the District of Columbia.* Charlottesville, Va.: Moseley & Tompkins, 1835.

Royall, Anne. *The Black Book.* vol. 1. Washington, D.C.: By the author, 1828.

Stearns, Charles. *The National Armories.* Springfield, Mass.: G. W. Wilson, 1852.

Thoreau, Henry David. *A Plea for Captain John Brown.* Boston: David R. Godine, 1969.

Washington, George. *The Writings of George Washington.* Edited by John C. Fitzpatrick. Vols. 33–39. Washington: Government Printing Office, 1940–1944.

Newspapers

Daily National Intelligencer (Washington, D.C.), 1835.
Springfield Republican (Springfield, Mass.), 1853, 1859.

SECONDARY SOURCES
Books

Barry, Joseph. *The Strange Story of Harper's Ferry.* 3d ed. Shepherdstown, W.Va.: Shepherdstown Register, 1959.

Boorstin, Daniel J. *The Americans: The National Experience.* New York: Random House, 1965.

Brooks, John G. *As Others See Us.* New York: Macmillan, 1908.

Brown, Stuart E. *The Guns of Harpers Ferry.* Berryville, Va.: Virginia Book Co., 1968.

Bushong, Millard K. *A History of Jefferson County, West Virginia.* Charlestown, W.Va.: Jefferson Publishing Co., 1941.

Calhoun, Daniel H. *The American Civil Engineer.* Cambridge, Mass.: The M.I.T. Press, 1960.

Clark, Victor S. *History of Manufactures in the United States.* Vol. 1. Washington: Carnegie Institution, 1929.

Cochran, Thomas C. *The Inner Revolution: Essays on the Social Sciences in History*. New York: Harper & Row, Torchbooks, 1964.

Cromwell, Giles. *The Virginia Manufactory of Arms*. Charlottesville: University Press of Virginia, 1975.

Cullum, G. W. *Biographical Register of the Officers and Graduates of the U.S. Military Academy*. Vol. 1. New York: Houghton Mifflin, 1891.

Cunliffe, Marcus. *George Washington: Man and Monument*. New York: The New American Library, Mentor Books, 1960.

Dew, Charles B. *Ironmaker to the Confederacy: Joseph R. Anderson and the Tredegar Iron Works*. New Haven: Yale University Press, 1966.

Deyrup, Felicia J. *Arms Makers of the Connecticut Valley: A Regional Study of the Economic Development of the Small Arms Industry, 1798–1870*. Smith College Studies in History. Vol. 33. Northampton, Mass.: Smith College, 1948.

Dillin, John G. W. *The Kentucky Rifle*. 4th ed. York, Pa.: Trimmer Printing, 1959.

Frisch, Michael H. *Town into City: Springfield, Massachusetts, and the Meaning of Community, 1840–1880*. Cambridge, Mass.: Harvard University Press, 1972.

Fuller, Claud E., and Richard D. Steuart. *Firearms of the Confederacy*. Huntington, W.Va.: Standard Publications, 1944.

Gardner, Robert E. *Small Arms Makers*. New York: Bonanza Books, 1963.

Gluckman, Arcadi, and Leroy D. Saterlee. *American Gun Makers*. Harrisburg, Pa.: The Stackpole Co., 1953.

Gould, William. *Portland in the Past*. Portland, Me.: B. Thurston & Co., 1886.

Hamlin, Talbot. *Benjamin Henry Latrobe*. New York: Oxford University Press, 1955.

Hicks, James E. *Notes on United States Ordnance*. Vol. 1. Mount Vernon, N.Y.: By the author, 1940.

Huntington, R.T. *Hall's Breechloaders*. York, Pa.: George Shumway, 1972.

Kauffman, Henry J. *Early American Gunsmiths, 1650–1850*. Harrisburg, Pa.: The Stackpole Co., 1952.

———. *The Pennsylvania-Kentucky Rifle*. Harrisburg, Pa.: Stackpole Books, 1960.

Kindig, Joe, Jr. *Thoughts on the Kentucky Rifle in Its Golden Age*. Wilmington, Del.: George N. Hyatt Publisher, 1960.

Layton, Edwin T., Jr., ed. *Technology and Social Change in America*. New York: Harper & Row, 1973.

Leggett, M. D. *Subject-Matter Index of Patents for Inventions Issued by the United States Patent Office, from 1790 to 1873*. Vols. 1–2. Washington: Government Printing Office, 1874.

Lovering Martin. *History of the Town of Holland, Massachusetts*. Rutland, Vt.: By the author, 1915.

Marx, Leo. *The Machine in the Garden*. New York: Oxford University Press, 1964.

Mesick, Jane L. *The English Traveller in America, 1785–1835*. New York: Columbia University Press, 1922.

Miller, Douglas T. *The Birth of Modern America, 1820–1850*. New York: Pegasus, 1970.

Miller, William, ed. *Men in Business*. New York: Harper & Row, Torchbooks, 1962.

North, S. N. D., and Ralph H. North. *Simeon North, First Official Pistol Maker of the United States*. Concord, N.H.: The Rumford Press, 1913.

Oates, Stephen B. *To Purge This Land with Blood: A Biography of John Brown*. New York: Harper & Row, 1970.

Pickering, Octavius, and Charles W. Upham. *The Life of Timothy Pickering*. Vol. 3. Boston: Little, Brown, 1873.

Pollard, Sidney. *The Genesis of Modern Management*. Cambridge, Mass.: Harvard University Press, 1965.

Quarles, Benjamin. *Allies for Freedom: Blacks and John Brown*. New York: Oxford University Press, 1974.

Quimby, Ian M. G., and Polly A. Earl, eds. *Technological Innovation and the Decorative Arts*. Winterthur Conference Report 1973. Charlottesville: University Press of Virginia, 1974.

Rosenberg, Nathan, ed. *The American System of Manufacturing*. Edinburgh, Scotland: Edinburgh University Press, 1969.

Rust, Ellsworth M. *Rust of Virginia: Genealogical and Biographical Sketches of the Descendents of William Rust, 1654–1940*. Washington, D.C.: By the author, 1940.

Sanford, Charles L., ed. *Quest for America*. New York: New York University Press, 1964.

Shanks, Henry T. *The Secession Movement in Virginia, 1847–1861*. Richmond, Va.: Garrett & Massie, 1934.

Smith, Henry Nash. *Virgin Land: The American West as Symbol and Myth*. Cambridge, Mass.: Harvard University Press, 1970.

Steiner, Bernard C. *The Life and Correspondence of James McHenry*. Cleveland: The Burrows Brothers Co., 1907.

Stockwell, George A. *The History of Worcester County, Massachusetts*. Boston: C. F. Jewett, 1879.

Thernstrom, Stephan. *Poverty and Progress: Social Mobility in a Nineteenth Century City*. Cambridge, Mass.: Harvard University Press, 1964.

Thompson, Edward P. *The Making of the English Working Class*. New York: Vintage Books, 1963.

Thorpe, Earl E. *Eros and Freedom in Southern Life and Thought*. Durham, N.C.: Seeman Printery, 1967.

Tindall, George B., ed. *The Pursuit of Southern History*. Baton Rouge: Louisiana State University Press, 1964.

Tuckerman, Henry T. *America and Her Commentators*. New York: Scribner's, 1864.

Villard, Oswald G. *John Brown, 1800–1859: A Biography of Fifty Years After*. Boston: Houghton Mifflin, 1911.

Waters, Asa H. *Biographical Sketch of Thomas Blanchard and His Inventions*. Worcester, Mass.: Lucius P. Goddard, 1878.

White, Leonard D. *The Jeffersonians: A Study in Administrative History, 1801–1829*. New York: The Free Press, paperback ed., 1965.

Bibliography

Wise, Barton H. *The Life of Henry A. Wise of Virginia, 1806–1876*. New York: Macmillan, 1899.
Woodbury, Robert S. *History of the Milling Machine*. Cambridge, Mass.: The M.I.T. Press, 1960.
——. *History of the Lathe*. Cambridge, Mass.: The M.I.T. Press, 1961.

Articles

Abbott, Jacob. "The Armory at Springfield." *Harper's New Monthly Magazine* 5 (July 1852):145–161.
Ames, Edward, and Nathan Rosenberg. "The Enfield Armory in Theory and History." *Economic Journal* 78 (1968):827–842.
Bacon-Foster, Corra. "The Story of Kalorama." *Records of the Columbia Historical Society* 13 (1910):98–118.
——. "Early Chapters in the Development of the Potomac Route to the West." *Records of the Columbia Historical Society* 15 (1912):96–244.
Banner, Lois W. "Religious Benevolence as Social Control: A Critique of an Interpretation." *Journal of American History* 60 (1973):23–41.
Battison, Edwin A. "Eli Whitney and the Milling Machine." *Smithsonian Journal of History* 1 (1966):9–34.
——. "A New Look at the 'Whitney' Milling Machine." *Technology and Culture* 14 (1973):592–598.
Buttrick, John. "The Inside Contract System." *Journal of Economic History* 12 (1952):205–221.
Daniels, George H. "The Big Questions in the History of American Technology." *Technology and Culture* 11(1970):1–35.
Durfee, William F. "The First Systematic Attempt at Interchangeability in Firearms." *Cassier's Magazine* 5 (1893–1894):469–477.
Fairbairn, Charlotte J., and C. Meade Patterson. "Captain Hall, Inventor." *The Gun Report* 5 (Oct. 1959):6–10.
Fitch, Charles H. "The Rise of a Mechanical Ideal." *Magazine of American History* 11 (June 1884):516–527.
Grampp, William D. "A Re-Examination of Jeffersonian Economics." *Southern Economic Journal* 12 (1946):263–282.
Griffin, Clifford S. "Religious Benevolence as Social Control, 1815–1860," *Mississippi Valley Historical Review* 44 (1957):423–444.
Gutman, Herbert G. "Work, Culture, and Society in Industrializing America, 1815–1919." *American Historical Review* 78 (1973):531–587.
Higham, John. "From Boundlessness to Consolidation: The Transformation of American Culture, 1848–1860." *William L. Clements Library* (1969; Bobbs-Merrill Reprint Series in American History, H-414):1–28.
Hubbard, Guy. "Development of Machine Tools in New England." *American Machinist* 60 (February 14, 1924):255–258.
Huntington, R. T. "Hall Rifles at Harpers Ferry." *Guns Magazine* 7 (April 1961):30–31.

Makers, G. "Old Machine Tools." *American Machinist* 32 (Aug. 12, 1909):291.

Mathews, Donald G. "The Second Great Awakening as an Organizing Process, 1780–1830: An Hypothesis." *American Quarterly* 31 (1969):23–43.

Meier, Hugo A. "Technology and Democracy, 1800–1860." *Mississippi Valley Historical Review* 43 (1957):618–640.

Montgomery, David. "The Working Class of the Preindustrial American City, 1780–1830." *Labor History* 9 (1968):1–22.

——. "The Shuttle and the Cross: Weavers and Artisans in the Kensington Riots of 1844." *Journal of Social History* 5 (1972):411–446.

Parkhurst, Edward G. "One of the Earliest Milling Machines." *American Machinist* 23 (March 8, 1900):217.

——. "More Early Milling Machines." *American Machinist* 23 (June 28, 1900):605–606.

Rosenberg, Nathan. "Technological Change in the Machine Tool Industry, 1840–1910." *Journal of Economic History* 23 (1963):414–443.

Sanford, Charles L. "The Intellectual Origins and New Worldliness of American Industry." *Journal of Economic History* 18 (1958):1–16.

Smith, Merritt R. "John H. Hall, Simeon North, and the Milling Machine: The Nature of Innovation among Antebellum Arms Makers." *Technology and Culture* 14 (1973):573–591.

——. "George Washington and the Establishment of the Harpers Ferry Armory." *Virginia Magazine of History and Biography* 81 (1973):415–436.

——. "The American Precision Museum." *Technology and Culture* 15 (1974):413–437.

Thompson, Edward P. "Time, Work-Discipline, and Industrial Capitalism." *Past and Present* 38 (1967):56–97.

Uselding, Paul J. "Henry Burden and the Question of Anglo-American Technological Transfer in the Nineteenth Century." *Journal of Economic History* 30 (1970):312–337.

——. "Technical Progress at the Springfield Armory, 1820–1850." *Explorations in Economic History* 9 (1972):291–316.

Waters, Asa H. "Thomas Blanchard, The Inventor." *Harper's New Monthly Magazine* 63 (July 1881):254–260.

Woodbury, Robert S. "The Legend of Eli Whitney and Interchangeable Parts." *Technology and Culture* 1 (1960):235–253.

Woodward, C. Vann. "The Southern Ethic in a Puritan World." *William and Mary Quarterly* 25 (1968):343–370.

Unpublished Studies

Cesari, Gene S. "American Arms-Making Machine Tool Development, 1789–1855." Ph.D. dissertation, University of Pennsylvania, 1970.

Peterson, Raymond G., Jr. "George Washington, Capitalistic Farmer." Ph.D. dissertation, The Ohio State University, 1970.

Smith, Philip R., Jr. "History of the Methodist Episcopal Church, 1818–1868, and the Free Church." Research report, Harpers Ferry National Historical Park, 1958.

——. "The Methodist Protestant Church and Odd Fellows Hall." Research report, Harpers Ferry National Historical Park, 1958.

——. "History of Lower Hall Island, 1796–1848." Research report, Harpers Ferry National Historical Park, 1959.

——. "St. Peter's Roman Catholic Church." Research report, Harpers Ferry National Historical Park, 1959.

——. "History of the Evangelical Lutheran Christ Church, Camp Hill, 1850–1868." Research report, Harpers Ferry National Historical Park, 1959.

——. "Protestant Episcopal Church." Research report, Harpers Ferry National Historical Park, 1959.

Snell, Charles W. "A Comprehensive History of the Construction, Maintenance, and Number of Armory Dwelling Houses, 1796–1869." Research report, Harpers Ferry National Historical Park, 1958.

——. "A Short History of the Island of Virginius, 1816–1870." Research report, Harpers Ferry National Historical Park, 1959.

——. "The Presbyterian Church." Research report, Harpers Ferry National Historical Park, 1959.

Uselding, Paul J. "Technological Change, Productivity Growth, and Linkages to Economic Development: 1820–1850." Paper delivered at the Eleutherian Mills-Hagley Foundation Conference on Technology in the Middle Atlantic Area in the Nineteenth Century, Greenville, Delaware, October 31, 1969.

Index

Index

Bedinger, Daniel, 239, 263
Bell, Colonel James, 205–206, 208, 229; *see also* Carrington committee
Bell, John, 267–269, 313
Benton, Nathan, 97
Blake, Eli W., 203, 223, 249
Blanc, Honoré, 88–89, 91, 232
Blanchard, Robert, 197, 245
Blanchard, Thomas, 124–138, 173, 230–232, 289, 326
Bolivar, Va., 263, 265
Bomford, Colonel George, 116, 135, 160, 166–170, 239; administrative reforms of, 141–142; as advocate of inter-armory co-operation, 111; background and early career of, 106, 153–154; and John H. Hall, 156, 185, 190–192, 199–202, 204–205, 208, 210–212, 214–216, 220, 256; and Roswell Lee, 171–173; clashes with Edward Lucas, Jr., 260–261; encourages mechanization, 103, 112–113, 117–118, 123–124, 167–168; as proponent of uniformity system, 106–109, 191, 220, 280; relations of, with the Junto, 155–156, 208, 332; retirement of, 273; and James Stubblefield, 153, 155–156, 160, 164, 171, 175, 178, 180–182; and superintendency controversy, 266–268, 271–272
Bookkeeping, practices of, 64, 178–179, 255, 274
Brantree, Samuel, 244
Breitenbaugh, Martin, 59, 93
Brindley, James, 39, 41–43
Brown, John, 304–310, 314, 322
Brown's raid, 305–307, 329; aftermath of, 307–310
Buckland, Cyrus, 248, 283, 285, 289
Burkhart, Philip, 248, 301
Burton, James H.: background of, 285n; on Hall's machinery, 236, 240; employed by Ames Manufacturing Company, 302; at Enfield armory, 302n, 312; at Harpers Ferry armory, 248, 261, 285, 287, 301, 326; mechanical practices of, 292, 313; at Richmond armory, 248, 312–313, 316, 320
Butt, Zadoc, 64, 299
Byington, Samuel, 302

Calhoun, John C., 156, 194–196, 198–200
Calhoun, William B., 268–269, 275
Cameron, Charles, 93
Carrington, James, 203–208, 242, 244; *see also* Carrington committee

Carrington committee, 208, 213, 226, 249; formation of, 173, 202–205, 242; report of, on Hall's machinery, 205–208, 221, 229–230, 234–237,240
Cass, Lewis, 213, 215–216
Chambers, William, 261–262
Change, technological: complexity of, 21–22, 283; historical stereotypes of, 22–23; impediments to, 67–68, 77, 84, 102–103, 109, 137, 139, 168, 279, 298, 326–335
Churches, *see* Religion
Civilian superintendents, restoration of, 299–301
Civil War, coming of, 313–319
Clowe, Henry W., 285, 301–303
Cocke, John H., 200–202, 208
Cocke, Philip St. George, 310–311
Cogsill, Elizur B., 244
Colt, Samuel, 19, 216, 307
Colt's Patent Fire Arms Company, 312, 326
Compton, Lenox, 244
Contracting, inside, 64, 134–136, 146, 239–240
Contractors, private arms, 113, 189, 204, 210, 249–250, 325–326
Cooper, George, 93
Copeland, Thomas, 155, 165–170, 174, 176–177, 179,.254
Cox, Ebenezer, 256
Craft traditions: challenged by industrialization, 20–21; persistence of, at Harpers Ferry, 62–68, 92–93, 182, 284, 334; reflected by Model 1803 rifle, 56
Craig, Colonel Henry K., 268, 271–276, 287–288, 297–298, 312, 315
Crawford, William H., 143
Creamer, Caspar, 93
Creamer, Daniel, 93

Dana, Daniel, 122–123
Dana and Olney lathe, 122–124, 127
Davis, Gideon, 204–205
Davis, Jefferson, 301
Dearborn, Henry, 53–54, 69–72, 74–75
Dexter, Samuel, 52
Discipline, *see* Work, discipline of
Dudley, Otis, 247
Dunn, Thomas B., 181, 253–256, 271, 303, 328

Eaton, John H., 177–178, 181, 253, 257
Education, 17, 158, 331–332
Election: of 1828, 253, 258; of 1839, 262–

Library of Congress Cataloging in Publication Data
(For library cataloging purposes only)

Smith, Merritt Roe, 1940-
   Harpers Ferry armory and the new technology.

   Bibliography: p.
   Includes index.
   1. Harpers Ferry, W. Va.—Armories—History.   2. Firearms indus-
try and trade—Harpers Ferry, W. Va.—History.   I. Title.
UF543.H37S63   338.4'7'62344250975499   76-28022
ISBN 0-8014-0984-5
ISBN 0-8014-9181-9 pbk.